The p53 Tumor Suppressor Pathway and Cancer

PROTEIN REVIEWS

Recent Volumes in this Series

VIRAL MEMBRANE PROTEINS: STRUCTURE, FUNCTION, AND DRUG DESIGN
Edited by Wolfgang B. Fischer

THE p53 TUMOR SUPPRESSOR PATHWAY AND CANCER
Edited by Gerard P. Zambetti

PROTEOMICS AND PROTEIN-PROTEIN INTERACTIONS: BIOLOGY, CHEMISTRY, BIOINFORMATICS, AND DRUG DESIGN
Edited by Gabriel Waksman

A Continuation Order Plan is available for this series. A continuation order will bring delivery of each new volume immediately upon publication. Volumes are billed only upon actual shipment. For further information please contact the publisher.

The p53 Tumor Suppressor Pathway and Cancer

Edited by

Gerard P. Zambetti

St. Jude Children's Research Hospital
Memphis, Tennessee

 Springer

Library of Congress Cataloging-in-Publication Data

Zambetti, Gerard P.
　　The p53 tumor suppressor pathway and cancer/Gerard P. Zambetti.
　　　　p.　cm. (Protein reviews)
　　Includes bibliographical references and index.
　　ISBN 0-387-24135-3
　　　　1. p53 antioncogene.　2. p53 protein.　I. Title.　II. Series.

　　RC268.44.P16Z35 2005
　　616.99′4042—dc22

　　　　　　　　　　　　　　　　　　　　　　　　　　　　　　　　2004065387

A C.I.P. Catalogue record for this book is available from the Library of Congress.

ISBN-10: 0-387-24135-3
ISBN-13: 978-0387-24135-7

Printed in Singapore　(BS/DH)

9　8　7　6　5　4　3　2　1

springeronline.com

Preface

The current year (2004) marks the Silver Anniversary of the discovery of the p53 tumor suppressor. The emerging field first considered p53 as a viral antigen and then as an oncogene that cooperates with activated *ras* in transforming primary cells in culture. Fueling the concept of p53 acting as a transforming factor, p53 expression was markedly elevated in various transformed and tumorigenic cell lines when compared to normal cells. In a simple twist of fate, most of the studies conducted in those early years inadvertently relied on a point mutant of p53 that had been cloned from a normal mouse genomic library. A bona fide wild-type p53 cDNA was subsequently isolated, ironically, from a mouse teratocarcinoma cell line. A decade after its discovery, p53 was shown to be a tumor suppressor that protects against cancer. It is now recognized that approximately half of all human tumors arise due to mutations within the p53 gene. As remarkable as this number may seem, it significantly underrepresents how often the p53 pathway is targeted during tumorigenesis. It is my personal view, as well as many in the p53 field, that the p53-signaling pathway is corrupted in nearly 100% of tumors. If you are interested in understanding cancer and how it develops, you must begin by studying p53 and its pathway.

After demonstrating that p53 functions as a tumor suppressor the field exploded and p53 became a major focus of scientists around the world. Indeed, there were approximately 300 published studies on p53 by 1990 and at last count there are more than 30,000 publications. The amount of information on p53 is truly overwhelming and in a real sense has created subspecialties within the field. It is quite difficult, if not impossible, to be well versed and up-to-date on all aspects of p53. It is for this reason that we have decided to consolidate the most important, landmark findings in one place, hence the purpose of writing this book.

The thought behind putting this book together was to assemble a group of outstanding scientist who significantly contributed to our understanding of how p53 functions in tumor suppression. By all means this book does not cover all aspects of p53; rather, it is meant to provide the necessary information to bring a novice up to speed with the field. This is no small feat as approximately 1,000 manuscripts have been published on p53 within just the first three months of this year and there are no signs of this pace letting up (ISI Web of Knowledge).

The book has been structured to first provide an overview and a historical perspective into how the p53 field became what it is today and where it may be heading, as conveyed by Dr. Arnold Levine, the codiscoverer of p53 (Chapter 1). Much has been learned about how p53 functions from its molecular structure and Dr. Thanos Halazonetis has reviewed the latest NMR and crystallography data (Chapter 2). Within this chapter it is important to note that the mutations observed in human cancer do not happen randomly and that many of these occur at the sites where p53 contacts DNA or at critical junctures that disrupt DNA binding. The studies on p53 binding to DNA leads to the chapters on how p53 functions as a tumor suppressor by regulating gene expression (Chapters 3 and 4). Clearly p53 recognizes specific DNA sequences and activates a series of downstream target genes. This area is reviewed by Dr. Wafik El-Deiry who identified a p53 DNA binding consensus site and discovered p21^{Cip1} as a p53-regulated gene while working as a postdoctoral fellow in Dr. Bert Vogelstein's lab (see Chapter 3). It is also known that p53 selectively turns off the expression of down stream targets (transrepression), some of which are thought to be required for cell survival. Our current understanding of how p53 suppresses gene expression is extensively reviewed by Dr. Maureen Murphy in Chapter 4. Dr. Murphy moved this area of research from artifact-prone, overexpression assays to the identification of endogenous genes that are downregulated by p53 and are required for cell viability.

In Chapter 5, Drs. Ettore Appella and Carl Anderson summarize our understanding of how p53 becomes activated during cell stress, which occurs almost exclusively by posttranslational modification. The development of site-specific antibodies has been critical for studying p53 activation and Drs. Appella and Anderson lead the field in the generation of these reagents. Site-specific antibodies have been instrumental in determining how phosphorylation and acetylation is regulated and how these modifications control p53 function.

As indicated above, it is the p53-signaling pathway acting as a whole that suppresses tumorigenesis. Obviously, perturbations along the pathway could compromise p53 activity and consequently promote tumor development. In Chapter 6, Dr. Ute Moll summarizes what mutations have been observed in p53, how they can arise, and how the pathway may be corrupted without directly affecting p53 itself. It was Dr. Moll who first observed the mislocalization of wild-type p53 in the cytoplasm of primary breast cancer cells and subsequently in neuroblastoma cells. With p53 residing in the wrong subcellular compartment, it no longer functions efficiently to protect against cancer despite being "normal" and this contributes to tumorigenesis.

A prominent mechanism for regulating p53 levels and activity is the intricate negative feedback loop that exists between p53 and Mdm2, a protein that is known to bind p53 and to block its function in multiple ways. How Mdm2 is regulated and in turn how it negatively regulates p53 is described in excellent detail in Chapter 7 by Dr. Jamil Momand, who identified Mdm2 as a critical p53 interacting protein while working as a postdoctoral fellow in Dr. Levine's lab. Subsequent studies revealed the existence of a highly related protein termed, MdmX. What is known about how

MdmX is expressed and how it functionally interacts with p53 and Mdm2 is also presented in this chapter. Pay close attention to the elegant genetic studies examining the phenotypes of Mdm2 and MdmX knockout mice and how these responses are influenced by p53 status.

Recent work has uncovered highly related p53 family members, specifically p63 and p73, suggesting that p53 may not act alone in suppressing tumor growth. Drs. Elsa Flores and Tyler Jacks review the literature surrounding the p53 family of proteins and present some of their latest studies on the consequence of p63 and p73 loss on cell growth and survival (Chapter 8). Intuitively, the functional interaction that exists between the family members would suggest an important role for p63 and p73 in tumor suppression. The consequence of deleting p63 and p73 in knockout mice on tumor susceptibility remains to be seen and should be enlightening.

The original observations of p53 acting as an oncogene during the time when most studies were carried out using the mutant form are not incorrect. Clearly, loss of p53 compromises tumor suppressor function as demonstrated by the finding that 100% of p53-knockout mice develop malignant tumors, usually within several months of age. Nevertheless, most tumors associated with defects in p53 express a full-length missense p53 protein. It is important to keep in mind that the missense protein is usually expressed at high levels in the nucleus and is not biochemically inert. The consequence that overexpression of mutant p53 has on tumor cell growth and survival is not completely understood. What is recognized however is that mutant p53 has the capability of making matters worse and can actively promote tumorigenesis through a gain-of-function mechanism. In Chapter 9, Drs. Alex Sigal and Varda Rotter discuss what is known about mutant p53 tumor promoting properties. Dr. Rotter is most appropriate to review this area as it was her seminal observations in the 1980s that provided the first insight into the tumorigenic properties of mutant.

Lastly, in Chapter 10, Dr. Andrei Gudkov summarizes the current state of affairs concerning the development of compounds that can restore tumor suppressor function to mutant p53. Considering that half of all human cancers express a mutant form of p53, the identification of a small molecule that could correct the biochemical defect could have a huge clinical impact. Although I was skeptical that this could ever happen, recent work by Wiman and coworkers provide a strong indication that this indeed can be possible. These exciting findings as well as other important studies are reviewed in this chapter. In addition, there are certain circumstances where p53 activation can actually be detrimental, such as during stroke, chemotherapy, or neurological degeneration. Therefore, compounds that inhibit wild-type p53 activity can also be desirable. Dr. Gudkov has been instrumental on this front and has recently identified a compound called pifithrin, which is also discussed here. My intent was to conclude the book with this chapter to provide some hope that there are reasonable ways to combat human cancers and other diseases associated with p53 mutations or perturbations in its pathway.

There is some redundancy between the chapters and this is unavoidable as each "subspecialty" overlaps to a degree with one another. Although these areas do overlap on occasion, the personal views, which are spun on the literature as discussed within each chapter, provides a rich and much broader understanding of the topic.

Dr. Arnie Levine once said that science moves forward in waves. He said that uncovering the true function of p53 as a tumor suppressor in 1989 was one such wave and that there would be others. Soon thereafter, Dr. Momand identified Mdm2 as a p53 binding protein. It did not take long to prove that Mdm2 was an oncogene because it blocks p53 function. From this simple observation a subfield was established leading to the demonstration that Mdm2 is frequently amplified and overexpressed in human cancers. What the future holds for p53 and the genes that are tied into this extremely important pathway should bring yet more exciting waves!

Gerard P. Zambetti

Contents

1

The p53 Network

Arnold J. Levine, Jill Bargonetti, Gareth L. Bond,
Josephine Hoh, Kenan Onel, Michael Overholtzer,
Archontoula Stoffel, Angelica K. Teresky,
Christine A. Walsh, and Shengkan Jin

SUMMARY

Cancer arises through a series of mutations in selected oncogenes, tumor suppressor genes, or genes involved in DNA repair or replication. The tumor suppressor gene products frequently monitor the efficiency of cellular duplication by populating checkpoints in the process of cell division. When defective, the tumor suppressor genes can lead to inherited predispositions in the development of cancers. Almost every human cancer contains mutations in the tumor suppressor pathways of p53, retinoblastoma (Rb), or both. Each of these pathways receives a complex set of signals and reports from the extracellular and intracellular environments of a cell and in response regulate "go-no go" decisions in the cell cycle. This chapter will review some of the origins of research into the p53 gene and its protein. This will form a basis for understanding the other chapters of this book and provide a foundation

A. J. LEVINE • Institute for Advanced Study, Einstein Drive, Princeton, NJ 08540, and The Cancer Institute of New Jersey, 195 Little Albany Street, New Brunswick, NJ 08903   J. BARGONETTI • Hunter College, Department of Biological Sciences, 695 Park Avenue, New York, NY 10021   G. L. BOND, A. K. TERESKY, AND S. JIN • The Cancer Institute of New Jersey, 195 Little Albany Street, New Brunswick, NJ 08903   JOSEPHINE HOH • Yale University, Department of Epidemiology / Public Health New Haven, CT 06520   K. ONEL • Memorial Sloan-Kettering Cancer Center, 1275 York Avenue, New York, NY 10021   M. OVERHOLTZER • Harvard Medical School, Cell Biology, 240 Longwood Avenue, Boston, MA 02115   A. STOFFEL AND C. A. WALSH • Rockefeller University, 1230 York Avenue, New York, NY 10021

The p53 Tumor Suppressor Pathway and Cancer, edited by Zambetti.
Springer Science+Business Media, New York, 2005.

upon which new facts are built. It also points to important future directions for this field.

1.1. HISTORICAL PERSPECTIVES

Cancer in human beings arises because of a series of mutations in selected oncogenes, tumor suppressor genes or genes involved in DNA repair or replication. The oncogene products participate in signal transduction pathways that regulate a variety of processes during cell division and growth. The tumor suppressor gene products frequently negatively regulate selected steps in these pathways and monitor the efficiency of cellular duplication by populating checkpoints in the process of cell division. DNA repair and replication is carefully monitored for fidelity in the transmission of information from one generation to another. Unrepaired DNA damage will increase the mutation rate many fold if replication is attempted on a damaged template.

Over the past 25 years more than 100 oncogenes, 20–30 tumor suppressor genes, and hundreds of genes involved in DNA repair and replication have been identified and shown to play a role in the origins of cancer in animals and humans (Hanahan and Weinberg, 2000). The tumor suppressor genes and some of the DNA repair genes, when defective, can lead to inherited predispositions in the development of cancers (Lander and Weinberg, 2000). What is poorly understood at present is why only a subset of all the oncogenes, that have been identified in animal cancer models (Liu et al., 2001), are found to have mutations in human cancers and why these mutations in oncogenes, tumor suppressor genes, and DNA repair genes often have a tissue or cell type specificity even when these gene products are expressed in many cell types. Inherited mutations in tumor suppressor genes commonly result in cancers of specific cell types, while spontaneous mutations in that same tumor suppressor gene can be found in many additional cancer cell types. For example, inherited mutations in mismatch repair genes predispose the host to colon cancer but the frequency of very few other cancers are elevated even when these defective gene products are expressed in other tissues. The most common explanation is that only the rate limiting steps in signal transduction pathways are targets for mutational impact resulting in cancers and just which step is rate limiting is different in different cell or tissue types. Alternatively, many steps in a pathway may be duplicated or backed up in a tissue specific fashion so that only a subset of mutations alter a pathway in the development of a cancer. These ideas have little or no experimental proof at present.

In cancers of humans, there are several genes that are commonly defective or are amplified in many cancer types and as such, these gene products ought to be good targets for therapy. Among the oncogenes, *myc*, *ras*, *bcl-2*, and the *EGF* receptor family are commonly observed in many tumor types. Almost every human cancer contains mutations in the p53 pathway, the retinoblastoma (Rb) pathway, or both of these pathways. The p53 and Rb pathways are interconnected at several levels (Vogelstein et al., 2000). Each of these pathways receive a complex set of signals and reports from the extracellular and intracellular environment of a cell and in response regulate "go-no go" decisions at the G1 to S phase transition of the cell cycle. Our present day understanding of these events owes a great deal to the study of viruses that cause cancer in animals. The retroviruses have led to the identification of many of the oncogenes that make up the signal transduction pathways. The small DNA tumor viruses led to the discovery of p53 (Lane and Crawford, 1979; Linzer and Levine, 1979) and helped to elucidate the functions of Rb (Whyte et al., 1989). The tools of genetics and the construction of signal transduction pathways from the study of the developmental biology of flies, worms, and cellular processes in yeast have shown the conservation of these ancient pathways and their centrality in understanding cancer (Hahn and Weinberg, 2002). For the purposes of this chapter and this book it will be useful to review some of the origins of research into the p53 gene and its protein. This will form a basis for understanding the other chapters of this book and provide a foundation upon which new facts are built. Because the functions of the p53 and Rb pathways are intertwined and an understanding of why viruses have targeted the p53 function for destruction requires an understanding of Rb function, both the p53 and Rb pathways will be described in detail. This in turn will make it clear why the small DNA tumor viruses choose Rb to inhibit and why in turn this alarms the p53 checkpoint. Viruses cannot duplicate in cells undergoing p53 mediated apoptosis and so the virus survives by inactivating p53 function and as a result has the ability to cause a cancer under certain circumstances. In this way the focus upon viruses that cause cancer in model systems uncovered the very genes that play a role in human cancers.

1.2. THE SMALL DNA TUMOR VIRUSES UNCOVER p53

During the decade of the 1960s and the 1970s the small DNA tumor viruses, SV40, the adenovirus and later (the 1980–90s) the papilloma viruses (HPV) were employed to focus upon the question of which genes that the virus contained were responsible for causing the cancers they produced. A detailed genetic analysis of these viruses demonstrated that in each case 2–3 genes were required for the establishment and the maintenance of transformation of cells in culture and for the formation of tumors in animals (Levine, 1993). These same genes were expressed in the cells of the tumor and the proteins encoded by these genes were recognized as foreign antigens by the host. These proteins came to be called tumor antigens and the antibodies directed against them were useful tools to measure the presence of these proteins in cells and extracts. These tumor antigens were shown to bind to and coimmunoprecipitate with the p53 (Lane and Crawford, 1979; Linzer and Levine, 1979) and Rb proteins (Whyte et al., 1989) in the cell (Table 1.1). Through the 1980s and into the early 1990s, the SV40 large tumor antigen (T-antigen) was shown to bind to the p53 and Rb proteins using separate domains in the T-antigen (Pipas and Levine, 2001). The E1A protein of the adenoviruses bound to Rb (Whyte et al., 1989) and the E1B-55KD protein bound to p53 (Sarnow et al., 1982). Finally, the E6 protein of the

Table 1.1. The interactions of the viral tumor antigens
with the p53 and Rb proteins.

Viruses	Viral proteins	Cellular targets
Polyoma, SV40	**T-antigen**	**p53, Rb**
Papilloma viruses	**E6, E7**	**p53, Rb**
Adenoviruses	**E1A, E1B**	**p53, Rb**

HPVs bound to p53 (Werness et al., 1990) while the E7 protein bound to Rb (Dyson et al., 1989; Munger et al., 1989) (Table 1.1). In human cancers somatic mutations in both Rb and p53 genes were described and at least in some cases these mutations appeared to be inactivating the protein or loss of function mutations (Dyson et al., 1989; Munger et al., 1989). In addition, inherited mutations in the Rb or p53 gene were described and shown to cause cancers to occur at a young age (Friend et al., 1986; Malkin et al., 1990). The p53 protein was shown to be a transcription factor that recognized and bound to specific DNA sequences and activated the transcription of a gene regulated by these DNA responsive elements (p53 RE) (Farmer et al., 1992; Zambetti et al., 1992). Both the binding of the viral oncoproteins (T-antigen, E1B-55K, E6) to p53 and the mutations in the p53 gene blocked the binding of p53 to its RE and reduced p53 specific transcription. In the case of HPV E6 and p53, the E6 protein promoted the degradation of p53 (Scheffner et al., 1990) and so it became clear that the viruses and the cellular mutations resulted in a loss of p53 function. Consistent with this was the demonstration that the wild-type p53 gene and its protein blocked cellular transformation (Finlay et al., 1989), as shown later by inducing apoptosis in the presence of activating oncogenes (Yonish-Rouach et al., 1991). Thus, p53 was a tumor suppressor gene and inactivating mutations resulted in a cancer prone phenotype.

The Rb protein was shown to bind to and negatively regulate a critical transcription factor for entry from G1 to S phase in the cell cycle—the E2F-1 transcription factor (Dyson, 1998; Lees et al., 1993; Nevins, 2001; Nevins et al., 1991; Trimarchi and Lees, 2002). E2F-1 recognizes a specific set of DNA sequences, which regulate a number of genes that are required for the synthesis of substrate precursors for DNA synthesis and DNA replication. When E2F-1 resides on the E2F-1 RE on the DNA and Rb is bound to it, the complex represses transcription of these genes. The SV40 T-antigen, the adenovirus E1A protein, and the HPV E7 protein bind to Rb and remove it from E2F-1, derepressing the transcription of these genes and sending the cell into S phase. Indeed the step regulated by Rb, termed the restriction point in the cell cycle, is the "go-no go" step in the G1 to S phase transition. The small DNA tumor viruses do not have the genetic coding capacity to synthesize all of the components that are required to take a resting cell into S phase. Instead they block Rb function, activate E2F-1, and turn on the cellular genes for entry into S phase. When T-antigen, the E1A protein or E7 protein are expressed all the time in cells, they are committed to S phase at every cycle, bypassing the restriction point. However the cell has a critical checkpoint looking for inappropriate signaling for cell division. A transcriptionally activated p53 protein in the presence of overexpressed levels of E2F-1, results in apoptosis of the cell. Thus, p53 senses the abnormal E2F-1 levels and rolls in a program of cell death. To counter this, the viral oncogenes bind to p53 and block its function. Thus, the small DNA tumor viruses overcome the Rb restriction point to produce the substrates and enzymatic activities to replicate its own genome, and in so doing trigger the p53 check point, which must be inactivated if the virus is to successfully replicate itself. In devising these strategies the small DNA viruses became the small DNA tumor viruses (Levine, 1994).

1.3. THE Rb PATHWAY

Resting cells move into the cell cycle in response to extracellular signals in the form of growth factors that are secreted by other cells, often in response to environmental signals (wound healing, etc). These growth factors bind to specific receptors at the cell surface resulting in the dimerization of these receptors (Attisano and Wrana, 2002; Schlessinger, 2000). This brings together their cytoplasmic domains, which are protein kinases that add phosphate residues to tyrosines in these receptors. The dimeric receptors phosphorylate each other and the phospho-tyrosine residue attract and bind an adaptor molecule (Fig. 1.1) (Attisano and Wrana, 2002; Schlessinger, 2000). This in turn activates a protein kinase cascade or brings to the complex a nucleotide exchange factor which activates a G-protein (ras) and then a protein kinase cascade (MAP kinases) (Lee and Goodbourn, 2001). As a result of these signal transduction pathways being activated several processes are put in place that prepares the cell for entry into G1 and the start of the cell cycle. The transcription factors beta-catenin-TCF-4, Ets, and AP-1 (Tetsu and McCormick, 1999), each of which is activated by these signal transduction pathways, help to transcribe the cyclin D1 gene and the myc gene which are essential for entry into S phase. The rac and rho G-proteins organize the cytoskeleton for division (Ridley, 2001; Settleman, 2001). Activation of the AKT kinase pathway acts to block apoptotic responses to growth signals (Vivanco and Sawyers, 2002) (Fig. 1.1). Almost every step in these pathways is constructed with oncogene products many of which have been shown to cause cancer under the wrong circumstances or when improperly regulated genes are overexpressed.

One result of this movement of the cell into G1 is the progressive synthesis of cyclin D1, whose continued synthesis is dependent upon the occupation of the receptor with growth hormone and the stimulation of the signal transduction pathway (Fig. 1.1). The cyclin D1 protein, when properly phosphorylated, binds to the cyclin-dependent kinase (CDK)-4 or 6, activating it so that it can now phosphorylate Rb which resides on E2F which is in turn bound to the E2F RE on the DNA (Dyson, 1998; Nevins, 2001). Cyclin D phosphorylation of Rb helps to remove it from E2F, just as the viral tumor antigens do (Kato et al., 1993). This begins the transcription of E2F 1 regulated genes. One of these E2F regulated genes is cyclin E, which when

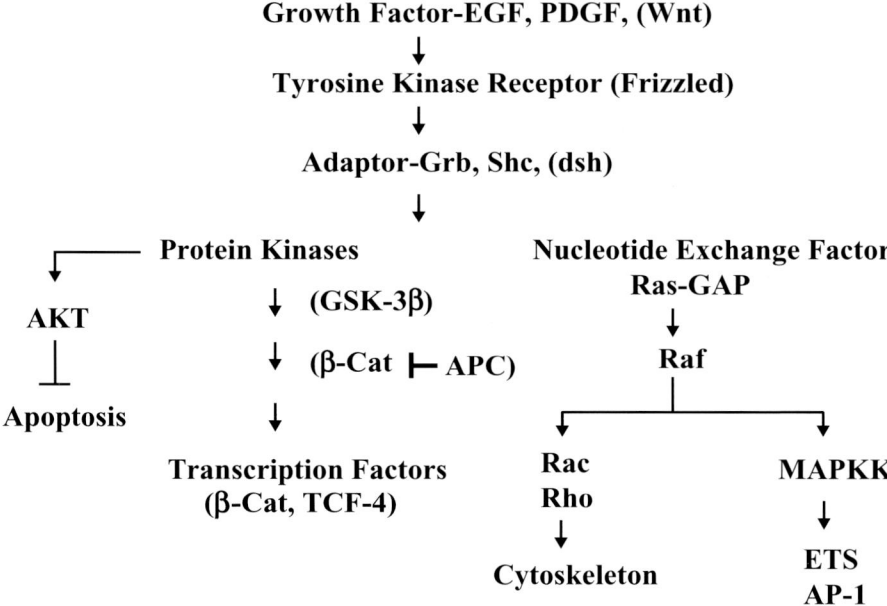

Figure 1.1. The signal transduction pathways that regulate entry into the cell cycle. A schematic review of the gene products that play a role in initiating cell division and guiding the cell through the G1 phase of the cell cycle and into Sphase. Virtually all of these genes and gene products can play a role in causing cancers when mutations activate these oncogenes.

synthesized binds to CDK-2 and enhances the phosphorylation of Rb, which results in the production of more cyclin E (DeGregori et al., 1995). This positive feed back loop does two things: (1) It makes the entry into S phase and the release from the restriction point autocatalytic and (2) The phosphorylation of Rb and the abrogation of the restriction point is no longer dependent upon the presence of a growth factor, which occupies the receptor. Indeed this represents an irreversible commitment to S phase that is independent of the extracellular and intracellular environment (Fig. 1.2). It is at this stage that p53 surveillance of the ribonucleoside triphosphate pools, hypoxia, the integrity of the DNA template, and oncogene activation occurs and the p53 G1 check point acts to arrest the cell cycle, kill the cell, or permit entry into S phase. The purpose of this checkpoint appears to be to assure that the cell can complete the cell cycle without additional nutrients and to insure DNA duplication fidelity, which would suffer if any critical component were not present in optimal amounts. When this p53 checkpoint fails and some component is not present in optimal levels as the DNA is replicated, the mutation rate rises dramatically. This remarkable coordination of multiple inputs from the cellular environment is the hallmark of a central node in a network (Vogelstein et al., 2000) and helps to explain why p53 has been called the guardian of the genome (Lane, 1992). It must integrate a wide variety of signals and respond in one of several ways.

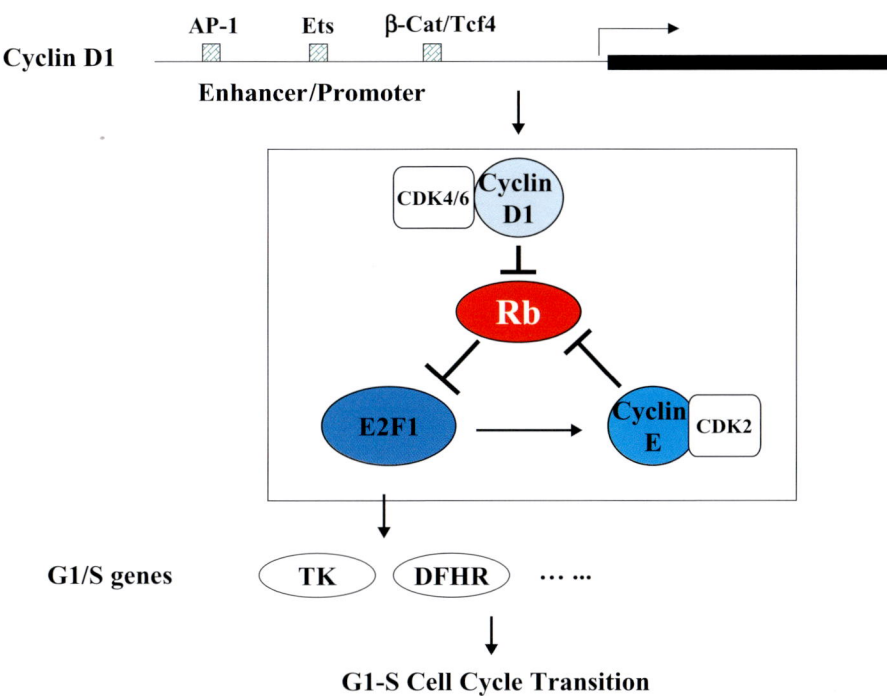

Figure 1.2. The cyclin E-cdk-2 positve feedback loop for Rb phosphorylation. The synthesis of cyclin E becomes autocatalytic and inactivates the restriction point of the cell cycle. This makes the rest of the cell cycle independent of growth hormones.

1.4. THE p53 PATHWAY

The p53 protein when synthesized in cells has a very short half-life, of 6 to 20 minutes (Reich et al., 1983). When first made, the protein is probably not active as a transcription factor and it may not bind efficiently to its RE in DNA. Modification of the protein by phosphorylation of serine and threonine residues at the amino-terminal domain and acetylation of lysine residues at the carboxy-terminal domain and stabilization of the protein (provide a longer half-life) probably result in a conformational change (for which there is no direct evidence at present) that permits binding to the p53 RE (Jayaraman and Prives, 1999; Prives and Manley, 2001). A wide variety of cellular stress signals appear to activate p53 (as measured by an increased half-life and transcriptional activity) such as DNA damage, hypoxia, ribonucleoside triphosphate pool depletion, mitotic spindle damage, nitric oxide signaling, and oncogene activation in a cell (Levine, 1997; Vogelstein et al., 2000) (Fig. 1.3). These signals act to stabilize p53 through mediators, some of whom are known. In response to gamma radiation, which makes single or double stranded breaks in DNA, the ATM protein kinase is activated and in the absence of this kinase p53 activation is delayed or

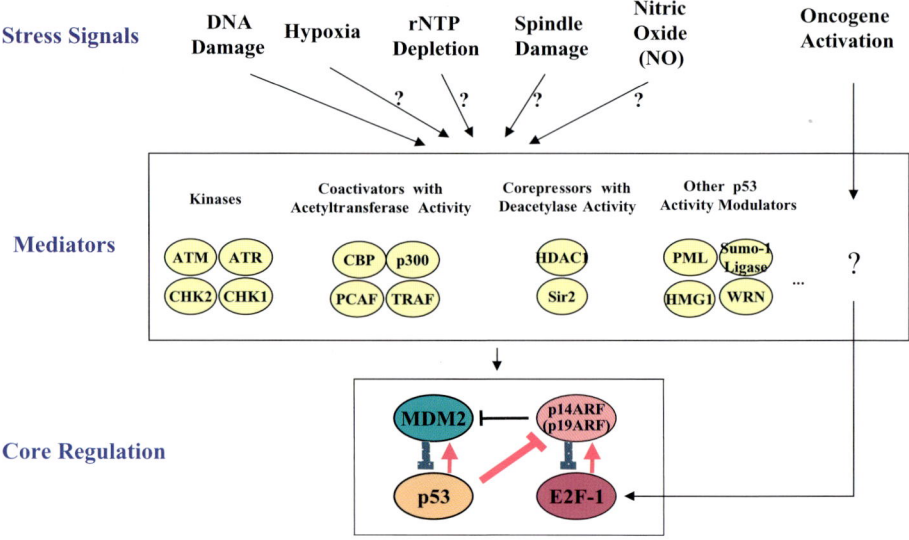

Figure 1.3. The upstream events of the p53 pathway. A variety of stress signals in the cell activate protein kinases and histone acetylases, which in turn modify the p53 and Mdm-2 proteins. This leads to increased levels of p53 and activation of p53 for DNA binding.

reduced (Herzog et al., 1998; Rotman and Shiloh, 1998). Similarly after UV irradiation the ATR kinase is induced and its absence alters the p53 response (Tibbetts et al., 1999). This suggests that different kinds of DNA damage, which are known to have different enzymatic repair mechanisms, are signaling to p53 with different kinases. Whether these kinases act directly upon p53 itself or activate other kinases is under study. The absence of the chk-2 kinase in humans gives rise to an inherited cancer syndrome just like a p53 mutation, suggesting that chk-2 may also play a role in this process (Bell et al., 1999).

Different types of stress signals appear to result in different patterns of phosphorylation of the p53 protein. There are 12 serines and 3 threonines in the amino-terminal 100 amino acids of the p53 protein and so the combinatorial number of possible charge changes resulting in different patterns of transcription in response to different stress signals is large. The histone acetyl-transferases (CBP/p300, PCAF, TRAF) acetylate the carboxy-terminal lysines in p53 (there are 12 lysines in the last 100 amino acids of p53) and neutralizes these positive charges (and phosphorylation of the penultimate serine) which enhances p53 binding to DNA in vitro (Jayaraman and Prives, 1999; Prives and Manley, 2001). It is not clear what mechanisms activate these enzymes or whether a specific combinatorial set of phosphorylated serines and acetylated lysines lead to different properties. At least some of the histone deacetylases (class 3 SIRT enzymes) can remove acetyl groups from the p53 protein suggesting a dynamic regulation of p53 (Luo et al., 2000, 2001; Vaziri et al., 2001). Similarly Wip-1 (PPM1D) is a protein phosphatase that can remove phosphate residues from a kinase that regulates p53 creating a feedback loop that is observed with a number of different p53-dependent gene products (Takekawa et al., 2000).

MDM-2 is a p53 inducible gene (Momand et al., 1992) whose protein product binds to p53 and acts as an E-3 ubiquitin ligase that adds ubiquitin to p53 and results in its degradation (Honda et al., 1997). This produces an autoregulatory loop where p53 results in the synthesis of MDM-2, which in turn degrades p53. This is a "fail-safe" to keep p53 levels in check and two experiments demonstrate this: (1) An MDM-2 knockout mouse is lethal upon implantation (possibly hypoxia?). The MDM-2 p53 double knockout mouse is viable (Jones et al., 1995; Montes de Oca Luna et al., 1995). (2) Some sarcomas have wild-type p53 genes but amplified copies of the MDM-2 gene, presumably inactivating p53 function (Cordon-Cardo et al., 1994). The p53–MDM-2 binding sites have been explored by extensive mutagenesis (Freedman et al., 1997; Lin et al., 1995) and X-ray crystallography (Kussie et al., 1996). MDM-2 forms a hydrophobic pocket into which an amino-terminal p53 amphipathic helix lies. Phosphorylation of p53 serine 20 clearly weakens the p53–MDM-2 complex and stabilizes the p53 protein. In addition phosphorylation modification of MDM-2 occurs. A p53 responsive gene, cyclin G1, binds to the PP2A phosphatase and acts upon MDM-2 to remove a phosphate. Cyclin G knockout mice are viable and have higher basal levels of p53 in all cells (removing the phosphate increases the MDM-2 activity and decreases p53 levels) (Okamoto et al., 2002). In addition, mice that contain a p53 protein that fails to bind to MDM-2 (a residue 22, 23 p53 mutant that is transcriptionally inactive) have higher basal levels of p53 demonstrating that MDM-2 regulates basal levels of p53 as well as activated or induced levels of p53 (Jimenez et al., 2000). Thus, p53 induces a set of activities, Wip-1, cyclin G, and MDM-2, which in turn alter p53 levels or function in the cell. These autocatalytic loops demonstrate the important role of protein modification and the fail-safe mechanisms put in place in the p53 pathway.

The MDM-2–p53 autoregulatory loop predicts that the levels of p53 and MDM-2 will oscillate in a cell and the two proteins should be out of phase with each other (Lev Bar-Or et al., 2000). This is in fact observed in some experiments and the oscillations can be modified or dampened by protein modifications; i.e. serine-20 phosphorylation should stabilize p53 at high levels (Fig. 1.5). Oscillating levels of p53 should also complicate the activation of different p53 responsive genes with different binding constants that result from different sequences in p53 REs. This should lead to changes in the levels of p53 responsive gene products with time, after p53 responses (Fig. 1.6). Yet another modulator of this process is the p14 or p19 ARF (alternate reading frame) gene (Kamijo et al., 1998; Quelle et al., 1995). The p19 ARF protein binds to MDM-2 and inhibits its ability to ubiquitinate p53 (Kamijo et al., 1998). Thus, the synthesis of this protein enhances p53 levels. The transcription of this gene is regulated by a number of oncogene transcription factors including the E2F-1 transcription factor (Bates et al., 1998). This explains why the inactivation of Rb by the adenovirus E1A gene product raises the level of p53 and sensitizes the cell to apoptotic signals. It also explains why the small DNA tumor viruses must inactivate p53 function to replicate themselves. There is some evidence that p19 ARF can regulate the levels of E2F-1,

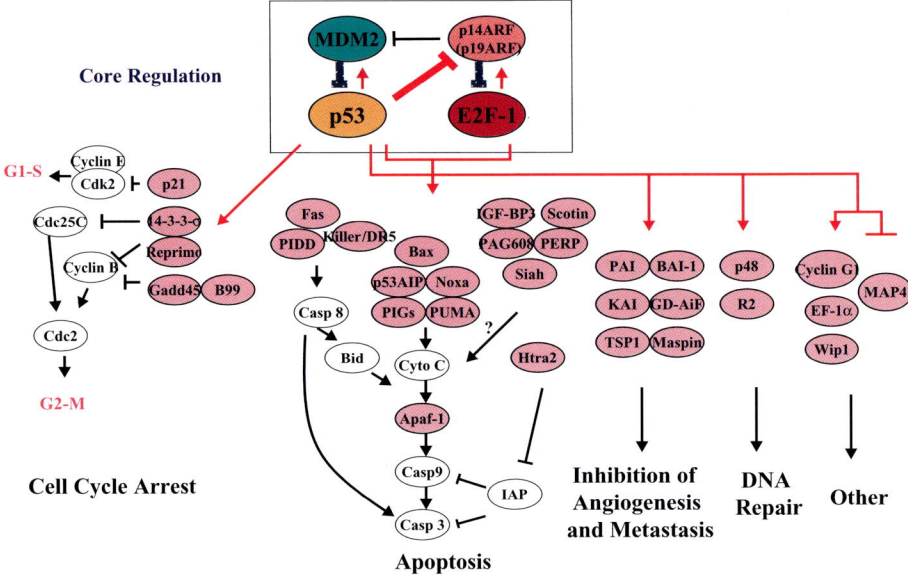

Figure 1.4. The down stream events of the p53 pathway that occur after p53 is activated. p53 activation allows it to bind to the p53 RE on DNA and regulate the transcription of p53 responsive genes, leading to cell cycle arrest or apoptosis. MDM-2, WIP-1, and Cyclin G form negative feedback loops upon p53 levels or activity while p19 or p14 ARF forms a positive feedback loop with p53 activity.

much like MDM-2 regulates p53 (Martelli et al., 2001) (Fig. 1.3), and this forms a joint oscillatory loop with these pairs of molecules. Finally, p53 knockout mice overproduce p19 ARF protein and p53 appears to negatively regulate p19 ARF, likely at the transcriptional level (Quelle et al., 1995; Stott et al., 1998). Clearly there are multiple inputs in the p53–MDM2 autoregulatory loop (Figs. 1.3 and 1.4).

Thus, several different types of stress acting within or upon cells are detected by protein complexes that contain protein kinases and histone acetylases. These activities modify the p53 protein and it appears that the specific serines or threonines that are phosphorylated or the specific lysines that are acetylated in the p53 protein produce combinations of protein modifications that depend upon the type of stress acting upon the cell (Jayaraman and Prives, 1999; Prives and Manley, 2001). Certainly different protein kinases are activated after different types of DNA damage. It is also the case that the nature of the transcriptional program regulated by p53 (which genes are transcribed) differs after different types of stress. There is likely a code, that we have yet to figure out, that relates the p53 protein modification to the transcriptional read-out regulated by this modified p53 protein. The nature of this transcriptional program is both quantitative and temporal. The p53–MDM-2 autoregulatory loop produces a p53 protein whose levels oscillate. The p53 RE sequence is very degenerate and binding constants of the p53 protein to a p53 RE can vary accordingly and the oscillating levels of p53 in a cell (Fig. 1.5) will thus impact gene expression patterns in a complex way.

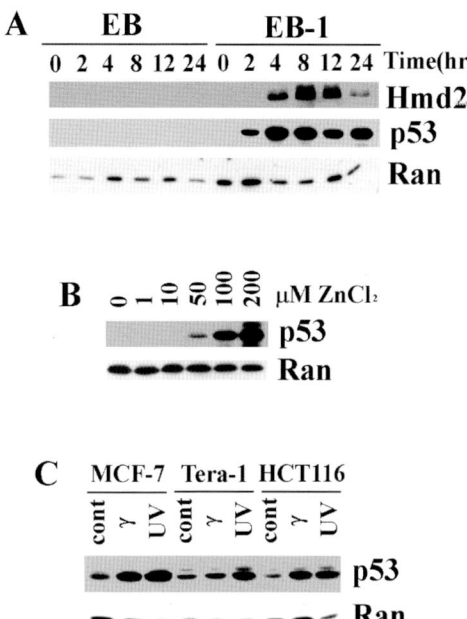

Figure 1.5. p53 levels oscillate over time after a p53 response. p53 levels are followed by immunoselection and Western blots as a function of time after the p53 protein is produced and activated. The p53 and MDM-2 levels oscillate out of phase over 24 hours.

About 31 genes have been shown to be regulated by p53 and to have p53 REs that bind the p53 protein (Jin and Levine, 2001; Vogelstein et al., 2000) (Fig. 1.4). These genes fall into several categories based upon their functions. Several of these proteins mediate cell cycle arrest by p53. The p21 protein binds to cyclin E–cdk2 and blocks it from phosphorylating the Rb protein which is required for entry into S phase (el-Deiry et al., 1993; Harper et al., 1993) (Fig. 1.2). The 14-3-3 sigma protein (Hermeking et al., 1997) binds to CDC-25C, a phosphatase essential for activating CDC-2 (cyclin B–CDC-2 is required for the G2–M phase transition). The 14-3-3-CDC25c complex localizes CDC-25c in the cytoplasm where it cannot act upon CDC-2, blocking cells in G-2 (Peng et al., 1997). The great majority of p53 responsive genes act in the proapoptotic pathway. p53 mediated apoptosis is triggered along at least three pathways; First, cytochrome *c* release from mitochondria binds to apaf-1 resulting in apaf-1 cleavage and activation (Soengas et al., 1999). This in turn cleaves caspase-9 which cleaves caspase-3 leading irreversibly to apoptosis. Here, the p53 inducible genes bax, noxa, and puma all enhance cytochrome *c* release from mitochondria. Second, the Fas ligand is produced in a p53 response and it plus trail engages the Killer/Dr5 receptor activating caspase-8, which in turn cleaves caspase-3 (el-Deiry, 1998; Wu et al., 2000). Third, p53 also induces a serine active site protease, HTRA-2, which binds to and cleaves a number of inhibitors of apoptosis (C-IAP's) that

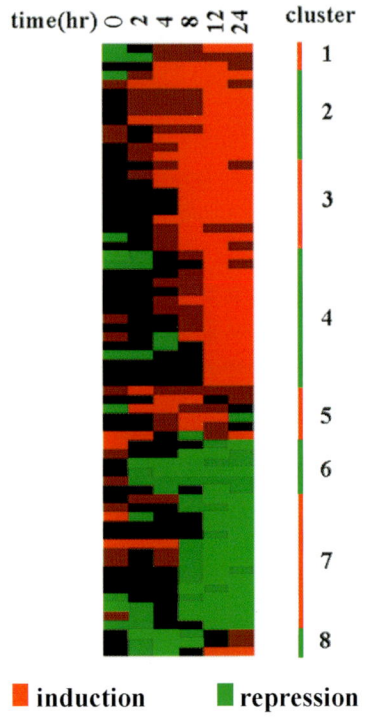

Figure 1.6. A representation of transcript levels of 70 genes whose m-RNA levels change with time after a p53 response. p53 is activated at time zero and 70 selected genes are followed for their transcript levels using Affymetrix chips.

normally block caspase activation (Jin et al., 2003). Thus, the p53 proapoptotic pathway acts upon many fronts that collectively execute programmed cell death. Another p53 inducible gene, IGF-BP-3 (Buckbinder et al., 1995), blocks the insulin-like growth factor outside of a cell so that it will not engage its receptor and activate the antiapoptotic kinase AKT. p53 responsive genes include many secreted proteins that impact upon adjacent cells that are not undergoing a p53 response to stress. This is likely responsible for the so called "by-stander" effect observed when adenoviruses containing a p53 cDNA infect tumor cells and adjacent uninfected cells appear to die (Qazilbash et al., 1997). Among the p53 inducible secreted proteins are those that are antiangiogenic (thrombospondin) (Dameron et al., 1994), and products which block the proteases that alter the extracellular matrix (PAI) (Kunz et al., 1995). Repairing DNA damage in the G0 or G1 phase of the cell cycle can run into a serious limitation in the deoxyribonucleoside triphosphate pool size required to repair damaged DNA. p53 induces a ribonucleotide reductase (Tanaka et al., 2000) to convert ribo-UDP to deoxy-UDP and enhance the pools of d-TTP. Perhaps most interesting in the list of p53 responsive genes are those gene products that, when made, modify the p53 response in specific ways. Included here are MDM-2, Wip-1, and cyclin G (Okamoto et al., 2002; Takekawa et al., 2000) each of which is made in a p53-dependent fashion and once synthesized negatively regulate p53 activity or levels.

The p53 transcriptional response has been studied by using DNA chips to monitor mRNA levels of many genes after p53 is activated (Zhao et al., 2000). In this case, there is no direct evidence as to which genes are regulated by p53 and which genes change their activities because they are regulated by another gene product that was altered by p53. When the p53 gene is activated for transcription in a cell by a zinc inducible promoter and p53 protein is made (Fig. 1.5) one can observe the oscillation of p53 and MDM-2 proteins out of phase in these cells. Many genes are turned on or off at different times after p53 induction (Fig. 1.6). Some downstream gene expression patterns are seen to oscillate off–on–off or on–off–on (Fig. 1.6). This is clearly a complex transcriptional response, which is further complicated by the fact that these cells are undergoing apoptosis. One can also use cell lines to explore the p53 transcriptional response to different DNA damaging agents. Several cell lines were treated with gamma irradiation or UV irradiation and these were compared to cells that were induced for p53 by the zinc inducible promoter. As a control, two cell lines that had mutant p53, and therefore did not respond to UV or gamma irradiation, were also examined (Fig. 1.7). Several interesting conclusions can be made from an examination of these results: (1) Several different cell lines exposed to the same stress, UV or gamma radiation, have slightly different responses. The cell or tissue type is a variable. (2) The same cells exposed to UV or gamma radiation have very different transcriptional responses. (3) Cells with mutant p53 do not show those mRNA inductions or repressions. (4) Inducing p53 in the absence of DNA damage (zinc-inducible promoter) gives yet a different mRNA profile response. (5) Some genes (p21, MDM-2, Cyclin G) are transcriptionally activated or repressed by all stress signals while others are activated by p53 only after a specific stress signal. By determining which p53 protein modifications result from which stress signals and which genes are specific to those stress signals the p53 transcriptional code may be deciphered.

1.5. DETECTING p53 RESPONSIVE ELEMENTS IN THE GENOME

With the sequence of the human and the mouse genome almost completed it should be possible to identify all of the genes regulated by p53 by screening the genome for p53 REs adjacent to a gene. This has not proven to be the case because of the degenerate nature of the p53 RE. There have been two approaches to identify the p53 RE. One is to incubate random oligonucleotide sequences (Funk et al., 1992) or random fragmented genomic DNA pieces (el-Deiry et al., 1992) with the purified p53 protein. Allowing the p53 protein to bind to its unique sequence then permits the selection of the oligonucleotide by immunoprecipitation of the associated p53

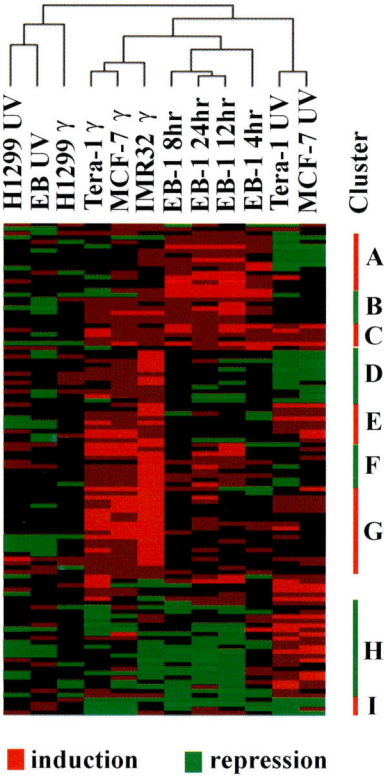

■ **induction**   ■ **repression**

Figure 1.7. A representation of the transcript levels of selected genes in several cell lines that are altered by UV or gamma radiation or induction of p53 without irradiation. Several cell lines were irradiated with UV light, gamma radiation or induced for high p53 levels without radiation. Affymetrix chips were employed to measure the levels of transcripts made by these cells at different times or in different cell lines. The results show that UV, gamma radiation, and p53 induction without radiation produce some transcripts in common and others unique to the inducing agent. The same stress signal in different cell lines did give rise to several different transcripts that are cell type specific. Cells with a mutant p53 gene failed to produce these transcripts.

protein. The oligonucleotide is then amplified by PCR and the selection is repeated until the pool of sequences is enriched and stable. In this way a consensus p53 binding sequence was obtained and it was determined to be RRRCWWGYYY where R is a purine, Y is a pyrimidine, and W is A or T. The crystal structure of the p53 DNA binding domain bound to this oligonuclotide (Cho et al., 1994) showed close contacts to the CWWG core and confirmed that the p53 protein recognized this sequence. The second approach to identifying the p53 RE was to recognize p53 responsive genes experimentally and then to examine the enhancer DNA sequences for elements that regulate p53 responses using mutational analysis to alter the p53 response. Here degenerate (containing several substitutions) RRRCWWGYYY elements were found and shown to be essential to regulation by p53 in studies employing mutagenesis. In addition every gene that was regulated by p53 had at least a second and sometimes several degenerate p53 RE sequences, which were separated by a spacer sequence. Some p53 responsive genes (MDM-2) had several quite degenerate p53REs that were not found by algorithms that required perfect or near matches to the consensus sequence (Barak et al., 1994a, b; Zauberman et al., 1995). The p53 protein bound less well to the p53 degenerate RE sequences in gel shift experiments and yet these were the same sequences found to regulate the p53 response in genes in vivo. Either essential cofactors are missing in the gel shift experiments or RE's with different nucleotide sequences, and as such lower binding constants, are regulated differently by p53 in a cell. This could explain the complex pattern of gene regulation seen in Figures 1.6 and 1.7. With oscillating levels of p53, genes with p53 RE's that bind p53 with different constants and the impact of different protein modifications upon binding of p53 to its RE one might reconstruct the program of transcription.

In an attempt to do this J Hoh and her colleagues (Hoh et al., 2002) have devised an algorithm that uses a filter and a scoring procedure to detect p53 REs in the genome. Table 1.2 presents the filter and scoring algorithm that was derived from the oligonucleotide sequences selected in vitro by p53 protein. A very similar scoring algorithm was obtained when p53 REs from p53 responsive genes were used. The algorithm scanned 10Kb around a gene of interest to look for a p53 RE and measure the distance to additional p53 REs that might regulate the gene. Then each p53 RE was assigned a score with the RRRCWWGYYY spacer RRRCWWGYYY given 100% (a perfect match). Genes that passed the filter with a score of 100% and no spacer (spacer length of zero bases) were found in the human genome. Orthologs of these genes were found in the mouse genome and they too had scores of 100% for p53 REs. Sixteen of these genes which had never before been identified as p53 responsive genes were then tested in two different cell lines to see if they were regulated by p53 using etoposide

Table 1.2. Weight and filter matrices.

	5'–R	R	R	C	W	W	G	Y	Y	Y	R	R	R	C	W	W	G	Y	Y	Y-3'
Weight																				
A	10	6	11	0	11	2.5	0	0.5	0	3	6	2	11.5	0	10	4	0	0.5	1	2
C	3	1	1	19	1	0.5	0	11.5	16	6	2	0	0.5	19	2	0	0	7.5	11	8
G	3	12	8	0	0	0	20	0	0	0	6.5	17	8	0	2	1	20	0	0	1
T	4	1	0	1	7	17	0	8	4	10	5.5	1	0	1	5	15	0	12	8	9
Filter[a]																				
A	1	1	1	0	1	1	0	1	0	1	1	1	1	0	1	1	0	1	1	1
C	1	1	1	1	1	1	0	1	1	1	1	1	1	1	1	1	0	1	1	1
G	1	1	1	0	0	0	1	0	1	1	1	1	1	0	1	1	1	1	1	1
T	1	1	1	1	1	1	0	1	1	1	1	1	0	1	1	1	0	1	1	1

[a]0 = filtered; 1 = nonfiltered

treatment (causing DNA damage) to activate p53 or a temperature sensitive p53 mutant in a cell line in culture. Ten of these genes were shown to be transcriptionally regulated by p53 by at least threefold changes in mRNA levels. Given the cell and tissue specificity and the stress signal specificity of the p53 response a greater experimental diversity will be needed to examine all these genes for their p53 regulation. Based upon the genes that we know are p53 responsive a score of 93 or better is a reasonable predictor for success but several p53 responsive genes (MDM-2) do not score that high.

It is possible to plot the p53 score of a 10 base pair RE as a function of its position (given as the base pair number) in the enhancer–promoter sequence regulating the gene. This should not only reflect the binding constant (a high score) of a p53 RE but should also include the spacer length between REs and the number of REs regulating a gene. When this is done for the p21 gene, which has two REs closely spaced with very high RE scores compared to the MDM-2 gene with four or five putative REs with poor scores and variable spacing, one can begin to correlate this with the observation that the p21 gene is transcribed under p53 control rapidly and at high levels after a p53 response while the MDM-2 gene is slower to produce m-RNA which is made in lower amounts, at least initially. The MDM-2 protein is made in larger amounts after some time (Fig. 1.6) and the multiple binding sites with poor binding constants could give a cooperative nature to this synthesis. The p53 response might want an effector like p21 to be made rapidly and the negative regulator (MDM-2) to be produced with a delay so as not to prematurely shut down the response. Thus, the regulatory complexities observed in the p53 transcriptional program shown in Figures 1.6 and 1.7, will need to be explained by both p53 enhancer–promoter strength but also by protein modification and stability. Protein kinases and phosphatases, protein acetylases and deacetylases, and ubiqutin ligases and deubiqutinases will all impact the outcome of a p53 response.

1.6. BREAKING THE p53 CODE

The previous discussion has identified three variables that play a critical role in the p53 response to a stress signal. These are: (1) protein modification, (2) the binding constants of the p53 protein to p53 REs, and (3) proteins that bind to p53 and regulate the p53 response. While it will be useful to review the state of knowledge about these three variables, it is important to understand the limitations upon the evidence in support of these observations. It is very difficult to prove a particular kinase phosphorylates a specific serine residue on a substrate in vivo and when this happens what the outcome is in a cell. Typically kinase activities are tested on substrates in vitro, or a kinase cDNA will be put in a cell and the kinase overexpressed in vivo, or a knockout mutation of a kinase may be employed to examine substrate alterations. All of these approaches are subject to problems. Knockout mice can accommodate and express some new genes in response to the absence of the kinase. Many kinases have duplicated or redundant kinase activities and so their absence may delay or not effect a substrate phosphorylation. Many proteins may bind to p53 and even modify its activity when those proteins are overproduced by a transfected cDNA in a cell but these proteins may have no physiological impact upon p53 when they are at normal concentrations in a cell. Even determining a protein–DNA binding constant in vitro (a gel shift, footprint or DNA binding assay) may have little reflection upon events in vivo on a chromatin template which may differ in cells with different states of differentiation. The endeavor to break the p53 code will result in false leads and speculative answers but where observations can be confirmed using genetics and diverse approaches we may move closer to an understanding of these processes.

Protein modifications of the p53 protein take place on serines and threonines (phosphorylation) and lysines (ubiquitin, sumoylation, and acetylation). There are a number of serines, threonines and lysines that are in conserved positions in the mouse and human p53 protein, and some residues that are unique to either species. For the sake of simplicity we will use the human protein and label each amino acid by its position, 1–393, from the amino to the carboxy-terminal end of the protein. After DNA damage the amino-terminal serines-9, 15, 20, and 46 and Threonine-18 are phosphorylated to various extents and in various combinations (Jayaraman and Prives, 1999). After gamma radiation (single or double stranded breaks) the ATM kinase is activated and loss of this kinase in mouse or human cells reduces or delays phosphorylation on all of those serine and threonine residues (Herzog et al., 1998; Rotman and Shiloh, 1998). In some experiments the Chk-2 kinase is essential for this pattern of phosphorylation (Chehab et al., 2000; Shieh et al., 2000). After this DNA damage lysines 328 and 382 are acetylated and the levels and timing of this is also ATM dependent. In cells and in mice missing the ATM kinase there is usually less apoptosis after gamma radiation (Herzog et al., 1998). Similarly, after UV irradiation the ATR kinase (ATM related) is activated along with a different DNA repair process (excision repair) (Tibbetts et al., 1999) and ATR then plays a role in modifying the p53 protein. The phosphorylation of serine 46 has been correlated with the apoptotic response and with the p53 induced transcription of the p53AIP-1 gene (Oda et al., 2000). AIP-1 is a mitochondrial associated protein that may play a role in cytochrome c release and apoptosis. ATM and ATR kinases also likely play a role in p53 independent pathways of cell cycle arrest, which of course complicates the interpretation of these results. WIP-1 is a p53 inducible gene that is a protein phosphatase and removes phosphate residues from a protein kinase that regulates p53 serines 33 and 46 (Takekawa et al., 2000), and has been implicated in reversing cell cycle arrest and premature senescence. The MAP kinase pathway (Ras) has been implicated in p53 serine 33 and 46 phosphorylation, and WIP-1 (PPM1D) gene amplifications have been reported in some breast cancers (Bulavin et al., 2002). Serine-15 phosphorylation will result in the movement of p53 from the cytoplasm into the nucleus and ATM/ATR activity is associated with this process.

The acetylation of p53 lysine residues is carried out by the histone acetyltransferases (p300/CBP, PCAF) (Gu and Roeder, 1997; Liu et al., 1999) and deacetylation by a class I histone deacetylase HDAC1 (Luo et al., 2000) and a class III histone deacetylase called SIRT1 enzyme (Luo et al., 2001; Vaziri et al., 2001). Some of these same lysines are the substrates for ubiquitination and sumoylation. MDM-2 is the ubiquitin ligase (E3) for p53. A p53 deubiquitinase has been described (Li et al., 2002), called HAUSP, that can regulate p53 levels in a cell. A HAUSP cDNA overexpression and an RNA-i under expression of HAUSP has been shown to increase and decrease the levels of p53, respectively (Li et al., 2002). The MDM-2 ubiqutin ligase activity is inhibited by the p19 ARF protein. MDM-2 is phosphorylated at threonine 216 by a cyclin A-CDK-2 kinase and the cyclin G-PP2A phosphatase removes this phosphate from threonine 216 (Okamoto et al., 2002). The cyclin G knock out mouse has higher levels of p53 in its cells (Okamoto et al., 2002), indicating that the phosphorylation of threonine 216 by cyclin A-CDK-2 makes MDM-2 less active and cyclin-G makes MDM-2 more active.

ASSP-1 and ASSP-2 are proteins that interact with p53 (ASSP-2 was first called p53 BP-2 found to bind to p53 in a two hybrid screen in yeast) and enhance its ability to promote apoptosis with no impact upon G1 arrest. ASSP-2 increases the binding of p53 to the bax p53 RE (gel shift experiment) and over expression of ASSP-2 in cells enhances the level of bax and pig-3 genes (proapoptotic genes).

The p53 protein has a proline rich region between residues 58–98, with several PXXP sequence motifs. Deletion of this region results in a reduced efficiency of apoptosis by p53 (Walker and Levine, 1996). Pin-1 is a proline isomerase that acts upon p53 as a substrate. Cells from Pin-1 knock out mice have very poor p53 responses and fail to activate p53 efficiently. Pin-1 activity on p53 requires the phosphorylation of serines-53, 81, and 315. This is a good example of how protein modification leads to protein–protein interactions.

1.7. CONCLUSIONS

This review demonstrates the roles of protein modification, protein–protein interaction, and protein–DNA binding upon p53 signaling and responses. It also points to important future directions for this field. First, we need to understand the mechanisms of the stress signals that activates p53 and the impact of protein modifications upon p53. Second, we need to understand how p53 protein modifications lets it select a subset of genes for an appropriate response to a selected stress signal. Third, we need to fill in the genes and proteins that make the p53 pathway functional. For each of these genes we will need to understand its function. Fourth, we will need to identify which of these genes are central to the origins of cancer and the responses of cancers to therapy. Finally, although this review has been focused upon p53 and the cancer phenotype, there are very likely other pathologies impacted by p53 activities. A central node in a network of pathways that integrates stress signals and responses must play a role in a large number of physiological processes. An overactive p53 protein can compromise stem cell regeneration and have an effect upon the rate of aging. An uncontrolled p53 protein can induce apoptosis in neuronal cells and trigger neurodegenerative diseases. Responses of normal and abnormal cells to therapeutic agents will reflect polymorphisms in p53 responsive genes responding to stress signals that result from a course of therapy. When we understand the differences in the p53 pathways of different species of animals, it will provide insights into the evolutionary processes of stress responses and some of the ways in which we are similar or unique. There is a great deal more to be learned.

REFERENCES

Attisano, L., and Wrana, J. L. (2002). Signal transduction by the TGF-beta superfamily. *Science* 296:1646–1647.

Barak, Y., Gottlieb, E., Juven-Gershon, T., and Oren, M. (1994a). Regulation of mdm2 expression by p53: alternative promoters produce transcripts with nonidentical translation potential. *Genes Dev* 8:1739–1749.

Barak, Y., Lupo, A., Zauberman, A., Juven, T., Aloni-Grinstein, R., Gottlieb, E., Rotter, V., and Oren, M. (1994b). Targets for transcriptional activation by wild-type p53: endogenous retroviral LTR, immunoglobulin-like promoter, and an internal promoter of the mdm2 gene. *Cold Spring Harb Symp Quant Biol* 59:225–235.

Bates, S., Phillips, A. C., Clark, P. A., Stott, F., Peters, G., Ludwig, R. L., and Vousden, K. H. (1998). p14ARF links the tumour suppressors RB and p53. *Nature* 395:124–125.

Bell, D. W., Varley, J. M., Szydlo, T. E., Kang, D. H., Wahrer, D. C., Shannon, K. E., Lubratovich, M., Verselis, S. J., Isselbacher, K. J., Fraumeni, J. F., et al. (1999). Heterozygous germ line hCHK2 mutations in Li-Fraumeni syndrome. *Science* 286:2528–2531.

Buckbinder, L., Talbott, R., Velasco-Miguel, S., Takenaka, I., Faha, B., Seizinger, B. R., and Kley, N. (1995). Induction of the growth inhibitor IGF-binding protein 3 by p53. *Nature* 377:646–649.

Bulavin, D. V., Demidov, O. N., Saito, S., Kauraniemi, P., Phillips, C., Amundson, S. A., Ambrosino, C., Sauter, G., Nebreda, A. R., Anderson, C. W., et al. (2002). Amplification of PPM1D in human tumors abrogates p53 tumor-suppressor activity. *Nat Genet* 31:210–215.

Chehab, N. H., Malikzay, A., Appel, M., and Halazonetis, T. D. (2000). Chk2/hCds1 functions as a DNA damage checkpoint in G(1) by stabilizing p53. *Genes Dev* 14:278–288.

Cho, Y., Gorina, S., Jeffrey, P. D., and Pavletich, N. P. (1994). Crystal structure of a p53 tumor suppressor-DNA complex: understanding tumorigenic mutations. *Science* 265:346–355.

Cordon-Cardo, C., Latres, E., Drobnjak, M., Oliva, M. R., Pollack, D., Woodruff, J. M., Marechal, V., Chen, J., Brennan, M. F., and Levine, A. J. (1994). Molecular abnormalities of mdm2 and p53 genes in adult soft tissue sarcomas. *Cancer Res* 54:794–799.

Dameron, K. M., Volpert, O. V., Tainsky, M. A., and Bouck, N. (1994). Control of angiogenesis in fibroblasts by p53 regulation of thrombospondin-1. *Science* 265:1582–1584.

DeGregori, J., Kowalik, T., and Nevins, J. R. (1995). Cellular targets for activation by the E2F1 transcription factor include DNA synthesis- and G1/S-regulatory genes. *Mol Cell Biol* 15:4215–4224.

Dyson, N. (1998). The regulation of E2F by pRB-family proteins. *Genes Dev* 12:2245–2262.

Dyson, N., Howley, P. M., Munger, K., and Harlow, E. (1989). The human papilloma virus-16 E7 oncoprotein is able to bind to the retinoblastoma gene product. *Science* 243:934–937.

el-Deiry, W. S. (1998). Regulation of p53 downstream genes. *Semin Cancer Biol* 8:345–357.

el-Deiry, W. S., Kern, S. E., Pietenpol, J. A., Kinzler, K. W., and Vogelstein, B. (1992). Definition of a consensus binding site for p53. *Nat Genet* 1:45–49.

el-Deiry, W. S., Tokino, T., Velculescu, V. E., Levy, D. B., Parsons, R., Trent, J. M., Lin, D., Mercer, W. E., Kinzler, K. W., and Vogelstein, B. (1993). WAF1, a potential mediator of p53 tumor suppression. *Cell* 75:817–825.

Farmer, G., Bargonetti, J., Zhu, H., Friedman, P., Prywes, R., and Prives, C. (1992). Wild-type p53 activates transcription in vitro. *Nature* 358:83–86.

Finlay, C. A., Hinds, P. W., and Levine, A. J. (1989). The p53 proto-oncogene can act as a suppressor of transformation. *Cell* 57:1083–1093.

Freedman, D. A., Epstein, C. B., Roth, J. C., and Levine, A. J. (1997). A genetic approach to mapping the p53 binding site in the MDM2 protein. *Mol Med* 3:248–259.

Friend SH, Bernards R, Rogelj S, Weinberg RA, Rapaport JM, Albert DM, and TP, D. (1986). A human DNA segment with properties of the gene that predisposes to retinoblastoma and osteosarcoma. *Nature* 323:643–646.

Funk, W. D., Pak, D. T., Karas, R. H., Wright, W. E., and Shay, J. W. (1992). A transcriptionally active DNA-binding site for human p53 protein complexes. *Mol Cell Biol* 12:2866–2871.

Gu, W., and Roeder, R. G. (1997). Activation of p53 sequence-specific DNA binding by acetylation of the p53 C-terminal domain. *Cell* 90:595–606.

Hahn, W. C., and Weinberg, R. A. (2002). Modelling the molecular circuitry of cancer. *Nat Rev Cancer* 2:331–341.

Hanahan, D., and Weinberg, R. A. (2000). The hallmarks of cancer. *Cell* 100:57–70.

Harper, J. W., Adami, G. R., Wei, N., Keyomarsi, K., and Elledge, S. J. (1993). The p21 Cdk-interacting protein Cip1 is a potent inhibitor of G1 cyclin-dependent kinases. *Cell* 75:805–816.

Hermeking, H., Lengauer, C., Polyak, K., He, T. C., Zhang, L., Thiagalingam, S., Kinzler, K. W., and Vogelstein, B. (1997). 14-3-3 sigma is a p53-regulated inhibitor of G2/M progression. *Mol Cell* 1:3–11.

Herzog, K. H., Chong, M. J., Kapsetaki, M., Morgan, J. I., and McKinnon, P. J. (1998). Requirement for Atm in ionizing radiation-induced cell death in the developing central nervous system. *Science* 280, 1089–1091.

Hoh, J., Jin, S., Parrado, T., Edington, J., Levine, A. J., and Ott, J. (2002). The p53MH algorithm and its application in detecting p53-responsive genes. *Proc Natl Acad Sci USA* 99:8467–8472.

Honda, R., Tanaka, H., and Yasuda, H. (1997). Oncoprotein MDM2 is a ubiquitin ligase E3 for tumor suppressor p53. *FEBS Lett* 420:25–27.

Jayaraman, L., and Prives, C. (1999). Covalent and noncovalent modifiers of the p53 protein. *Cell Mol Life Sci* 55:76–87.

Jimenez, G. S., Nister, M., Stommel, J. M., Beeche, M., Barcarse, E. A., Zhang, X. Q., O'Gorman, S., and Wahl, G. M. (2000). A transactivation-deficient mouse model provides insights into Trp53 regulation and function. *Nat Genet* 26:37–43.

Jin, S., Kalkum, M., Overholtzer, M., Stoffel, A., Chait, B., and Levine, A. (2003). CIAP1 and the serine protease HTRA2 are involved in a novel p53-dependent apoptosis pathway in mammals. *Genes Dev* 17:359–367.

Jin, S., and Levine, A. J. (2001). The p53 functional circuit. *J Cell Sci* 114:4139–4140.

Jones, S. N., Roe, A. E., Donehower, L. A., and Bradley, A. (1995). Rescue of embryonic lethality in Mdm2-deficient mice by absence of p53. *Nature* 378:206–208.

Kamijo, T., Weber, J. D., Zambetti, G., Zindy, F., Roussel, M. F., and Sherr, C. J. (1998). Functional and physical interactions of the ARF tumor suppressor with p53 and Mdm2. *Proc Natl Acad Sci USA* 95:8292–8297.

Kato, J., Matsushime, H., Hiebert, S. W., Ewen, M. E., and Sherr, C. J. (1993). Direct binding of cyclin D to the retinoblastoma gene product (pRb) and pRb phosphorylation by the cyclin D-dependent kinase CDK4. *Genes Dev* 7:331–342.

Kunz, C., Pebler, S., Otte, J., and von der Ahe, D. (1995). Differential regulation of plasminogen activator and inhibitor gene transcription by the tumor suppressor p53. *Nucleic Acids Res* 23:3710–3717.

Kussie, P. H., Gorina, S., Marechal, V., Elenbaas, B., Moreau, J., Levine, A. J., and Pavletich, N. P. (1996). Structure of the MDM2 oncoprotein bound to the p53 tumor suppressor transactivation domain. *Science* 274:948–953.

Lander, E. S., and Weinberg, R. A. (2000). Genomics: journey to the center of biology. *Science* 287:1777–1782.

Lane, D. P. (1992). Cancer. p53, guardian of the genome. *Nature* 358:15–16.

Lane, D. P., and Crawford, L. V. (1979). T antigen is bound to a host protein in SV40-transformed cells. *Nature* 278:261–263.

Lee, M., and Goodbourn, S. (2001). Signalling from the cell surface to the nucleus. *Essays Biochem* 37:71–85.

Lees, J. A., Saito, M., Vidal, M., Valentine, M., Look, T., Harlow, E., Dyson, N., and Helin, K. (1993). The retinoblastoma protein binds to a family of E2F transcription factors. *Mol Cell Biol* 13:7813–7825.

Lev Bar-Or, R., Maya, R., Segel, L. A., Alon, U., Levine, A. J., and Oren, M. (2000). Generation of oscillations by the p53-Mdm2 feedback loop: a theoretical and experimental study. *Proc Natl Acad Sci USA* 97:11250–11255.

Levine, A. J. (1993). The oncogenes of the DNA tumor viruses. In Molecular Genetics of Nervous System Tumors, AJ Levine and HH Schmidek (eds), John Wiley & Sons, New York, pp. 137–143.

Levine, A. J. (1994). The origins of the small DNA tumor viruses. *Adv Cancer Res* 65:141–168.

Levine, A. J. (1997). p53, the cellular gatekeeper for growth and division. *Cell* 88:323–331.

Li, M., Chen, D., Shiloh, A., Luo, J., Nikolaev, A. Y., Qin, J., and Gu, W. (2002). Deubiquitination of p53 by HAUSP is an important pathway for p53 stabilization. *Nature* 416:648–653.

Lin, J., Teresky, A. K., and Levine, A. J. (1995). Two critical hydrophobic amino acids in the N-terminal domain of the p53 protein are required for the gain of function phenotypes of human p53 mutants. *Oncogene* 10:2387–2390.

Linzer DI, and AJ, L. (1979). Characterization of a 54K dalton cellular SV40 tumor antigen present in SV40-transformed cells and uninfected embryonal carcinoma cells. *Cell* 17:43–52.

Liu, L., Scolnick, D. M., Trievel, R. C., Zhang, H. B., Marmorstein, R., Halazonetis, T. D., and Berger, S. L. (1999). p53 sites acetylated in vitro by PCAF and p300 are acetylated in vivo in response to DNA damage. *Mol Cell Biol* 19:1202–1209.

Liu, T., Yan, H., Kuismanen, S., Percesepe, A., Bisgaard, M. L., Pedroni, M., Benatti, P., Kinzler, K. W., Vogelstein, B., Ponz de Leon, M., et al. (2001). The role of hPMS1 and hPMS2 in predisposing to colorectal cancer. *Cancer Res* 61:7798–7802.

Luo, J., Nikolaev, A. Y., Imai, S., Chen, D., Su, F., Shiloh, A., Guarente, L., and Gu, W. (2001). Negative control of p53 by Sir2alpha promotes cell survival under stress. *Cell* 107:137–148.

Luo, J., Su, F., Chen, D., Shiloh, A., and Gu, W. (2000). Deacetylation of p53 modulates its effect on cell growth and apoptosis. *Nature* 408:377–381.

Malkin, D., Li, F. P., Strong, L. C., Fraumeni, J. F., Jr., Nelson, C. E., Kim, D. H., Kassel, J., Gryka, M. A., Bischoff, F. Z., Tainsky, M. A., and et al. (1990). Germ line p53 mutations in a familial syndrome of breast cancer, sarcomas, and other neoplasms. *Science* 250:1233–1238.

Martelli, F., Hamilton, T., Silver, D. P., Sharpless, N. E., Bardeesy, N., Rokas, M., DePinho, R. A., Livingston, D. M., and Grossman, S. R. (2001). p19ARF targets certain E2F species for degradation. *Proc Natl Acad Sci USA* 98, 4455–4460.

Momand, J., Zambetti, G. P., Olson, D. C., George, D., and Levine, A. J. (1992). The mdm-2 oncogene product forms a complex with the p53 protein and inhibits p53-mediated transactivation. *Cell* 69:1237–1245.

Montes de Oca Luna, R., Wagner, D. S., and Lozano, G. (1995). Rescue of early embryonic lethality in mdm2-deficient mice by deletion of p53. *Nature* 378:203–206.

Munger, K., Werness, B. A., Dyson, N., Phelps, W. C., Harlow, E., and Howley, P. M. (1989). Complex formation of human papillomavirus E7 proteins with the retinoblastoma tumor suppressor gene product. *EMBO J* 8:4099–4105.

Nevins, J. R. (2001). The Rb/E2F pathway and cancer. *Hum Mol Genet* 10:699–703.

Nevins, J. R., Chellappan, S. P., Mudryj, M., Hiebert, S., Devoto, S., Horowitz, J., Hunter, T., and Pines, J. (1991). E2F transcription factor is a target for the RB protein and the cyclin A protein. *Cold Spring Harb Symp Quant Biol* 56:157–162.

Oda, K., Arakawa, H., Tanaka, T., Matsuda, K., Tanikawa, C., Mori, T., Nishimori, H., Tamai, K., Tokino, T., Nakamura, Y., and Taya, Y. (2000). p53AIP1, a potential mediator of p53-dependent apoptosis, and its regulation by Ser-46-phosphorylated p53. *Cell* 102:849–862.

Okamoto, K., Li, H., Jensen, M. R., Zhang, T., Taya, Y., Thorgeirsson, S. S., and Prives, C. (2002). Cyclin G recruits PP2A to dephosphorylate Mdm2. *Mol Cell* 9:761–771.

Peng, C. Y., Graves, P. R., Thoma, R. S., Wu, Z., Shaw, A. S., and Piwnica-Worms, H. (1997). Mitotic and G2 checkpoint control: regulation of 14-3-3 protein binding by phosphorylation of Cdc25C on serine-216. *Science* 277:1501–1505.

Pipas, J. M., and Levine, A. J. (2001). Role of T antigen interactions with p53 in tumorigenesis. *Semin Cancer Biol* 11:23–30.

Prives, C., and Manley, J. L. (2001). Why is p53 acetylated? *Cell* 107:815–818.

Qazilbash, M. H., Xiao, X., Seth, P., Cowan, K. H., and Walsh, C. E. (1997). Cancer gene therapy using a novel adeno-associated virus vector expressing human wild-type p53. *Gene Ther* 4:675–682.

Quelle, D. E., Zindy, F., Ashmun, R. A., and Sherr, C. J. (1995). Alternative reading frames of the INK4a tumor suppressor gene encode two unrelated proteins capable of inducing cell cycle arrest. *Cell* 83:993–1000.

Reich, N. C., Oren, M., and Levine, A. J. (1983). Two distinct mechanisms regulate the levels of a cellular tumor antigen, p53. *Mol Cell Biol* 3:2143–2150.

Ridley, A. J. (2001). Rho family proteins: coordinating cell responses. *Trends Cell Biol* 11:471–477.

Rotman, G., and Shiloh, Y. (1998). ATM: from gene to function. *Hum Mol Genet* 7:1555–1563.

Sarnow, P., Ho, Y. S., Williams, J., and Levine, A. J. (1982). Adenovirus E1b-58kd tumor antigen and SV40 large tumor antigen are physically associated with the same 54 kd cellular protein in transformed cells. *Cell* 28:387–394.

Scheffner, M., Werness, B. A., Huibregtse, J. M., Levine, A. J., and Howley, P. M. (1990). The E6 oncoprotein encoded by human papillomavirus types 16 and 18 promotes the degradation of p53. *Cell* 63:1129–1136.

Schlessinger, J. (2000). Cell signaling by receptor tyrosine kinases. *Cell* 103:211–225.

Settleman, J. (2001). Rac 'n Rho: the music that shapes a developing embryo. *Dev Cell* 1:321–331.

Shieh, S. Y., Ahn, J., Tamai, K., Taya, Y., and Prives, C. (2000). The human homologs of checkpoint kinases Chk1 and Cds1 (Chk2) phosphorylate p53 at multiple DNA damage-inducible sites. *Genes Dev* 14:289–300.

Soengas, M. S., Alarcon, R. M., Yoshida, H., Giaccia, A. J., Hakem, R., Mak, T. W., and Lowe, S. W. (1999). Apaf-1 and caspase-9 in p53-dependent apoptosis and tumor inhibition. *Science* 284:156–159.

Stott, F. J., Bates, S., James, M. C., McConnell, B. B., Starborg, M., Brookes, S., Palmero, I., Ryan, K., Hara, E., Vousden, K. H., and Peters, G. (1998). The alternative product from the human CDKN2A locus, p14(ARF), participates in a regulatory feedback loop with p53 and MDM2. *EMBO J* 17:5001–5014.

Takekawa, M., Adachi, M., Nakahata, A., Nakayama, I., Itoh, F., Tsukuda, H., Taya, Y., and Imai, K. (2000). p53-inducible wip1 phosphatase mediates a negative feedback regulation of p38 MAPK-p53 signaling in response to UV radiation. *EMBO J* 19:6517–6526.

Tanaka, H., Arakawa, H., Yamaguchi, T., Shiraishi, K., Fukuda, S., Matsui, K., Takei, Y., and Nakamura, Y. (2000). A ribonucleotide reductase gene involved in a p53-dependent cell-cycle checkpoint for DNA damage. *Nature* 404:42–49.

Tetsu, O., and McCormick, F. (1999). Beta-catenin regulates expression of cyclin D1 in colon carcinoma cells. *Nature* 398:422–426.

Tibbetts, R. S., Brumbaugh, K. M., Williams, J. M., Sarkaria, J. N., Cliby, W. A., Shieh, S. Y., Taya, Y., Prives, C., and Abraham, R. T. (1999). A role for ATR in the DNA damage-induced phosphorylation of p53. *Genes Dev* 13:152–157.

Trimarchi, J. M., and Lees, J. A. (2002). Sibling rivalry in the E2F family. *Nat Rev Mol Cell Biol* 3:11–20.

Vaziri, H., Dessain, S. K., Ng Eaton, E., Imai, S. I., Frye, R. A., Pandita, T. K., Guarente, L., and Weinberg, R. A. (2001). hSIR2(SIRT1) functions as an NAD-dependent p53 deacetylase. *Cell* 107:149–159.

Vivanco, I., and Sawyers, C. L. (2002). The phosphatidylinositol 3-Kinase AKT pathway in human cancer. *Nat Rev Cancer* 2:489–501.

Vogelstein, B., Lane, D., and Levine, A. J. (2000). Surfing the p53 network. *Nature* 408:307–310.

Walker, K. K., and Levine, A. J. (1996). Identification of a novel p53 functional domain that is necessary for efficient growth suppression. *Proc Natl Acad Sci USA* 93:15335–15340.

Werness, B. A., Levine, A. J., and Howley, P. M. (1990). Association of human papillomavirus types 16 and 18 E6 proteins with p53. *Science* 248:76–79.

Whyte, P., Williamson, N. M., and Harlow, E. (1989). Cellular targets for transformation by the adenovirus E1A proteins. *Cell* 56:67–75.

Wu, G. S., Kim, K., and el-Deiry, W. S. (2000). KILLER/DR5, a novel DNA-damage inducible death receptor gene, links the p53-tumor suppressor to caspase activation and apoptotic death. *Adv Exp Med Biol* 465:143–151.

Yonish-Rouach, E., Resnitzky, D., Lotem, J., Sachs, L., Kimchi, A., and Oren, M. (1991). Wild-type p53 induces apoptosis of myeloid leukaemic cells that is inhibited by interleukin-6. *Nature* 352:345–347.

Zambetti, G. P., Bargonetti, J., Walker, K., Prives, C., and Levine, A. J. (1992). Wild-type p53 mediates positive regulation of gene expression through a specific DNA sequence element. *Genes Dev* 6:1143–1152.

Zauberman, A., Flusberg, D., Haupt, Y., Barak, Y., and Oren, M. (1995). A functional p53-responsive intronic promoter is contained within the human mdm2 gene. *Nucleic Acids Res* 23:2584–2592.

Zhao, R., Gish, K., Murphy, M., Yin, Y., Notterman, D., Hoffman, W. H., Tom, E., Mack, D. H., and Levine, A. J. (2000). Analysis of p53-regulated gene expression patterns using oligonucleotide arrays. *Genes Dev* 14:981–993.

2

The Three-Dimensional Structure of p53

Elena S. Stavridi, Yentram Huyen, Emily A. Sheston, and Thanos D. Halazonetis

Protein function is completely dependent on three-dimensional structure. Yet, molecular biologists often pay little attention to structural data. In a field, like p53, where numerous functions and protein-protein interactions have been proposed, the structural information can serve as a sieve to distinguish between credible models and models that are less likely to be physiologically relevant. Structural information can also help design meaningful experiments. In this chapter we will present the structural data that are available for p53 and consider their implications for p53 function. As will become evident, there are still major gaps in our knowledge of p53 structure.

2.1. p53 DOMAINS AND REGIONS

The human p53 protein is 393 amino acids long. Protein domains, defined as independently folding units of a protein, typically have a size of between 40 and 200 amino acids (Koonin et al., 2002). This suggests that p53 contains more than one protein domain, a prediction that has been confirmed by structural and functional studies (Vogelstein et al., 2000). Three domains are recognized in p53 (Fig. 2.1). A transactivation domain (residues 1–70), a sequence-specific DNA binding domain (residues 94–293) and a tetramerization domain (residues 324–355). These domains are flanked

E. S. STAVRIDI AND E. A. SHESTON • Wistar Institute, Philadelphia, PA 19104-4268   Y. HUYEN • Wistar Institute, Philadelphia, PA 19104-4268, and Graduate Group in Biomedical Sciences   T. D. HALAZONETIS • Wistar Institute, Philadelphia, PA 19104-4268, and Department of Pathology and Laboratory Medicine, University of Pennsylvania, Philadelphia, PA 19104.

The p53 Tumor Suppressor Pathway and Cancer, edited by Zambetti.
Springer Science+Business Media, New York, 2005.

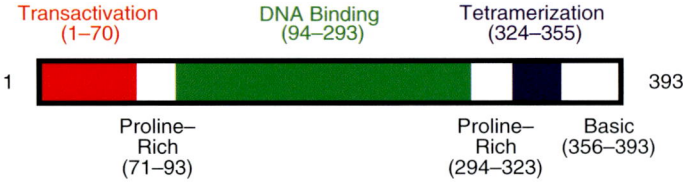

Figure 2.1. Domains of human p53. Linear diagram of human p53 showing its three major domains, the proline-rich regions and the C-terminal basic region. The codon numbers indicate the boundaries of the various domains and regions.

by linker regions. A proline-rich region (residues 71–93) links the transactivation and sequence-specific DNA binding domains; a second proline-rich region (residues 294–323) links the sequence-specific DNA binding and tetramerization domains; and a basic region (residues 356–393) forms the very C-terminus of the protein.

2.1.1. The Transactivation Domain

Functional studies indicate that the N-terminus of p53 (residues 1-70) can activate transcription either in the context of full-length p53 or when grafted to heterologous proteins (Fields and Jang, 1990; Raycroft et al., 1990; Lin et al., 1994; Candau et al., 1997). Accordingly, the N-terminus is referred to as the transactivation domain. Technically, the term domain to describe this region of p53 is inaccurate, since the N-terminus of p53 is not an independently folding unit. A better term would have been transactivating region; however, the term transactivating domain has been used so extensively that it is unlikely to be replaced.

The transactivation domain of p53 serves at least two roles. It activates transcription and it also regulates p53 function and stability. It does so by interacting with transcription factors, such as p300 and CBP, and also with Mdm2, a ubiquitin protein ligase that targets p53 for degradation in the proteasome (Avantaggiati et al., 1997; Gu et al., 1997; Lill et al., 1997; Scolnick et al., 1997; Lin et al., 1994; Honda et al., 1997). p300/CBP and Mdm2 have overlapping binding sites within the N-terminus of p53. In response to DNA damage phosphorylation of p53 N-terminal residues decreases the affinity of p53 for Mdm2 and, concomitantly, increases its affinity for p300/CBP (Shieh et al., 1997; Lambert et al., 1998; Chehab et al., 1999). This leads to increased p53 protein levels and increased p53 transcriptional activity.

A three-dimensional structure of the N-terminus of p53 bound to p300, CBP or any other transcription factor is not yet available. However, a three-dimensional structure of an N-terminal p53 peptide bound to the N-terminus of Mdm2 has been determined (Kussie et al., 1996) and explains how phosphorylation of residues Thr18 (see footnote for list of amino acid codes) and Ser20 of p53 decrease its affinity for Mdm2. The structure comprises residues 25–109 of human Mdm2 bound to a

---

Single and three-letter amino acid codes: A, Ala, alanine; C, Cys, cysteine; D, Asp, aspartic acid; E, Glu, glutamic acid; F, Phe, phenylalanine; G, Gly, glycine; H, His, histidine; I, Ile, isoleucine; K, Lys, lysine; L, Leu, leucine; M, Met, methionine; N, Asn, asparagine; P, Pro, proline; Q, Gln, glutamine; R, Arg, arginine; S, Ser, serine; T, Thr, threonine; V, Val, valine; W, Trp, tryptophan; Y, Tyr, tyrosine.

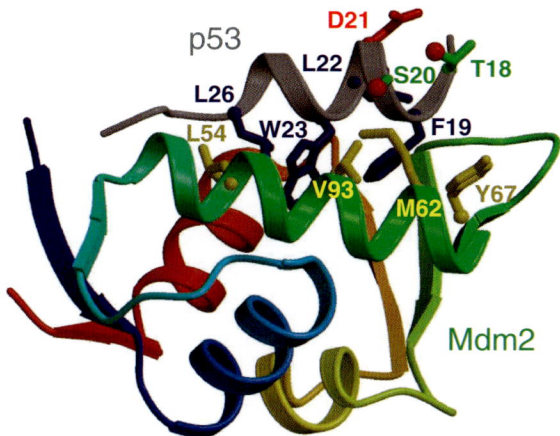

Figure 2.2. Three-dimensional structure of residues 17–29 of human p53 bound to the N-terminus of Mdm2. The hydroxyl groups of Thr18 and Ser20 of human p53 are colored red.

15-residue p53 peptide (Fig. 2.2). The region of Mdm2 that binds to p53 comprises an independently-folding domain consisting of a four helix bundle flanked on either side by β-sheets. The p53 peptide corresponds to amino acids 15–29 of the full-length protein. The N-terminal end of this peptide was not structured and was therefore invisible in the solved structure. The rest of the peptide corresponding to residues 17–29 of full-length p53 was structured and visible; most of this segment folds as an α-helix that interacts with a deep cleft on the Mdm2 surface. The interaction between the two proteins is mediated by hydrophobic interactions; residues Phe19, Trp23 and Leu26 of p53 interact with multiple conserved Mdm2 hydrophobic residues, including Leu54, Met62, Tyr67 and Val93, whose side chains line the cleft on the Mdm2 surface. Thus, the amphipathic nature of the p53 helix (hydrophobic on the side that interacts with Mdm2 and hydrophilic on the side exposed to solvent) is critical for the interaction.

The p53-Mdm2 structure explains how various phosphorylation events, induced in response to DNA damage, decrease the affinity between the two proteins (Chehab et al., 1999; Craig et al., 1999; Sakaguchi et al., 2000). The hydroxyl group of Ser20 is close to the side chain of Met62 of Mdm2; thus, Ser20 phosphorylation would lead to a steric clash, as well as position a negatively charged phosphate group next to the Met hydrophobic side chain. Phosphorylation of Thr18 is also predicted to weaken the interaction between p53 and Mdm2. The hydroxyl group of the Thr18 side chain forms a hydrogen bond with the carboxyl group of Asp21; this hydrogen bond stabilizes the amphipathic α-helix and would be disrupted by phosphorylation of Thr18. The structure further suggests that phosphorylation of Ser15 is unlikely to directly affect the interaction between p53 and Mdm2, because Ser15 is not involved in the p53-Mdm2 interface. However, Ser15 phosphorylation may have an effect *in vivo*, because it enhances the affinity of p53 for p300 and CBP, which compete with Mdm2 for binding to p53 (Shieh et al., 1997; Lambert et al., 1998).

The p53-Mdm2 structure further explains how amino acid substitutions that replace Leu22 and Trp23 of p53 with Gln and Ser, respectively, abolish the interaction between p53 and Mdm2 (Lin et. al., 1994). *In vivo* these two substitutions make p53 refractory to Mdm2-dependent regulation (Chehab et al., 1999). The same substitutions also compromise the transcriptional activity of p53 (Lin et al., 1994) suggesting that the interaction of p53 with p300 and CBP may also involve an amphipathic helix from p53 binding to a hydrophobic cleft on the surface of these transcription factors.

The p53 amphipathic helix that interacts with Mdm2 (residues 18–26) spans only a small part of the p53 transactivation domain (residues 1–70). In some transactivation assays substitutions targeting Leu22 and Trp23 compromise, but do not abolish p53 transcriptional activity. Complete loss of transcriptional activity requires additional substitutions targeting residues Trp53 and Phe54 of p53 (Candau et al., 1997), suggesting that the p53 transactivation domain encompasses multiple elements with which it can interact with transcription factors and other proteins (such as Mdm2).

While the structure of the p53-Mdm2 complex displays a specific p53 conformation, in the absence of Mdm2 the p53 transactivation domain adopts mostly an unstructured conformation (Botuyan et al., 1997; Lee et al., 2000). The α–helical fold is stabilized when p53 binds to Mdm2, because Mdm2 provides a hydrophobic cleft that favors partitioning of the p53 hydrophobic and hydrophilic residues on distinct surfaces (the surfaces towards Mdm2 and solvent, respectively). The spacing of the hydrophobic residues on the p53 primary sequence is such that they partition on one surface when they adopt an α–helical fold. Ligand-receptor interactions, in which binding is associated with conformational changes, are referred to as induced-fit (Koshland, 1958). The conformational flexibility of the p53 transactivation domain may have been selected to allow p53 to interact with multiple proteins, since each p53-protein complex may stabilize a specific p53 conformation suitable for that specific interaction. Further, because of the conformational flexibility, the p53 transactivation domain residues can adapt to the active sites of various enzymes (kinases, acetylases, etc.), which explains why many post-translational modifications of p53 map to the transactivation domain (Vogelstein et al., 2000). As discussed above, these modifications can favor or disfavor specific p53-protein interactions and therefore regulate p53 function.

2.1.2. The DNA Binding Domain

2.1.2.1. Overall Structure and DNA Contacts

Probably the most interesting domain of the p53 tumor suppressor protein is its sequence-specific DNA binding domain, which encompasses amino acids 94–293 of the full-length protein and which is targeted by the vast majority of cancer-associated *p53* mutations (Vogelstein et al., 2000). The structure of the p53 DNA binding domain has been solved in complex with DNA and also in the absence of DNA (Cho et al., 1994; Zhao et al., 2001).

The scaffold of the p53 DNA binding domain is a β–sandwich formed by two antiparallel β–sheets that pack against each other (Fig. 2.3). One end of the β–sandwich is formed by evolutionarily weakly conserved loops that join β–strands from the opposing β–sheets. In contrast, the other end of the β–sandwich is characterized by the presence of conserved secondary structure elements interspersed within the loops

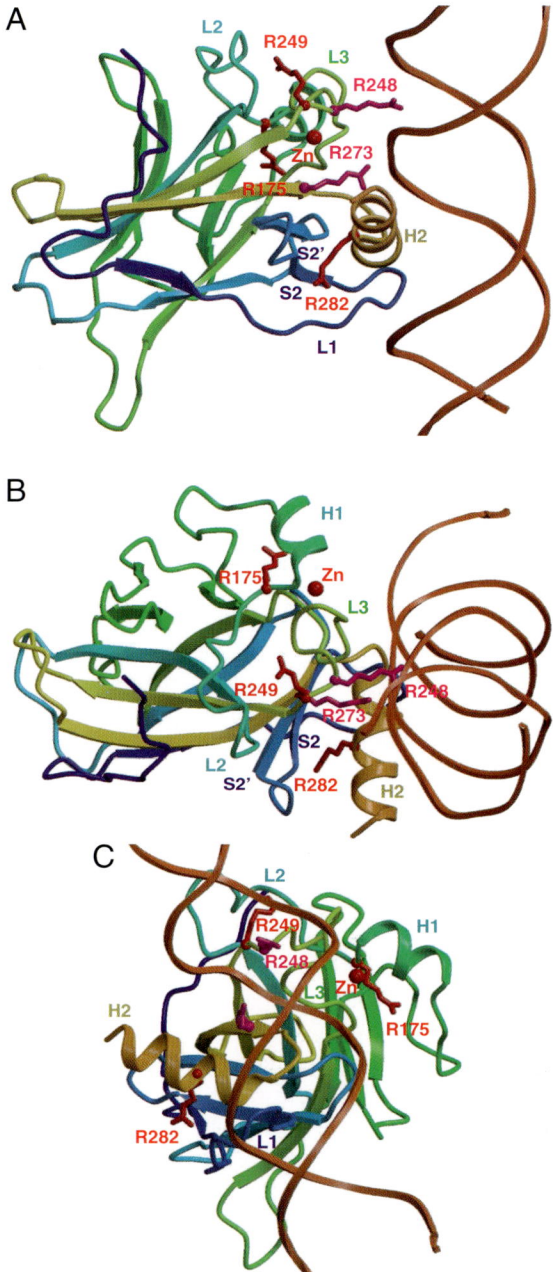

Figure 2.3. Three-dimensional structure of the DNA binding domain of human p53 in complex with DNA. Select secondary structure elements are marked (H1, H2, L1, L2, and L3). The side chains of residues that are frequently substituted in human cancer are shown. Of the residues corresponding to mutation hotspots, Arg175, Arg249, and Arg282 (red side chains) stabilize the native structure; whereas, Arg248 and Arg273 (purple side chains) contact DNA. The DNA backbone is shown as a ribbon. A, B, and C are orthogonal views.

29

that join the β–sheets. The first such element is a β–hairpin formed by two antiparallel β–strands (S2 and S2′). The second such element is a short α–helix (H1) that contains two Zn-chelating residues, which together with two other Zn-chelating residues contributed by loop L3, coordinate a Zn atom. The third element is a conserved α–helix (H2) that extends to the C-terminal end of the DNA binding domain. The presence of conserved secondary structure elements decorating one end of the β–sandwich clearly suggests that this is the functionally important part of the domain. Indeed, this is the side of the domain that contacts DNA, as revealed by the structure of the DNA binding domain in complex with DNA (Fig. 2.3).

The DNA sequence recognized by p53 consists of four tandem copies of the pentamer consensus sequence GGGCA arranged head-to-tail (El-Deiry et al., 1992). Two such pentamers comprising a p53 half-site are shown in Fig. 2.4A. To facilitate describing the structure of p53 bound to DNA, we assigned a number to the position of each nucleotide in the pentamer. The complementary nucleotides are indicated by the same number and an apostrophe (Fig. 2.4A). Each pentamer within the 20-nucleotide binding site is recognized predominantly by a single DNA binding domain (since p53 is a homotetramer). The contacts with the pentamer sequence are mediated by loops L1 and L3, strand S10, helix H2 and the loop that connects S10 to H2 (hereafter referred to as loop L4). Loops L1, L4, strand S10 and helix H2 contact the major groove of DNA (Fig. 2.4). From loop L1, the side chain of Lys120 makes a sequence-specific contact with guanine G4 and the amide nitrogen of the same residue contacts the phosphate backbone. From S10, the side chain of Arg273 contacts the phosphate backbone. From loop L4, the amide nitrogen of Ala276 contacts the phosphate backbone and the side chain of Cys277 makes a specific contact with cytosine C3′. From H2, the side chain of Arg280 makes a specific contact with guanine G2′, whereas the side chain of Arg283 contacts the phosphate backbone. Loop L3 interacts with the minor groove of DNA. The side chain of Ser241 contacts the phosphate backbone, whereas the side chain of Arg248 makes four contacts with DNA backbone sugar and phosphate atoms. The ability of Arg248 to make multiple contacts with the DNA backbone requires significant compression of the minor groove. In turn, this might explain the preference for adenine at position 1 of the consensus binding site, since adenine:thymidine pairs favor compression of the minor groove of double-stranded DNA (Yoon et al., 1988).

Taking into account the preference for adenine:thymidine bases at position 1 of the GGGCA pentamer to allow compression of the minor groove, it becomes evident that the structure explains binding specificity for positions 1, 2, 3 and 4 of the pentamer. This is in fact consistent with mutagenesis data suggesting that the nucleotide at position 5 contributes very little to the sequence-specificity of p53 DNA binding (El-Deiry et al., 1992; Halazonetis et al., 1993). Additional mutagenesis data further supports the p53–DNA structure solved by N. Pavletich (Cho et al., 1994). Extensive mutagenesis of the mouse p53 DNA binding domain identified the same residues as being important for DNA binding as the crystal structure (Halazonetis and Kandil, 1993). Further, mutagenesis of Lys120 changes the specificity of p53 for the nucleotide at position 4 of the pentamer (Freeman et al., 1994), whereas mutagenesis of Arg273 decreases the affinity of p53 for DNA, but does not change sequence-specificity (Wieczorek et al., 1996), consistent with the structure, which

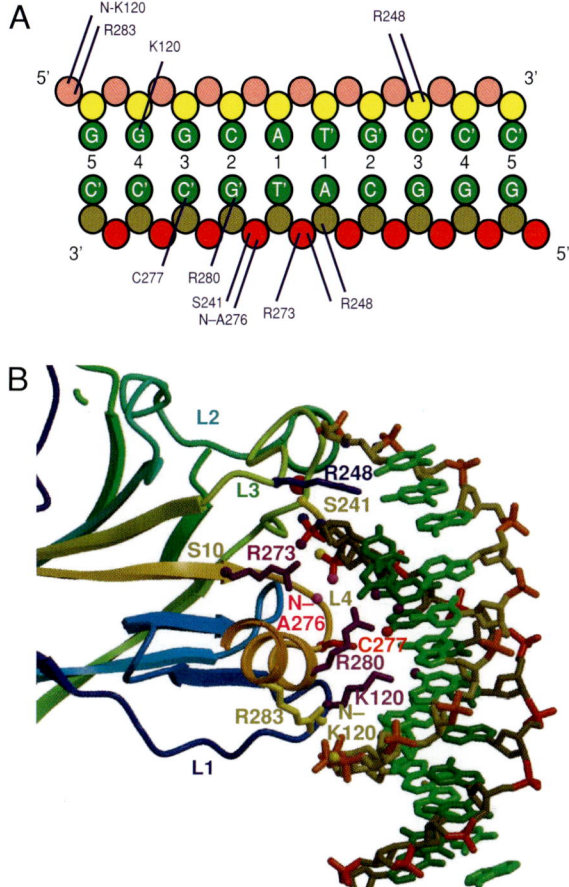

Figure 2.4. Human p53–DNA contacts. **A**: Diagram of a p53 half-site consisting of two head-to-tail pentamer repeats. Two tandem half-sites constitute a complete p53 binding site. The nucleotides in the GGGCA pentamer are numbered 5, 4, 3, 2, and 1, respectively. The consensus pentamer has pyrimidine at position 1, cytosine at position 2, and purines at positions 3, 4, and 5. The lines indicate contacts of p53 residues with specific atom groups of the p53 DNA binding site. The green circles represent the bases, the yellow circles represent the sugars and the red circles the phosphate groups. Contacts mediated by amide nitrogens are indicated by the p53 residue number preceded by an N- prefix. **B**: Three-dimensional structure of the p53 DNA binding domain in complex with DNA highlighting the p53–DNA contacts. The orientation is similar to the one shown in Figure 2.3A. The p53 residues (side chains or amide nitrogen) are shown in the same color as the DNA atoms they contact. Select secondary structure elements are marked (L1, L2, L3, L4, and S10).

shows Lys120 in contact with the base at position 4 and Arg273 in contact with the DNA phosphate backbone. In fact, the Arg273 substitution can be rescued by substituting Thr284 in helix H2 with Arg. Thr284 does not contact DNA in the solved p53–DNA structure, but its substitution with Arg is predicted to allow a new contact to be established with the DNA backbone (Wieczorek et al., 1996).

2.1.2.2. Mapping Cancer-Associated Mutations on the DNA Binding Domain

Most of the tumor-associated *p53* mutations target the sequence-specific DNA binding domain (Vogelstein et al., 2000). Certain residues within this domain are targeted much more frequently than others (Hollstein et al., 1991). The structure of the p53 DNA binding domain bound to DNA provides an explanation: the so-called "hotspot" residues contribute critically to DNA binding activity. Generally, p53 mutants have been divided into two classes (Cho et al., 1994). Class I mutants substitute residues that directly contact DNA, such as Arg248 and Arg273, whereas Class II mutants substitute residues, such as Arg175, Arg249 and Arg282, that stabilize the native structure of the p53 DNA binding domain (Fig. 2.3).

The nature of the amino acid substitutions explains the different properties of Class I and II mutants. Since Class I mutants target surface residues, the proteins encoded by these mutants are typically able to adopt the native fold. In contrast, Class II mutants, by definition, cannot adopt the native fold and are either partially (i.e. locally) or completely unfolded. Accordingly Class II mutants tend to be sequestered in the cytoplasm in complex with protein chaperones, such as the heat shock protein hsc70 (Finlay et al., 1988). Furthermore, Class II mutants react with antibodies, such as PAb240, that do not react with wild-type p53. These antibodies recognize epitopes that are buried in native p53, but become exposed in the unfolded p53 mutants (Gannon et al., 1990). The realization that Class II tumor-derived p53 mutants are actually unfolded proteins needs to be considered when interpreting their various activities. Unfolding exposes "sticky" hydrophobic residues through which Class II mutants can bind non-specifically to a wide array of proteins.

There are at least three reasons to explain why the tumor-associated mutations target predominantly the DNA binding domain of p53. The first reason is that mutations in the DNA binding domain generate dominant negative mutants. This is because p53 binds DNA as a homotetramer and tetramers containing both mutant and wild-type subunits are defective in DNA binding (Bargonetti et al., 1992; Halazonetis and Kandil, 1993). A second reason is the low melting temperature of the sequence-specific DNA binding domain of p53, which for human p53 is just a few degrees above 37°C (Bullock et al., 1997). The low melting temperature of the p53 DNA binding domain means that practically every non-conservative substitution in its core will decrease the melting temperature below 37°C preventing the protein from folding. The third reason is that many of the functionally important residues of this domain are arginines. Arginine codons are particularly susceptible to mutagenesis, as they contain CG dinucleotides, which, when damaged, are repaired with lower fidelity than other dinucleotides (Pfeifer et al., 2002).

2.1.3. The Oligomerization Domain

2.1.3.1. Overall Structure

The oligomerization domain of p53 resides within its C-terminus between residues 324 to 355. Its three-dimensional structure has been solved by X-ray

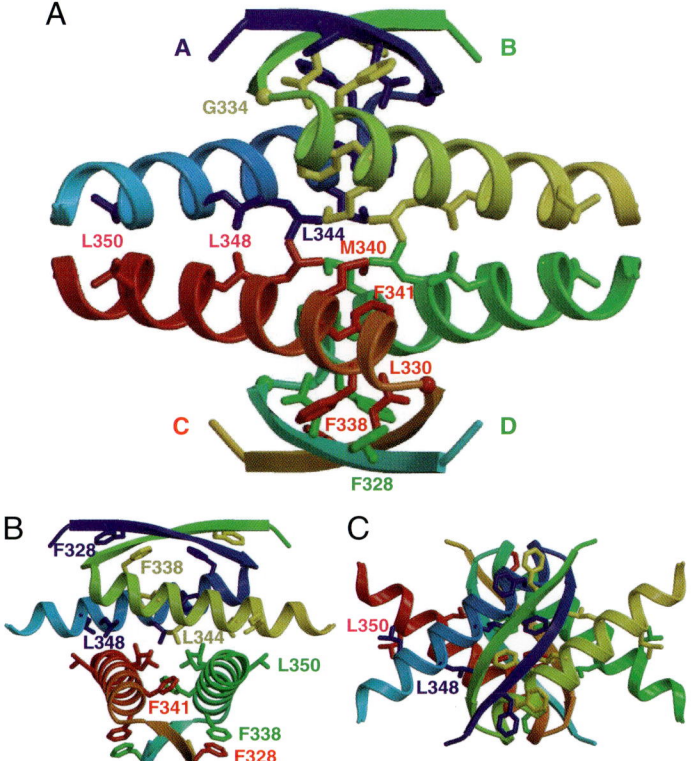

Figure 2.5. Three-dimensional structure of the human p53 oligomerization domain. The four subunits are labeled A, B, C, and D. The side chains of hydrophobic residues that mediate important inter-subunit interactions are shown. A, B, and C represent three different views.

crystallography and NMR spectroscopy (Lee et al., 1994; Jeffrey et al., 1995; Clore et al., 1995). The four monomer subunits adopt identical conformations and are related to each other by three perpendicular axes of symmetry (dihedral symmetry). Each monomer subunit of the p53 oligomerization domain consists of a β–strand (residues 326–333) and an α–helix (residues 335–354) that together form a V-like shape (Fig. 2.5). A conserved glycine residue (G334) allows a tight turn to be formed between these two secondary structure elements.

The interactions between the four subunits are extensive and are primarily hydrophobic in nature. The subunits are typically labeled A, B, C and D. In the dimer formed by subunits A and B (or the symmetrically equivalent subunits C and D) the β–strands pack antiparallel to each other and are stabilized by inter-subunit hydrophobic interactions between Phe328 and Phe338 and between the two Leu330 residues (Fig. 2.5). Because of the V-like shape of each subunit, the α–helices of subunits A and B also pack antiparallel to each other. Packing of the helices against the β–sheet is stabilized primarily by hydrophobic interactions mediated by Phe341.

In the dimer formed by subunits A and C (or the symmetrically equivalent subunits B and D) there are no interactions between the β–strands, but there are extensive interactions between the C-termini of the α–helices mediated primarily by Leu348 and Leu350 (Fig. 2.5). Interestingly, in this part of the structure the helices pack parallel to each other. Thus, the α–helices of the p53 oligomerization domain have the capacity to form both antiparallel (between subunits A and B) and parallel (between subunits A and C) interactions.

The central part of the α–helices is involved in stabilizing the interaction between the A-B and C-D dimers. Residues Met340 and Leu344 from all four subunits interact with each other at the very center of the domain and are responsible for the tetramer stoichiometry (Fig. 2.5).

2.1.3.2. Mapping Cancer-Associated Mutations on the Oligomerization Domain

The frequency of mutations targeting the p53 oligomerization domain in human cancer is at least 100-fold lower than the frequency of mutations targeting the DNA binding domain (Hollstein et al., 1991). There are several explanations for this observation. First, the p53 oligomerization domain is thermodynamically very stable, since its melting temperature is about 70°C (Johnson et al., 1995). Given this high melting temperature, most single amino acid substitutions cannot lead to unfolding of the domain. In fact, the most common cancer-associated substitution in the p53 oligomerization domain targets Gly334; this residue is critical for the tight turn between the β–strand and α–helix and its substitution with practically every other residue is sufficient to unfold the domain (Lee et al., 1994). The second reason for the low frequency of mutations targeting the oligomerization domain is that inactivating the oligomerization domain does not lead to dominant negative mutants (Shaulian et al., 1992). In contrast, p53 mutants with inactive DNA binding domains retain the ability to form hetero-oligomers with wild-type p53 and, consequently, have dominant negative activity.

2.1.3.3. Determinants of Oligomerization Stoichiometry

The structure of the p53 oligomerization domain allows us to visualize the residues that mediate inter-subunit interactions and therefore the residues that determine the stoichiometry of this domain as a tetramer. Because of the dihedral symmetry it is possible to disrupt interactions between subunits A and C (and the equivalent B and D) without affecting the interactions between subunits A and B (or the equivalent C and D). This has the effect of dissociating a p53 tetramer into A-B and C-D dimers. Amino acid substitutions that have this effect include Leu344 to Ala and a double substitution of Met340 to Gln and Leu344 to Arg (Waterman et al., 1995; Davison et al., 2001).

A more interesting way to change the stoichiometry of the p53 oligomerization domain from tetramer to dimer involves substitution of two hydrophobic residues, Phe341 and Leu344, at the core of the domain with other hydrophobic residues of different side chain size (McCoy et al., 1997). Whenever the side chain of the residue at position 344 becomes larger than the side chain of the residue at position 341, the stoichiometry of the domain switches from tetramer to dimer (Fig. 2.6). In this case

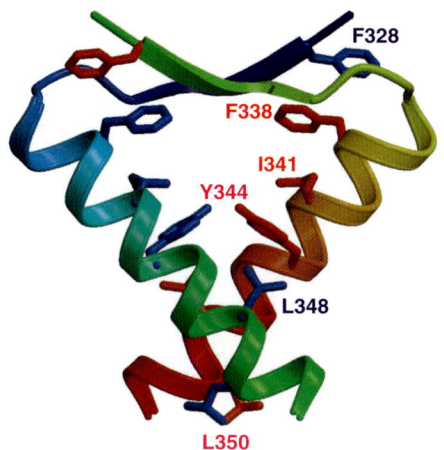

Figure 2.6. Three-dimensional structure of a mutant p53 oligomerization domain with dimeric stoichiometry. In this mutant residues Phe341 and Leu344 are substituted with Ile and Tyr, respectively.

the change in stoichiometry is accompanied by a change in the fold of the domain; the dimer retains the antiparallel packing of the β–strands (compare Fig. 2.6 to Fig. 2.5B), but the α–helices pack parallel to each other, as the overall shape of each subunit changes from V-like to L-like. Interestingly, the packing of the C-termini of the α–helices in this mutant p53 dimer is practically identical to the way the C-termini of the α–helices of subunits A and C pack in the wild-type structure (compare Fig. 2.6 to Fig. 2.5C). To our knowledge this is the only example of designed amino acid substitutions that change the fold of a protein domain. One implication of these findings is that the side chain size of hydrophobic residues is an important determinant of protein fold.

2.1.3.4. Determinants of Oligomerization Specificity

p53 belongs to a family of proteins with conserved sequence-specific DNA binding and oligomerization domains. While under certain conditions p53 can hetero-oligomerize with its family members, under physiological conditions the majority of p53 is thought to exist as homotetramers (Davison et al., 1999; Mateu et al., 1999). In turn, this implies oligomerization specificity. Specificity in the interaction between various subunits is generally thought to be mediated by charged and polar residues, which are capable of forming electrostatic interactions and hydrogen bonds. The p53 oligomerization domain has very few electrostatic interactions between the subunits. In fact, the oligomerization specificity of p53 can be changed by amino acid substitutions that target its hydrophobic residues. Substitutions that weaken existing strong hydrophobic interactions between the subunits coupled with substitutions that enhance existing weak hydrophobic interactions create variant p53 oligomerization domains that form homotetramers, but interact very weakly with the wild-type domain (Mateu et al., 1999; Stavridi et al., 1999). These results suggest that hydrophobic interactions are critical determinants of both the fold and oligomerization specificity of the p53 oligomerization domain.

2.1.4. The N-terminal Proline-Rich Region

As shown in Fig. 2.1 the p53 domains discussed above are separated from each other by short linker regions. The most N-terminal of these regions, spans residues 71–93 and links the transactivation and DNA binding domains. This region has ten prolines and is often referred to as the polyproline region.

Functional experiments suggest that the polyproline region contributes to the functional activity of p53 and especially to its ability to induce apoptosis (Walker et al., 1996; Sakamuro et al., 1997; Ruaro et al., 1997; Venot et al., 1998). There are at least two ways to explain the functional significance of this region. Regions rich in prolines are often ligands for proteins containing SH3 domains (Yu et al., 1994). Thus the polyproline region may mediate an interaction between p53 and a protein containing an SH3 domain. While definitive evidence for a protein that binds to the p53 polyproline region is not yet available, 53BP2, an SH3 domain-containing protein, is known to bind p53 (Iwabuchi et al., 1994). The interaction between 53BP2 and p53 will be described further below in this Chapter. A second way by which the polyproline region could affect p53 function might be indirect through an effect on p53 folding. Proline-rich regions have conformationally-constrained backbones and may affect the kinetics of protein folding by acting as folding barriers separating two protein domains. As discussed above the DNA binding domain of p53 has a very low melting temperature; it is therefore possible that the polyproline region may be needed to facilitate its folding. Such a role has not yet been experimentally demonstrated for p53, but in other proteins polyproline regions have been shown to facilitate folding (Kusano et al., 2001; Wang et al., 2002).

2.1.5. The C-terminal Proline-Rich Region

The C-terminal proline-rich region spans residues 294–323 and links the DNA binding and tetramerization domains. This region is not described in the literature as being proline-rich, but has seven prolines. The primary function of this region is to accomodate the different symmetries of the p53 DNA binding and oligomerization domains in p53 tetramers bound to DNA (discussed below). However, as discussed above, a second function of the C-terminal proline-rich region might be to facilitate folding of the p53 DNA binding domain. As in the case of the N-terminal proline-rich region, such a function needs to be experimentally documented for p53.

2.1.6. The C-terminal Basic Region

The C-terminal basic region of p53 spans residues 356–393. As its name indicates this region is rich in basic amino acids. Like the N-terminal transactivation domain, this region is conformationally flexible (Ayed et al., 2001) and is subject to multiple post-translational modifications that regulate p53 function (Appella and Anderson,

2000; Brooks and Gu, 2003; Xu, 2003). Because of its basic nature, this region resembles histone tails. As such it appears to be modified by the same enzymes that modify histone tails and is subjected to similar post-translational modifications (phosphorylation and acetylation). The functional significance of these modifications is beyond the scope of this chapter. *In vitro*, they enhance the sequence-specific DNA binding activity of wild-type p53 (Appella and Anderson, 2000; Brooks and Gu, 2003; Xu, 2003) and a similar effect can be achieved by deleting this region or masking it with a monoclonal antibody (Hupp et al., 1992; Halazonetis et al., 1993). It is unclear, however, whether these modifications enhance the sequence-specific DNA binding activity of p53 *in vivo*. Instead, *in vivo* the C-terminal post-translational modifications may enhance the transcriptional activity of p53 by facilitating the recruitment of transcriptional coactivators to p53-target genes; these coactivators, many of which are histone acetyltransferases, can modify both p53 and histones leading to "opening up" of the chromatin structure, which is thought to be required for efficient transcription (Barlev et al., 2001; Espinosa and Emerson, 2001; Wang et al., 2001).

The structure of the p53 C-terminal basic region has been solved in complex with two proteins: Sir2, a protein deacetylase (Avalos et al., 2002), and S100B, a calcium-binding protein (Rustandi et al., 2000). Since the p53 C-terminal basic region on its own is conformationally flexible its interaction with Sir2 and S100B is governed by the "induced-fit" model. Thus, the p53 C-terminal basic region adopts conformations that fit the ligand-binding sites in Sir2 and S100B. As shown below these two conformations are completely different from each other.

Sir2 is a member of the evolutionarily highly conserved family of sirtuin NAD+-dependent deacetylases. This family of proteins is implicated in a variety of functions related to DNA, such as transcriptional silencing, DNA repair and chromosomal stability (Guarente, 2000; Denu, 2003). Members of this family catalyze the removal of acetyl groups from the ε-amino group of lysines in reactions that also yield nicotinamide and O-acetyl-ADP-ribose (Sauve et al., 2001) as products. Acetylated histones are physiological substrates of yeast Sir2 and their deacetylation leads to transcriptional silencing. As mentioned above, the C-terminal basic region of p53 is acetylated *in vivo* by the same enzymes that acetylate histone tails. This observation suggested that Sir2 proteins might also deacetylate p53, a prediction now confirmed by three laboratories (Luo et al., 2001; Vaziri et al., 2001; Langley et al., 2002).

The structure of a peptide corresponding to residues 372–389 of human p53 with an acetylated lysine at position 382 was solved in complex with an archaeal Sir2 protein (Avalos et al., 2002). In this complex, residues 379–387 of p53 were structured and the p53 peptide adopts the secondary structure of a β–strand, which together with two β–strands of Sir2 (strands 7 and 9) participates in a three-strand β–sheet (Fig. 2.7). Remarkably, other than for the side chain of acetylated Lys382, the contacts between Sir2 and the p53 peptide are not sequence-specific; rather the interaction is mediated by backbone atom hydrogen bonds of the kind present in all β–sheets. Thus, the structure suggests that Sir2 enzymes can deacetylate practically any acetylated peptide, a prediction confirmed by enzymatic assays involving several Sir2 proteins, including human SIRT2 (Avalos et al., 2002). One would have to assume

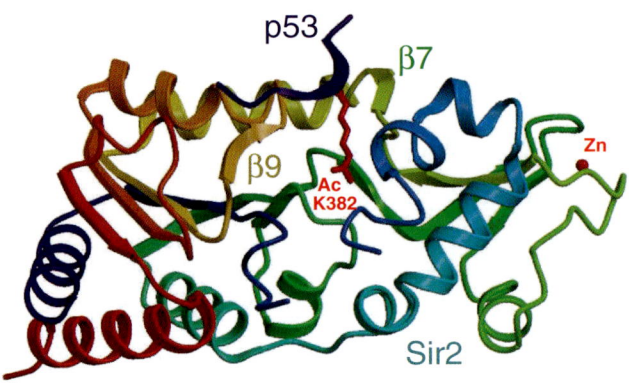

Figure 2.7. Three-dimensional structure of archaeal Sir2 in complex with a peptide corresponding to residues 379–387 of human p53 with an acetylated lysine at position 382 (Ac-K382). The side chain of this residue is shown in red.

that substrate specificity for Sir2 enzymes *in vivo* will be mediated by interactions between the substrate and enzyme that are further away from the enzyme active site; such interactions are not evident in structures containing short peptides as substrates. Thus, a structure of a longer p53 polypeptide bound to a human Sir2 protein is needed to understand how p53 is targeted by human Sir2 *in vivo*.

S100B belongs to the so-called EF-hand family of calcium binding proteins. This family includes well-known proteins such as calmodulin and troponin and its members change conformation in response to changes in the intracellular concentration of calcium ions (Yap et al., 1999; Lewit-Bentley and Rety, 2000; Donato, 2001). Many EF-hand proteins, including S100B, are involved in various cytoskeletal functions, such as contractile activity of muscle cells and in the dynamics of actin filaments. The three-dimensional structure of S100B has been solved in the unbound and calcium-bound forms and shows the protein as a homodimer with each subunit consisting of four α–helices. Upon calcium binding helix 3 changes orientation exposing a patch of hydrophobic residues that serves as binding site for various proteins (Drohat et al., 1996; Kilby et al., 1996; Matsumura et al., 1998). S100B binds to the actin capping protein CapZ and the structure of calcium-bound S100B has been solved in complex with a 12-residue peptide corresponding to residues 265–276 of CapZ (Inman et al., 2002). In this structure the CapZ peptide adopts an α–helical conformation and its interaction with S100B involves several CapZ hydrophobic residues, including Ile269, Trp271, Ile274 and Leu275 (Fig. 2.8A). Trp271 is located in a deep hydrophobic cleft of S100B and is conserved in the majority of S100B binding peptides.

S100B was discovered to bind p53 as part of studies examining the phosphorylation of the p53 C-terminal basic region by the calcium and phospholipid-dependent protein kinase PKC. As a result of these studies binding between p53 and S100B was demonstrated *in vitro* and was shown to involve the C-terminal basic region of p53 (Baudier et al., 1992). The structure of S100B bound to a p53 peptide containing residues 367-388 of human p53 was subsequently solved by NMR spectroscopy (Rustandi et al., 2000). In this structure the p53 peptide adopts an α–helical conformation

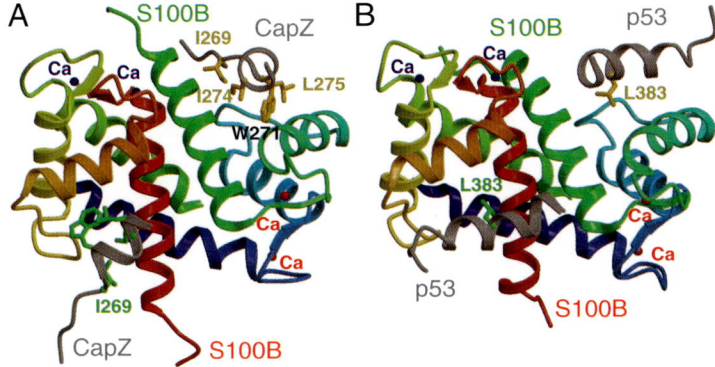

Figure 2.8. Three-dimensional structure of S100B in complex with a CapZ (A) or a human p53 (B) peptide. S100B is a symmetric dimer, but the view shown is not along an axis of symmetry, so the two S100B subunits and their bound peptides are viewed from different angles.

(Fig. 2.8B). However, the orientation of the α–helix is different from the orientation of the CapZ helix in the S100B-CapZ structure (compare Fig. 2.8A to Fig. 2.8B). Interestingly, only a single p53 hydrophobic residue, Leu383, is present in the S100B hydrophobic cleft and p53 lacks the tryptophan residue that is present in the consensus sequence of S100B-binding peptides. Thus, this structure does not provide strong support for the hypothesis that S100B is a physiologically important regulator of p53 function *in vivo*. Further studies are needed to address this question; such studies can now be designed with help from the S100B-p53 structural data.

2.2. MODELS FOR THE STRUCTURE OF FULL-LENGTH p53 HOMOTETRAMERS

As described above the three-dimensional information regarding p53 derives from analysis of isolated domains and not from analysis of full-length protein. In principle, it should be straightforward to model the structure of a full-length p53 tetramer bound to DNA using the available structures of the isolated domains. However, more careful analysis reveals that a model of a p53 tetramer cannot be easily derived from the available structures. There are two problems that need to be considered. The first problem relates to the discrepancy between the symmetry of the various p53 subunits relative to each other in a p53 homotetramer and the symmetry of the p53 DNA binding sites in DNA. This discrepancy, which is explained below, suggests that p53 undergoes global changes in its conformation as it binds DNA. The second problem relates to whether four p53 sequence-specific DNA binding domains can bind DNA without steric clashes, when the crystal structure of the p53 monomeric DNA binding domain bound to DNA is used to model the structure of a p53 tetramer bound to DNA. As described below, computer modeling reveals steric clashes implying that binding of p53 tetramers to DNA is associated with local conformational changes either in the DNA or in the p53 DNA binding domains or in both.

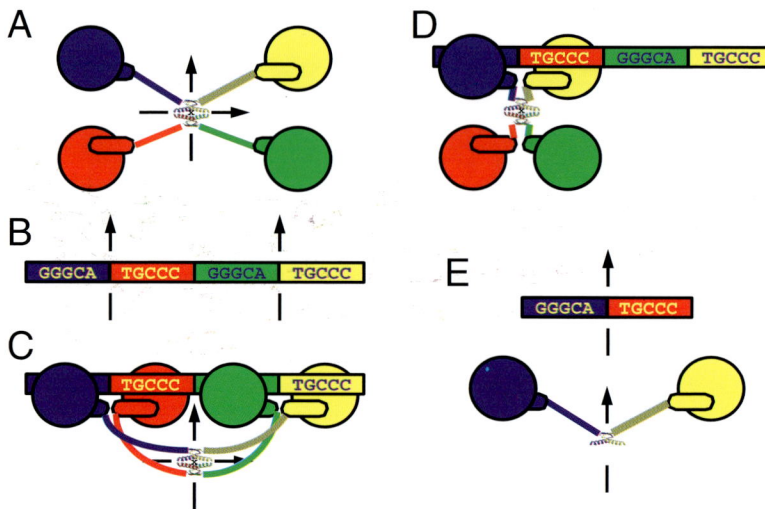

Figure 2.9. Symmetry-related p53 conformational changes associated with DNA binding. **A**: Dihedral symmetry (three perpendicular axes of symmetry) of a p53 homotetramer in the absence of DNA. **B**: Cyclic-translation symmetry of pentamers comprising a p53 binding site. **C**: Change in the conformation of the C-terminal proline-rich region allows the DNA binding domains to adopt cyclic-translation symmetry, while the oligomerization domain maintains dihedral symmetry. **D**: Deletions that shorten the C-terminal proline-rich region prevent p53 from binding to a full site, but allow binding to half-sites. **E**: Both p53 dimers and DNA half-sites exhibit cyclic symmetry (single axis of symmetry). The DNA binding domains of p53 are shown as spheres with a cylinder representing helix H2. Arrows indicate axes of symmetry. Axes of symmetry perpendicular to the page are indicated by a letter x and are found at the center of the p53 oligomerization domain.

2.2.1. Global (Symmetry-Related) p53 Conformational Changes Associated with DNA Binding

The need for global p53 conformational changes associated with DNA binding arises because the p53 homotetramer and its DNA binding site have different symmetries. The symmetry of p53 is dictated by the symmetry of its oligomerization domain. As mentioned above this domain has dihedral symmetry, which means that the four subunits are related to each other by three axes of symmetry that are perpendicular to each other (Fig. 2.5). Assuming that all four p53 subunits adopt the same conformation (the one that is energetically most favored), then the entire p53 tetramer, including the DNA binding domains, will adopt dihedral symmetry (Fig. 2.9A).

The p53 DNA binding site consists of four tandem pentamers arranged head-to-tail. The symmetry relating these pentamers can be described as cyclic-translation: within each half-site the repeats are related by cyclic symmetry via a single rotation axis and the two half-sites are related to each other by translation (Fig. 2.9B).

Because p53 and its DNA binding site have different symmetries, DNA binding must be associated with a conformational switch that allows the DNA binding domains to adopt the cyclic-translation symmetry of the DNA binding site. The proline-rich region that links the DNA binding domain and oligomerization domains is critical for resolving the symmetry discrepancy (Waterman et al., 1995). Thanks to its length and conformational flexibility, this region can adopt distinct conformations in each subunit, such that the DNA binding domains exhibit cyclic-translation symmetry, while the tetramerization domain retains dihedral symmetry (Fig. 2.9C). In support of this model, deletions within the C-terminal proline-rich region prevent p53 from binding to full-length DNA sites, although binding to DNA half-sites is still possible (Waterman et al., 1995; Fig. 2.9D).

The extent to which the DNA binding domains of full-length tetrameric p53 adopt a stable dihedrally-symmetric state in the absence of DNA (Fig. 2.9A) is not known. Based on *in vitro* observations that deletion of the C-terminal basic region of p53 enhances DNA binding, it was proposed that this region participates in inter-subunit interactions that stabilize the dihedrally-symmetric state of p53 (Hupp et al., 1995; Waterman et al., 1995). To address this model, the conformations of p53 molecules that have or do not have the C-terminal basic region were compared by NMR spectroscopy (Ayed et al., 2001). One of the p53 fragments that was examined included the sequence-specific DNA binding domain, the C-terminal proline-rich region and the oligomerization domain, while the other included in addition the C-terminal basic region. Both p53 fragments had amino acid substitutions within the oligomerization domain that disrupted the interactions between the A-B and C-D dimers; thus these p53 fragments assembled as dimers. Dimeric stoichiometry was necessary, because the sensitivity of NMR spectroscopy experiments decreases as protein size increases. The conformations of these two p53 fragments were compared, but not determined. This is possible, because the magnetic resonance frequencies of atoms are dependent on their local environment. Thus, similar magnetic resonance frequencies indicate similar conformations. The analysis revealed that the two p53 fragments adopted similar conformations, suggesting that the C-terminal basic region does not affect the conformation or symmetry of p53. However, these experiments do not completely rule out the possibility that the C-terminal basic region stabilizes a dihedrally-symmetric state of p53 for the following reasons. First, the C-terminal basic region of p53 may stabilize the dihedrally-symmetric state by interacting with the N-terminus, which was not present in the p53 constructs examined by NMR spectroscopy. Second, the interactions that stabilize the dihedrally symmetric state may occur only in the context of p53 tetramers and not in the context of p53 dimers. Third, p53 dimers do not need to resolve differences in p53 and DNA symmetries to bind DNA half-sites, since p53 dimers and DNA half-sites both have cyclic symmetry (Fig. 2.9E). Thus, the use of dimeric p53 proteins, while necessary for technical reasons, may not have been optimal to study the effects of the C-terminal basic region on p53 conformation.

2.2.2. Local (Sequence-Specific DNA Binding Domain and/or DNA) Conformational Changes Associated with p53 DNA Binding

The structure of the human p53 DNA binding domain was solved from crystals that contained in the asymmetric unit three DNA binding domains and one oligonucleotide containing two tandem pentamers (Cho et al., 1994). Only one of the three

DNA binding domains was bound to DNA in a sequence-specific manner, yet the structures of all three domains were virtually identical suggesting that binding of isolated DNA binding domains to DNA is not associated with any significant conformational change in the p53 protein structure (Cho et al., 1994). This conclusion is further supported by the structure of the mouse p53 DNA binding domain solved in the absence of DNA (Zhao et al., 2001); the mouse p53 structure is essentially identical to that of the human p53 DNA binding domain bound to DNA.

The absence of significant conformational changes associated with binding of isolated human p53 DNA binding domains to DNA does not preclude the possibility of conformational changes in the p53 DNA binding domain, when full-length p53 binds DNA as a homotetramer. In fact, attempts to model four p53 sequence-specific DNA binding domains bound to linear B-form DNA using the available crystallographic data reveals steric clashes between the four p53 sequence-specific DNA binding domains (Nagaich et al., 1997). One way to resolve the steric clashes is by bending the p53 DNA site and experimental data indeed suggests that p53 bends its target DNA upon binding (Balagurumoorthy et al., 1995; Cherny et al., 1999; Nagaich et al., 1999). However, it is also possible that the steric clashes are resolved by changes in the conformation of the p53 DNA binding domains themselves in a way that changes slightly their position relative to DNA. It is hard to predict whether the steric clashes are resolved by DNA bending, changes in the conformation of the p53 DNA binding domain or both. Interestingly, NMR analysis of isolated human p53 DNA binding domains examined in the absence and presence of specific p53 DNA sites reveals that DNA binding is accompanied by changes in p53 magnetic resonance frequencies of residues that are far away from the bound DNA (Klein et al., 2001; Rippin et al., 2002). These frequency shifts might be indicative of conformational changes in the p53 DNA binding domain, although other interpretations cannot be excluded. The definitive way to understand how p53 homotetramers resolve the steric clash problem would be to solve the structure of a p53 homotetramer bound to DNA. This has proven to be technically very difficult.

2.3. STRUCTURES OF p53 WITH 53BP1 AND 53BP2

53BP1 and 53BP2 (p53 binding proteins 1 and 2, respectively) were identified by a yeast two-hybrid screen for proteins that bind to p53 (Iwabuchi et al., 1994). Both 53BP1 and 53BP2 bind to overlapping surfaces of the p53 sequence-specific DNA binding domain, yet utilize different structural motifs to do so. Functional data link p53 to 53BP1 and 53BP2, but it still remains to be proven whether the functional interactions are due to direct protein-protein interactions.

2.3.1. The p53-53BP1 Interaction

53BP1 localizes rapidly to sites of DNA double-strand breaks after exposure of cells to ionizing radiation, where it is thought to activate ATM (Schultz et al., 2000; Mochan et al., 2003). Thus, 53BP1 functions in the same DNA damage checkpoint pathway as p53. Orthologs of 53BP1 are present in all eukaryotes (Weinert and Hartwell, 1988; Willson et al., 1997). The common structural feature among these orthologs are highly conserved C-terminal BRCT repeats. Deletion mapping analysis indicates that the interaction between p53 and 53BP1 is mediated by the sequence-specific DNA binding domain of p53 and the BRCT repeats of 53BP1 (Iwabuchi et al., 1994).

BRCT repeats are protein-protein interaction motifs. They are present in proteins that function in the cellular response to DNA damage (Bork et al., 1997). 53BP1 and its orthologs have two tandem BRCT repeats connected by a short linker region. The structure of the two tandem BRCT repeats of 53BP1 in complex with the sequence-specific DNA binding domain of 53 has been solved by X-ray crystallography (Joo et al., 2002; Derbyshire et al., 2002).

In the three-dimensional structure, the two BRCT repeats and the linker region between the repeats pack tightly against each other to form a single globular domain (Fig. 2.10A). Each BRCT repeat consists of a β–sheet surrounded by α–helices on

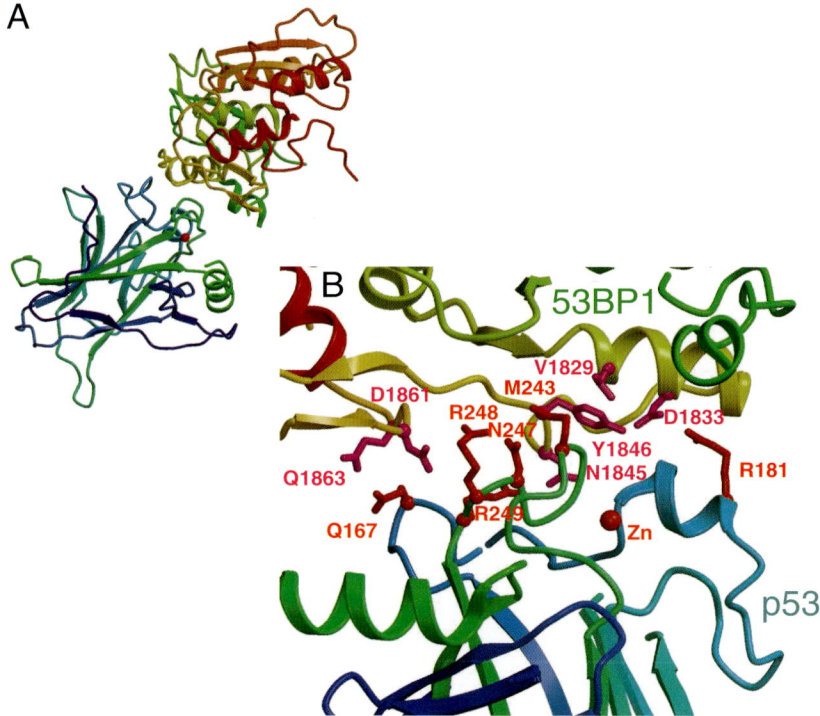

Figure 2.10. Three-dimensional structure of the p53 DNA binding domain in complex with the BRCT repeats of 53BP1. **A**: View of the protein domains with the p53 DNA binding domain shown in the same orientation as in Figure 2.3A. The evolutionarily conserved cleft at the interface of the two BRCT repeats of 53BP1 maps to the helices colored in red. **B**: p53–53BP1 interface showing the side chains of residues involved in the protein–protein interaction. The orientation is not the same as that shown in panel A.

either side. The interface between the two repeats is formed by conserved hydrophobic and charged interactions that allow helix 2 from the first repeat to form a three-helix bundle with helices 1 and 3 from the second repeat. This leads to a cleft on the surface of 53BP1 at the interface of the two BRCT repeats. The surface of this cleft is highly conserved in evolution. In BRCA1, another DNA damage checkpoint protein with two C-terminal tandem BRCT repeats, this cleft is targeted by most cancer-associated mutations (Williams et al., 2001; Joo et al., 2002).

In the p53-53BP1 structure the DNA binding domain of p53 does not contact the evolutionarily conserved cleft at the interface of the two BRCT repeats (Fig. 2.10A). Instead, the p53 DNA binding domain contacts a surface of 53BP1 formed by the N-terminal BRCT repeat and the linker region between the two BRCT repeats. From the p53 side, the interaction with 53BP1 is mediated primarily by the L3 loop and to a lesser extent by the L2 loop and H1 helix. Specifically, from loop L3 of p53, Met243 contacts Val1829 and Tyr1846 of 53BP1; Asn247 contacts the 53BP1 main chain; Arg248 contacts Asp1861; and Arg249 contacts Asn1845 (Fig. 2.10B). From loop L2 of p53, Gln167 contacts Gln1863 and from helix H1, Arg181 contacts Asp1833 of 53BP1 (Fig. 2.10B). The p53 residues that are key to its interaction with 53BP1, Arg248 and Arg249, are highly conserved in evolution; however, the evolutionary conservation may reflect the importance of these residues for sequence-specific DNA binding (Cho et al., 1994). Of the 53BP1 residues that are key to the interaction with p53 several are conserved in Xenopus and C. elegans. Nevertheless, as mentioned above, the p53 binding surface of 53BP1 is less well-conserved in evolution than the cleft at the interface between the two BRCT repeats.

The functional significance of the p53-53BP1 physical interaction is as yet unclear. Comparison of the surfaces of p53 involved in DNA and 53BP1 binding indicates significant overlap such that it would be impossible for a p53 sequence-specific DNA binding domain to contact simultaneously both DNA and 53BP1. This could imply that p53 interacts with 53BP1 transiently prior to or during the process of its activation by DNA damage. A transient interaction between p53 and 53BP1 might explain why an interaction between endogenous p53 and 53BP1 proteins has not yet been reported in human cells. Furthermore, a transient interaction might allow 53BP1 to recruit p53 to activated ATM and Chk2, where p53 is phosphorylated; activated p53 could then be released from 53BP1 and relocalize to the promoters of its target genes. The availability of the 53BP1-p53 structure will allow this putative model to be tested by facilitating the construction of p53 mutants that lose the capacity to interact with 53BP1, while retaining DNA binding activity. These p53 mutants can then be tested for activation in response to irradiation.

2.3.2. The p53-53BP2 Interaction

53BP2 was the second protein identified in the original yeast two-hybrid screen for proteins that interact with p53 (Iwabuchi et al., 1994). 53BP2 turns out to be the C-terminal fragment of a full-length protein termed ASPP2. In turn, ASPP2 belongs to a family of proteins that in humans includes 3 members, ASPP1, ASPP2 and iASPP (Samuels-Lev et al., 2001; Bergamaschi et al., 2003). All three proteins are characterized by the presence of 4 tandem ankyrin repeats followed by an SH3 domain in their C-terminus. In addition, they all bind the DNA binding domain of human p53, but have different activities; ASPP1 and ASPP2 promote the apoptotic activity of p53, whereas iASPP inhibits p53 activity. In *C. elegans* there is only one member of this family that most closely resembles human iASPP. Suppression of *C. elegans* iASPP leads to apoptosis in germ cells, which can be rescued by suppression of p53. Thus, in *C. elegans*, which lack Mdm2, the activity of p53 is curtailed by iASPP (Bergamaschi et al., 2003).

The structure of the C-terminus of ASPP2 (53BP2) has been solved in complex with the sequence-specific DNA binding domain of p53 (Gorina and Pavletich, 1996). The 53BP2 C-terminus contains four ankyrin repeats (Michaely and Bennett, 1992) and an SH3 domain (Cohen et al., 1995) (Fig. 2.11A). Each ankyrin repeat consists of a β–hairpin followed by two antiparallel β–helices arranged perpendicular to the

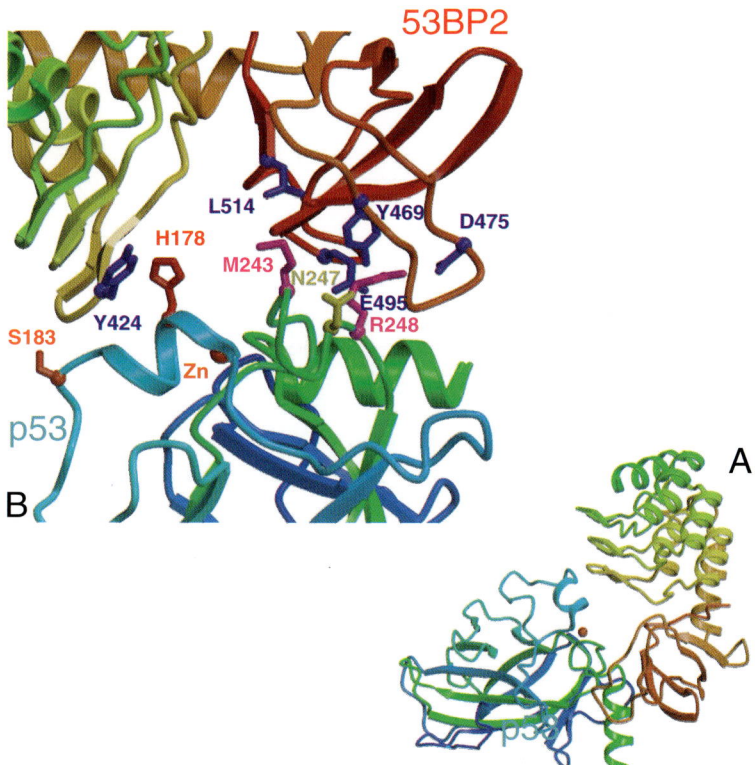

Figure 2.11. Three-dimensional structure of the p53 DNA binding domain in complex with the ankyrin and SH3 domains of 53BP2 (ASPP2). **A**: View of the protein domains with the p53 DNA binding domain shown in the same orientation as in Figure 2.3B. **B**: Interface of p53 with 53BP2 showing the side chains of residues involved in the protein–protein interaction.

β–hairpin creating the shape of the letter L. The four ankyrin repeats are tightly packed against each other, such that the β–hairpins form a continuous β–sheet and the β–helices form helix bundles. The C-terminal ankyrin repeat packs against the SH3 domain, which adopts the same β–strand-rich fold as previously described SH3 domains in other proteins.

The interaction between 53BP2 and p53 involves loops L2 and L3 and helix H1 of p53. In that regard, the interaction surface in p53 overlaps pretty well with the 53BP1 interaction surface and partially with the sequence-specific DNA binding surface. From the 53BP2 side, the interaction involves the SH3 domain, which interacts with loop L3 of p53, and the fourth ankyrin repeat, which interacts with loop L2 and helix H1 of p53.

The 53BP2 SH3 domain contains the characteristic peptide-binding groove present in SH3 domains (Fig. 2.11B). This groove is occupied by the L3 loop of p53, which makes electrostatic, hydrogen bond and hydrophobic interactions with 53BP2. Specifically, Arg248 of human p53 interacts with Asp475 and Glu495 of 53BP2; Asn247 of p53 interacts with Tyr469 of 53BP2 and Met243 of p53 interacts with Leu514 of 53BP2. The latter contact appears to be important for the specificity of the p53/53BP2 interaction. Most SH3 domains have a tyrosine at the position corresponding to Leu514 of ASPP1, which would not be able, due to its bulky side chain, to accomodate the Met243 side chain of human p53.

The fourth ankyrin repeat of 53BP2, via its β–hairpin, interacts with the L2 loop and helix H1 of p53 (Fig. 2.11B). This interaction involves hydrogen bonds mediated by residues Ser183 and His178 of p53 with backbone amides of 53BP2. The interaction is facilitated by the presence of a tyrosine insertion in the β–hairpin of 53BP2; this tyrosine (Tyr424) allows steric complementarity between p53 and 53BP2 and most likely contributes to the specificity of the p53/53BP2 interaction.

As with the interaction of p53 with 53BP1, the genetic evidence in human cells and in *C. elegans* suggests that ASPP proteins and p53 functionally interact (Samuels-Lev et al., 2001; Bergamaschi et al., 2003). However, the significance of the direct protein-protein interaction between p53 and ASPP2 (53BP2) remains to be established. Analysis of p53 mutants that fail to bind 53BP2, but retain DNA and 53BP1-binding activities, will help address this question.

2.4. STRUCTURE OF THE p73 C-TERMINAL SAM DOMAIN

As discussed elsewhere in this volume, p53 is a member of a protein family that in humans includes two other members, p63 and p73 (Yang and McKeon, 2000). The similarity between p53, p63 and p73 at the amino acid sequence level is high and all three proteins have similar domain organizations with a DNA binding domain at the center of the protein and an oligomerization domain at the C-terminus. One difference, however, is the presence of a C-terminal extension in certain splice variants of p63 and p73 that contains a Sterile Alpha Motif (SAM) domain (Thanos and Bowie, 1999).

SAM domains are protein-protein interaction motifs, primarily found in signaling molecules and transcription factors involved in developmental regulation (Schultz

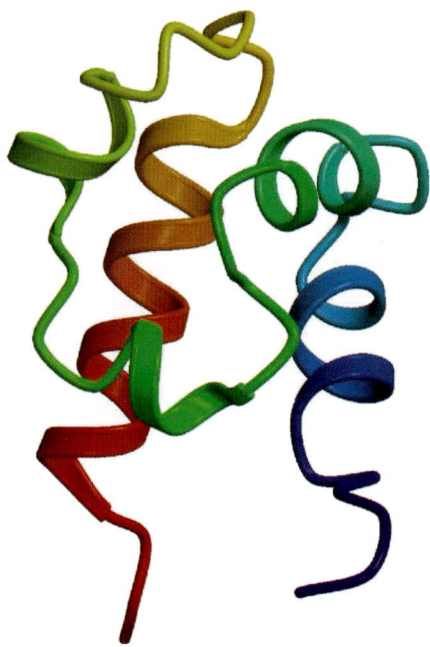

Figure 2.12. Three-dimensional structure of the SAM domain of p73.

et al., 1997). The structure of the SAM domain of human p73 was determined by NMR spectroscopy (Chi et al., 1999) and consists of 5 α–helices that form a tight globular structure (Fig. 2.12). The overall structure is very similar to that reported for the SAM domain of the ephrin receptor tyrosine kinase with the only difference being that whereas the ephrin receptor SAM domain homo-oligomerizes, the p73 SAM domain is a monomer (Smalla et al., 1999; Stapleton et al., 1999; Thanos et al., 1999). It is very likely that the SAM domains of p63 and p73 mediate protein-protein interactions with, as yet, unknown proteins. The identification of these proteins will greatly facilitate the understanding of the function of the p63 and p73 SAM domains.

2.5. CONCLUSIONS AND FUTURE DIRECTIONS

The p53 tumor suppressor is one of the best studied human proteins. As described in this chapter, significant progress has been made towards understanding its three-dimensional structure. The structural information has helped our understanding of p53 function and regulation and has also explained how tumor-associated mutations inactivate p53. The challenge now is to extend the structural studies to visualize multidomain fragments of p53 and even full-length p53. Structures of p53 with important partner molecules, such as transcriptional coactivators, are also needed. In the end the structural information is likely to form the basis for pharmacologic rescue of p53 function in human cancer.

ACKNOWLEDGMENTS

The authors thank Nikola Pavletich, Phil Jeffrey, Cheryl Arrowsmith, Stanley Opella, Mark McCoy, Rodney Harrington and Victor Zhurkin for helpful discussions.

REFERENCES

Appella, E., and Anderson, C. W. (2000). Signaling to p53: breaking the posttranslational modification code. Pathol. Biol. 48: 227–245.

Avalos, J. L., Celic, I., Muhammad, S., Cosgrove, M. S., Boeke, J. D., and Wolberger, C. (2002). Structure of a Sir2 enzyme bound to an acetylated p53 peptide. Mol. Cell 10: 523–535.

Avantaggiati, M. L., Ogryzko, V., Gardner, K., Giordano, A., Levine, A. S., and Kelly, K. (1997). Recruitment of p300/CBP in p53-dependent signal pathways. Cell 89: 1175–1184.

Ayed, A., Mulder, F. A., Yi, G. S., Lu, Y., Kay, L. E., and Arrowsmith, C. H. (2001). Latent and active p53 are identical in conformation. Nat. Struct. Biol. 8: 756–760.

Balagurumoorthy, P., Sakamoto, H., Lewis, M. S., Zambrano, N., Clore, G. M., Gronenborn, A. M., Appella, E., and Harrington, R. E. (1995). Four p53 DNA-binding domain peptides bind natural p53-response elements and bend the DNA. Proc. Natl. Acad. Sci. USA 92: 8591–8595.

Bargonetti, J., Reynisdottir, I., Friedman, P. N., and Prives, C. (1992). Site-specific binding of wild-type p53 to cellular DNA is inhibited by SV40 T antigen and mutant p53. Genes Dev. 6: 1886–1898.

Barlev, N. A., Liu, L., Chehab, N. H., Mansfield, K., Harris, K. G., Halazonetis, T. D., and Berger, S. L. (2001). Acetylation of p53 activates transcription through recruitment of coactivators/histone acetyltransferases. Mol. Cell 8: 1243–1254.

Baudier, J., Delphin, C., Grunwald, D., Khochbin, S., and Lawrence, J. J. (1992). Characterization of the tumor suppressor protein p53 as a protein kinase C substrate and a S100b-binding protein. Proc. Natl. Acad. Sci. USA 89: 11627–11631.

Bergamaschi, D., Samuels, Y., O'Neil, N. J., Trigiante, G., Crook, T., Hsieh, J. K., O'Connor, D. J., Zhong, S., Campargue, I., Tomlinson, M. L., Kuwabara, P. E., and Lu, X. (2003). iASPP oncoprotein is a key inhibitor of p53 conserved from worm to human. Nat. Genet. 33: 162–167.

Bork, P., Hofmann, K., Bucher, P., Neuwald, A. F., Altschul, S. F., and Koonin, E. V. (1997). A superfamily of conserved domains in DNA damage-responsive cell cycle checkpoint proteins. FASEB J. 11: 68–76.

Botuyan, M. V., Momand, J., and Chen, Y. (1997). Solution conformation of an essential region of the p53 transactivation domain. Fold. Des. 2: 331–342.

Brooks, C. L., and Gu W. (2003). Ubiquitination, phosphorylation and acetylation: the molecular basis for p53 regulation. Curr. Opin. Cell Biol. 15: 164–171.

Bullock, A. N., Henckel, J., DeDecker, B. S., Johnson, C. M., Nikolova, P. V., Proctor, M. R., Lane, D. P., Fersht, A. R. (1997). Thermodynamic stability of wild-type and mutant p53 core domain. Proc. Natl. Acad. Sci. USA 94: 14338–14342.

Candau, R., Scolnick, D. M., Darpino, P., Ying, C. Y., Halazonetis, T. D., and Berger, S. L. (1997). Two tandem and independent sub-activation domains in the amino terminus of p53 require the adaptor complex for activity. Oncogene 15: 807–816.

Chehab, N. H., Malikzay, A., Stavridi, E. S., and Halazonetis, T. D. (1999). Phosphorylation of Ser-20 mediates stabilization of human p53 in response to DNA damage. Proc. Natl. Acad. Sci. USA 96: 13777–13782.

Cherny, D. I., Striker, G., Subramaniam, V., Jett, S. D., Palecek, E., and Jovin, T. M. (1999). DNA bending due to specific p53 and p53 core domain-DNA interactions visualized by electron microscopy. J. Mol. Biol. 294: 1015–1026.

Chi, S. W., Ayed, A., and Arrowsmith, C. H. (1999). Solution structure of a conserved C-terminal domain of p73 with structural homology to the SAM domain. EMBO J. 18: 4438–4445.

Cho, Y., Gorina, S., Jeffrey, P. D., and Pavletich, N. P. (1994). Crystal structure of a p53 tumor suppressor-DNA complex: understanding tumorigenic mutations. Science 265: 346–355.

Clore, G. M., Ernst, J., Clubb, R., Omichinski, J. G., Kennedy, W. M., Sakaguchi, K., Appella, E., and Gronenborn, A. M. (1995). Refined solution structure of the oligomerization domain of the tumour suppressor p53. Nat. Struct. Biol. 2: 321–333.

Cohen, G. B., Ren, R., and Baltimore, D. (1995). Modular binding domains in signal transduction proteins. Cell 80: 237–248.

Craig, A. L., Burch, L., Vojtesek, B., Mikutowska, J., Thompson, A., and Hupp, T. R. (1999). Novel phosphorylation sites of human tumour suppressor protein p53 at Ser20 and Thr18 that disrupt the binding of mdm2 (mouse double minute 2) protein are modified in human cancers. Biochem. J. 342: 133–141.

Davison, T. S., Nie, X., Ma, W., Lin, Y., Kay, C., Benchimol, S., and Arrowsmith, C. H. (2001). Structure and functionality of a designed p53 dimer. J. Mol. Biol. 307: 605–617.

Davison, T. S., Vagner, C., Kaghad, M., Ayed, A., Caput, D., and Arrowsmith, C. H. (1999). p73 and p63 are homotetramers capable of weak heterotypic interactions with each other but not with p53. J. Biol. Chem. 274: 18709–18714.

Denu, J. M. (2003). Linking chromatin function with metabolic networks: Sir2 family of NAD(+)-dependent deacetylases. Trends Biochem. Sci. 28: 41–48.

Derbyshire, D. J., Basu, B. P., Serpell, L. C., Joo, W. S., Date, T., Iwabuchi, K., and Doherty, A. J. (2002). Crystal structure of human 53BP1 BRCT domains bound to p53 tumour suppressor. EMBO J. 21: 3863–3872.

Donato R. (2001). S100: a multigenic family of calcium-modulated proteins of the EF-hand type with intracellular and extracellular functional roles. Int. J. Biochem. Cell Biol. 33: 637–668.

Drohat, A. C., Amburgey, J. C., Abildgaard, F., Starich, M. R., Baldisseri, D., and Weber, D. J. (1996). Solution structure of rat apo-S100B(beta beta) as determined by NMR spectroscopy. Biochemistry 35: 11577–11588.

El-Deiry, W. S., Kern, S. E., Pietenpol, J. A., Kinzler, K. W., and Vogelstein, B. (1992). Definition of a consensus binding site for p53. Nat. Genet. 1: 45–49.

Espinosa, J. M., and Emerson, B. M. (2001). Transcriptional regulation by p53 through intrinsic DNA/chromatin binding and site-directed cofactor recruitment. Mol. Cell 8: 57–69.

Fields, S., and Jang, S. K. (1990). Presence of a potent transcription activating sequence in the p53 protein. Science 249: 1046–1049.

Finlay, C. A., Hinds, P. W., Tan, T. H., Eliyahu, D., Oren, M., and Levine, A. J. (1988). Activating mutations for transformation by p53 produce a gene product that forms an hsc70-p53 complex with an altered half-life. Mol. Cell. Biol. 8: 531–539.

Freeman, J., Schmidt, S., Scharer, E., and Iggo, R. (1994). Mutation of conserved domain II alters the sequence specificity of DNA binding by the p53 protein. EMBO J. 13: 5393–5400.

Gannon, J. V., Greaves, R., Iggo, R., and Lane, D. P. (1990). Activating mutations in p53 produce a common conformational effect. A monoclonal antibody specific for the mutant form. EMBO J. 9: 1595–1602.

Gorina, S., and Pavletich, N. P. (1996). Structure of the p53 tumor suppressor bound to the ankyrin and SH3 domains of 53BP2. Science 274: 1001–1005.

Guarente, L. (2000). Sir2 links chromatin silencing, metabolism, and aging. Genes Dev. 14: 1021–1026.

Gu, W., Shi, X. L., and Roeder, R. G. (1997). Synergistic activation of transcription by CBP and p53. Nature 387: 819–823.

Halazonetis, T. D., and Kandil, A. N. (1993). Conformational shifts propagate from the oligomerization domain of p53 to its tetrameric DNA binding domain and restore DNA binding to select p53 mutants. EMBO J. 12: 5057–5064.

Halazonetis, T. D., Davis, L. J., and Kandil, A. N. (1993). Wild-type p53 adopts a 'mutant'-like conformation when bound to DNA. EMBO J. 12: 1021–1028.

Hollstein, M., Sidransky, D., Vogelstein, B., and Harris, C. C. (1991). p53 mutations in human cancers. Science 253: 49–53.

Honda, R., Tanaka, H., and Yasuda, H. (1997). Oncoprotein MDM2 is a ubiquitin ligase E3 for tumor suppressor p53. FEBS Lett. 420: 25–27.

Hupp, T. R., Meek, D. W., Midgley, C. A., and Lane, D. P. (1992). Regulation of the specific DNA binding function of p53. Cell 71: 875–886.

Hupp, T. R., Sparks, A., and Lane, D. P. (1995). Small peptides activate the latent sequence-specific DNA binding function of p53. Cell 83: 237–245.

Inman, K. G., Yang, R., Rustandi, R. R., Miller, K. E., Baldisseri, D.M., and Weber, D. J. (2002). Solution NMR structure of S100B bound to the high-affinity target peptide TRTK-12. J. Mol. Biol. 324: 1003–1014.

Iwabuchi, K., Bartel, P. L., Li, B., Marraccino, R., and Fields, S. (1994). Two cellular proteins that bind to wild-type but not mutant p53. Proc. Natl. Acad. Sci. USA 91: 6098–6102.

Jeffrey, P. D., Gorina, S., and Pavletich, N. P. (1995). Crystal structure of the tetramerization domain of the p53 tumor suppressor at 1.7 angstroms. Science 267: 1498–1502.

Johnson, C. R., Morin, P. E., Arrowsmith, C. H., and Freire, E. (1995). Thermodynamic analysis of the structural stability of the tetrameric oligomerization domain of p53 tumor suppressor. Biochemistry 34: 5309–5316.

Joo, W. S., Jeffrey, P. D., Cantor, S. B., Finnin, M. S., Livingston, D. M., and Pavletich, N. P. (2002). Structure of the 53BP1 BRCT region bound to p53 and its comparison to the Brca1 BRCT structure. Genes Dev. 16: 583–593.

Kilby, P. M., Van Eldik, L. J., and Roberts, G. C. (1996). The solution structure of the bovine S100B protein dimer in the calcium-free state. Structure 4: 1041–1052.

Klein, C., Planker, E., Diercks, T., Kessler, H., Kunkele, K. P., Lang, K., Hansen, S., and Schwaiger, M. (2001). NMR spectroscopy reveals the solution dimerization interface of p53 core domains bound to their consensus DNA. J. Biol. Chem. 276: 49020–49027.

Koonin, E. V., Wolf, Y. I., and Karev, G. P. (2002). The structure of the protein universe and genome evolution. Nature 420: 218–223.

Koshland, D. E., Jr. (1958). Application of a theory of enzyme specificity to protein synthesis. Proc. Natl. Acad. Sci. USA 44, 98–104.

Kusano, K., Sakaguchi, M., Kagawa, N., Waterman, M. R., and Omura, T. (2001). Microsomal p450s use specific proline-rich sequences for efficient folding, but not for maintenance of the folded structure. J. Biochem. 129: 259–269.

Kussie, P. H., Gorina, S., Marechal, V., Elenbaas, B., Moreau, J., Levine, A. J., and Pavletich, N. P. (1996). Structure of the MDM2 oncoprotein bound to the p53 tumor suppressor transactivation domain. Science 274: 948–953.

Lambert, P. F., Kashanchi, F., Radonovich, M. F., Shiekhattar, R., and Brady, J. N. (1998). Phosphorylation of p53 serine 15 increases interaction with CBP. J. Biol. Chem. 273: 33048–33053.

Langley, E., Pearson, M., Faretta, M., Bauer, U. M., Frye, R. A., Minucci, S., Pelicci, P. G., and Kouzarides, T. (2002). Human SIR2 deacetylates p53 and antagonizes PML/p53-induced cellular senescence. EMBO J. 21: 2383–2396.

Lee, H., Mok, K. H., Muhandiram, R., Park, K. H., Suk, J. E., Kim, D. H., Chang, J., Sung, Y. C., Choi, K. Y., and Han, K. H. (2000). Local structural elements in the mostly unstructured transcriptional activation domain of human p53. J. Biol. Chem. 275: 29426–29432.

Lee, W., Harvey, T. S., Yin, Y., Yau, P., Litchfield, D., and Arrowsmith, C. H. (1994). Solution structure of the tetrameric minimum transforming domain of p53. Nat. Struct. Biol. 1: 877–890.

Lewit-Bentley, A., and Rety, S. (2000). EF-hand calcium-binding proteins. Curr. Opin. Struct. Biol. 10: 637–643.

Lill, N. L., Grossman, S. R., Ginsberg, D., DeCaprio, J., and Livingston, D. M. (1997). Binding and modulation of p53 by p300/CBP coactivators. Nature 387: 823–827.

Lin, J., Chen, J., Elenbaas, B., and Levine, A. J. (1994). Several hydrophobic amino acids in the p53 amino-terminal domain are required for transcriptional activation, binding to mdm-2 and the adenovirus 5 E1B 55-kD protein. Genes Dev. 8: 1235–1246.

Luo, J., Nikolaev, A. Y., Imaim S., Chenm D., Su, F., Shiloh, A., Guarente, L., and Gu, W. (2001). Negative control of p53 by Sir2alpha promotes cell survival under stress. Cell 107: 137–148.

Mateu, M. G., and Fersht, A. R. (1999). Mutually compensatory mutations during evolution of the tetramer-ization domain of tumor suppressor p53 lead to impaired hetero-oligomerization. Proc. Natl. Acad. Sci. USA 96: 3595–3599.

Matsumura, H., Shiba, T., Inoue, T., Harada, S., and Kai, Y. (1998). A novel mode of target recognition suggested by the 2.0 A structure of holo S100B from bovine brain. Structure 6: 233–241.

McCoy, M., Stavridi, E. S., Waterman, J. L., Wieczorek, A. M., Opella, S. J., and Halazonetis, T. D. (1997). Hydrophobic side-chain size is a determinant of the three-dimensional structure of the p53 oligomerization domain. EMBO J. 16: 6230–6236.

Michaely, P., and Bennett, V. (1992). The ANK repeat: a ubiquitous motif involved in macromolecular recognition. Trends Cell Biol. 2: 127–129.

Mochan, T. A., Venere, M., DiTullio, R. A., Jr., and Halazonetis, T. D. (2003). 53BP1 and NFBD1/MDC1-Nbs1 function in parallel interacting pathways activating ataxia-telangiectasia mutated (ATM) in response to DNA damage. Cancer Res. 63: 8586–8591.

Nagaich, A. K., Zhurkin, V. B., Durell, S. R., Jernigan, R. L., Appella, E., and Harrington, R. E. (1999). p53-induced DNA bending and twisting: p53 tetramer binds on the outer side of a DNA loop and increases DNA twisting. Proc. Natl. Acad. Sci. USA 96: 1875–1880.

Nagaich, A. K., Zhurkin, V. B., Sakamoto, H., Gorin, A. A., Clore, G. M., Gronenborn, A. M., Appella, E., and Harrington, R. E. (1997). Architectural accommodation in the complex of four p53 DNA binding domain peptides with the p21/waf1/cip1 DNA response element. J. Biol. Chem. 272: 14830–14841.

Pfeifer, G. P., Denissenko, M. F., Olivier, M., Tretyakova, N., Hecht, S. S., and Hainaut, P. (2002). Tobacco smoke carcinogens, DNA damage and p53 mutations in smoking-associated cancers. Oncogene 21: 7435–7451.

Raycroft, L., Wu, H. Y., and Lozano, G. (1990). Transcriptional activation by wild-type but not transforming mutants of the p53 anti-oncogene. Science 249: 1049–1051.

Rippin, T. M., Freund, S. M., Veprintsev, D. B., and Fersht, A. R. (2002). Recognition of DNA by p53 core domain and location of intermolecular contacts of cooperative binding. J. Mol. Biol. 319: 351–358.

Ruaro, E. M., Collavin, L., Del Sal, G., Haffner, R., Oren, M., Levine, A. J., and Schneider, C. (1997). A proline-rich motif in p53 is required for transactivation-independent growth arrest as induced by Gas1. Proc. Natl. Acad. Sci. USA 94: 4675–4680.

Rustandi, R. R., Baldisseri, D. M., and Weber, D. J. (2000). Structure of the negative regulatory domain of p53 bound to S100B(betabeta). Nat. Struct. Biol. 7: 570–574.

Sakaguchi, K., Saito, S., Higashimoto, Y., Roy, S., Anderson, C. W., and Appella, E. (2000). Damage-mediated phosphorylation of human p53 threonine 18 through a cascade mediated by a casein 1-like kinase. Effect on Mdm2 binding. J. Biol. Chem. 275: 9278–9283.

Sakamuro, D., Sabbatini, P., White, E., and Prendergast, G. C. (1997). The polyproline region of p53 is required to activate apoptosis but not growth arrest. Oncogene 15: 887–898.

Samuels-Lev, Y., O'Connor, D. J., Bergamaschi, D., Trigiante, G., Hsieh, J. K., Zhong, S., Campargue, I., Naumovski, L., Crook, T., and Lu, X. (2001). ASPP proteins specifically stimulate the apoptotic function of p53. Mol. Cell 8: 781–794.

Sauve, A. A., Celic, I., Avalos, J., Deng, H., Boeke, J. D., and Schramm, V. L. (2001). Chemistry of gene silencing: the mechanism of NAD+-dependent deacetylation reactions. Biochemistry 40: 15456–15463.

Schultz, J., Ponting, C. P., Hofmann, K., and Bork, P. (1997). SAM as a protein interaction domain involved in developmental regulation. Protein Sci. 6: 249–253.

Schultz, L. B., Chehab, N. H., Malikzay, A., and Halazonetis, T. D. (2000). p53 binding protein 1 (53BP1) is an early participant in the cellular response to DNA double-strand breaks. J. Cell Biol. 151: 1381–1390.

Scolnick, D. M., Chehab, N. H., Stavridi, E. S., Lien, M. C., Caruso, L., Moran, E., Berger, S. L., and Halazonetis, T. D. (1997). CREB-binding protein and p300/CBP-associated factor are transcriptional coactivators of the p53 tumor suppressor protein. Cancer Res. 57: 3693–3696.

Shaulian, E., Zauberman, A., Ginsberg, D., and Oren, M. (1992). Identification of a minimal transforming domain of p53: negative dominance through abrogation of sequence-specific DNA binding. Mol. Cell. Biol. 12: 5581–5592.

Shieh, S. Y., Ikeda, M., Taya, Y., and Prives, C. (1997). DNA damage-induced phosphorylation of p53 alleviates inhibition by MDM2. Cell 91: 325–334.

Smalla, M., Schmieder, P., Kelly, M., TerLaak, A., Krause, G., Ball, L., Wahl, M., Bork, P., and Oschkinat, H. (1999). Solution structure of the receptor tyrosine kinase EphB2 SAM domain and identification of two distinct homotypic interaction sites. Protein Sci. 8: 1954–1961.

Stapleton, D., Balan, I., Pawson, T., and Sicheri, F. (1999). The crystal structure of an Eph receptor SAM domain reveals a mechanism for modular dimerization. Nat. Struct. Biol. 6: 44–49.

Stavridi, E. S., Chehab, N. H., Caruso, L. C., and Halazonetis, T. D. (1999). Change in oligomerization specificity of the p53 tetramerization domain by hydrophobic amino acid substitutions. Protein Sci. 8: 1773–1779.

Thanos, C. D., and Bowie, J. U. (1999). p53 Family members p63 and p73 are SAM domain-containing proteins. Protein Sci. 8: 1708–1710.

Thanos, C. D., Goodwill, K. E., and Bowie, J. U. (1999). Oligomeric structure of the human EphB2 receptor SAM domain. Science 83: 833–836.

Vaziri, H., Dessain, S. K., Ng, Eaton E., Imai, S. I., Frye, R. A., Pandita, T. K., Guarente, L., and Weinberg, R. A. (2001). hSIR2(SIRT1) functions as an NAD-dependent p53 deacetylase. Cell 107: 149–159.

Venot, C., Maratrat, M., Dureuil, C., Conseiller, E., Bracco, L., and Debussche, L. (1998). The requirement for the p53 proline-rich functional domain for mediation of apoptosis is correlated with specific PIG3 gene transactivation and with transcriptional repression. EMBO J. 17: 4668–4679.

Vogelstein, B., Lane, D., and Levine, A. J. (2000). Surfing the p53 network. Nature 408: 307–310.

Walker, K. K., and Levine, A. J. (1996). Identification of a novel p53 functional domain that is necessary for efficient growth suppression. Proc. Natl. Acad. Sci. USA 93: 15335–15340.

Wang, J., Tan, N. S., Ho, B., and Ding, J. L. (2002). Modular arrangement and secretion of a multidomain serine protease. Evidence for involvement of proline-rich region and N-glycans in the secretion pathway. J. Biol. Chem. 277: 36363–36372.

Wang, T., Kobayashi, T., Takimoto, R., Denes, A. E., Snyder, E. L., El-Deiry, W. S., and Brachmann, R. K. (2001). hADA3 is required for p53 activity. EMBO J. 20: 6404–6413.

Waterman, J. L., Shenk, J. L., and Halazonetis, T. D. (1995). The dihedral symmetry of the p53 tetramerization domain mandates a conformational switch upon DNA binding. EMBO J. 14: 512–519.

Weinert, T. A., and Hartwell, L. H. (1988). The RAD9 gene controls the cell cycle response to DNA damage in Saccharomyces cerevisiae. Science 241: 317–322.

Wieczorek, A. M., Waterman, J. L., Waterman, M. J., and Halazonetis, T. D. (1996). Structure-based rescue of common tumor-derived p53 mutants. Nat. Med. 2: 1143–1146.

Williams, R. S., Green, R., and Glover, J. N. (2001). Crystal structure of the BRCT repeat region from the breast cancer-associated protein BRCA1. Nat. Struct. Biol. 8: 838–842.

Willson, J., Wilson, S., Warr, N., and Watts, F. Z. (1997). Isolation and characterization of the Schizosaccharomyces pombe rhp9 gene: a gene required for the DNA damage checkpoint but not the replication checkpoint. Nucleic Acids Res. 25: 2138–2146.

Xu, Y. (2003). Regulation of p53 responses by post-translational modifications. Cell Death Differ. 10: 400–403.

Yang, A., and McKeon, F. (2000). P63 and P73: P53 mimics, menaces and more. Nat. Rev. Mol. Cell Biol. 1: 199–207.

Yap, K. L., Ames, J. B., Swindells, M. B., and Ikura, M. (1999). Diversity of conformational states and changes within the EF-hand protein superfamily. Proteins 37: 499–507.

Yoon, C., Prive, G. G., Goodsell, D. S., and Dickerson, R. E. (1988). Structure of an alternating-B DNA helix and its relationship to A-tract DNA. Proc. Natl. Acad. Sci. USA 85: 6332–6336.

Yu, H., Chen, J. K., Feng, S., Dalgarno, D. C., Brauer, A. W., and Schreiber, S. L. (1994). Structural basis for the binding of proline-rich peptides to SH3 domains. Cell 76: 933–945.

Zhao, K., Chai, X., Johnston, K., Clements, A., and Marmorstein, R. (2001). Crystal structure of the mouse p53 core DNA-binding domain at 2.7 A resolution. J. Biol. Chem. 276: 12120–12127.

3

Transcriptional Activation by p53: Mechanisms and Targeted Genes

Timothy MacLachlan and Wafik El-Deiry

SUMMARY

The best-characterized function of p53, the most renowned tumor suppressor, is that of transcriptional activation. Upon its first description as a regulator of gene expression, p53 was simply thought to bind to an element within the 5' UTR of a target gene, which would lead to transcription, by the appropriate machinery within the cell. Over the past 10 years however, this process has proven to be much more complex and tightly regulated than originally visualized. From the posttranslational modifications and proteins associated with p53 to the choice of a subset of genes that p53 possesses the capability to activate, the regulation of p53 transcriptional activity exists on several levels. This review will focus on the alterations of p53 protein that control activity of the protein, how genomic binding sites for p53 are presently being found and then finally, how the p53 target gene group is growing and being clustered into subsets of gene families.

T. MACLACHLAN AND W. EL-DEIRY • Laboratory of Molecular Oncology and Cell Cycle Regulation, Howard Hughes Medical Institute, Departments of Medicine, Genetics, and Pharmacology, and Abramson Cancer Center, University of Pennsylvania School of Medicine, Philadelphia, PA 19104, USA

The p53 Tumor Suppressor Pathway and Cancer, edited by Zambetti.
Springer Science+Business Media, New York, 2005.

3.1. INTRODUCTION

Although first thought of as an oncogene (DeLeo et al., 1979; Jenkins et al., 1984; Lane, 1984), the TP53 gene encodes for the most commonly mutated tumor suppressor proteins found in human cancer. Mouse models as well as studies in human systems in vitro have shown that p53 governs essential checkpoints in cellular growth (Vogelstein et al., 2000). Without proper attention being paid to these sentries, multiple deleterious effects on the homeostasis of the cell can occur, such as untimed DNA replication, chromosomal instability, and lack of engagement of the apoptosis machinery when all else fails. Although transcription independent mechanisms of tumor suppression are possible for p53 (Bennett et al., 1998; Caelles et al., 1994; Chen et al., 1996; Haupt et al., 1995; Wagner et al., 1994), the most well studied theory is that of a regulator of transcription (Farmer et al., 1992). The induction or repression of these targets alters the homeostasis of the cell to undergo one of several processes among which are apoptosis induction or growth arrest (el-Deiry, 1998). Some 10 years ago, a consensus genomic DNA binding site for p53 was defined that existed upstream of genes that were mostly known to be induced by p53 (el-Deiry et al., 1992). Shortly thereafter, the first few mRNAs found to be directly regulated by p53 were discovered, including the p21WAF1 cyclin-dependent kinase inhibitor and the Bax protein that affects mitochondrial membrane potential (el-Deiry et al., 1993; Miyashita and Reed, 1995). These inductions were preceded by a stabilization of p53 that was the result of processes affecting phosphorylation, and later found to also involve acetylation, ubiquitination, and sumoylation (Prives and Manley, 2001; Vousden, 2002). Interestingly, as the numbers of p53 target genes increased, it was found that these genes can be very generally classified into two separate categories— those arresting proliferation but allowing for cell repair and those that immediately shunt the cell toward a pathway leading to cell death. Many steps are thought to lead to the eventual decision of cell life or apoptosis by way of alteration of gene expression.

3.2. POSTTRANSLATIONAL MODIFICATIONS

3.2.1. Phosphorylation

A great deal of data presently exists on the phosphorylation of p53, primarily at serines 15, 20, and 46 (Vousden, 2002). These modifications play a role in the stabilization of the protein, but also could regulate transcriptional specificity. While phosphorylation of serine 15 and 20 have not been shown to affect DNA binding capacity, the status of serine 46 may determine the genes that p53 will control downstream. Via a yeast enhancer trap protocol, the p53-regulated apoptosis inducing protein (p53AIP1) was identified as a p53 target gene that is able to induce apoptosis (Oda et al., 2000b). What made this particular discovery different from all the other apoptosis regulators controlled by p53 was the finding that phosphorylation of serine 46 was required for activation of p53AIP1. Phosphorylation of Ser46 lagged behind other serines such as 15 and 20, suggesting that this modification was not involved in the initial process of stabilization, but at a later stage that could involve selective DNA binding. This was supported by the fact that activation of p53AIP1 took place after p21WAF1.

This serine is followed by a proline, which initially led to speculation that it was phosphorylated by proline-directed kinases such as CDKs and MAPKs. Indeed, Bulavin et al. (1999) found that UV induced apoptosis was abrogated in cells expressing a mutant form of p53 with an alanine in place of serine at position 46, and that this effect was found to lie in the MAP kinase p38 kinase pathway. The Wip1 phosphatase, also a target gene of p53, is involved in the p38 pathway by dephosphorylating at both the ser-46 and 33 residues affected (Takekawa et al., 2000). Soon thereafter, a kinase termed Homeodomain-interacting protein kinase 2 (HipK2) was found to be a kinase specific for serine 46 (D'Orazi et al., 2002; Hofmann et al., 2002; Kim et al., 2002). The physiological role of this family of kinases, originally discovered as associated proteins to the Nkx-1.2 homeoprotein, was previously unknown. A number of experiments from three independent groups established that HipK2 was able to specifically phosphorylate at serine 46 and as a result, enhance transcriptional activity from p53. The kinase activity of endogenous HipK2 is enhanced under conditions of apoptosis, and significantly increased the number of apoptotic cells when expressed exogenously. While one group found that HipK2 was only able to activate p53 transcription of apoptosis genes specifically (D'Orazi et al., 2002), such as PIG3, another found that it lacked specificity in p53 target gene activation (Hofmann et al., 2002). Interestingly, the association of p53 with this kinase was found in a day-11 mouse embryonic library, a time in which apoptosis may be necessary for cell deletion during development. Further studies will be interesting in determining what serine 46 phosphorylation of p53 does to the protein biochemically. One line of evidence points to association of HipK2 with CBP and that this grouping and activity facilitates acetylation of p53 at lysine 382.

A target gene of p53, p53DINP1, appears to be involved in the complex phosphorylating p53 at ser-46, as knockdown of the transcript abolishes phosphorylation following several types of DNA damage (Okamura et al., 2001).

3.2.2. Acetylation

For some time, acetylation of core histones has been a hallmark of gene activation (Berger, 2002). Key lysine residues are acetylated allowing access to the genome by transcriptional regulators such as RNA polymerase. Allis and colleagues provided the link between transcription factors and histone acetylation by showing that a known histone acetyltransferase in tetrahymena was actually a homologue of a yeast transcription factor (Brownell et al., 1996). This set off a flurry of activities identifying several components of the basal transcription machinery as well as transcription coactivators that possess histone acetyltransferase, as well as deacetylase, activity. Therefore, it was not surprising to find that when p53 bound to the p300/CBP protein, acetyltransferase activity of p300/CBP increased p53 transcriptional activity (Avantaggiati et al., 1997; Lill et al., 1997; Ogryzko et al., 1996; Sang and Giordano, 1997; Somasundaram and el-Deiry, 1997). What was quite surprising however was a paper by Gu and Roeder that showed that p300/CBP was not only acetylating histones, but also p53 itself (Gu and Roeder, 1997). Extreme C-terminal lysine residues were shown to be acetylated by p300/CBP as well as P/CAF, a HAT associated with p300/CBP in another report (Liu et al., 1999; Sakaguchi et al., 1998). p300/CBP acetylated lysines 372, 373, 381, and 382 in the basic region of p53, and P/CAF acetylated lysine 320 in a region between the DNA binding and tetramerization domains. In a mobility shift assay using synthetic oligonucleotides, acetylation enhanced DNA binding, leading to the conclusion that this was an essential step in p53 activation. Several reports followed that acetylation of p53 at these residues was induced by a number of stimuli, including those that induced apoptosis (Gottifredi et al., 2001; Ito et al., 2001; Liu et al., 1999; Pearson et al., 2000; Sakaguchi et al., 1998).

Recently however, the notion that acetylation affects DNA binding has come into question. While the assay used by Gu and Roeder to determine protein-DNA association has been an effective means to look at such interactions for some time, more physiologically relevant techniques have been developed to look at DNA binding by proteins. First, Espinosa and Emerson (2001) constructed chromatin containing the p21WAF1 promoter and showed that, acetylated or unacetylated, p53 bound with the same affinity. Next, Barlev et al. (2001) used the chromatin immunoprecipitation method to show that p53 mutants lacking the critical lysines at the C-terminus were just as effective in binding the endogenous p21 promoter as wild type. Therefore, it appears that acetylation does not affect binding to promoter regions. It is also unclear whether acetylation affects transcriptional activity in any respect. Experiments done by these two separate groups have seen no effect on transcriptional activation by p53 dependent on the status of the C-terminal lysines, while a third group has shown slight induction in the presence of acetylated lysines (Barlev et al., 2001; Espinosa and Emerson, 2001; Nakamura et al., 2000). One consistency between these three studies is the fact that transcriptional activation was measured each time by use of the p21WAF1 promoter. A very attractive possibility is that acetylation could modify the target specificity of p53—growth arrest and DNA repair, or cell death genes—which would ultimately affect the outcome of activation of p53 in that case. This theory seems to have borne out for a relative of p53, p73 (Costanzo et al., 2002). Using p53–/– cells, Costanzo et al. showed that inhibitors of p300 HAT activity reduced apoptosis induction. It was found that the substrate for p300 in this case was p73. Mutants of p73 that could not be acetylated were not able to bind to the promoter region of p53AIP1 in vivo, while their binding capacity for the p21WAF1 promoter was unchanged. It remains to be seen if the same will be true for p53 as well.

Two other proteins in addition to p300/CBP and P/CAF that have induced acetylation of p53 are hADA3 and p33ING2 (Nagashima et al., 2001; Wang et al., 2001). Using a yeast dissociator assay, hADA3 was identified as a protein that interfered with p53 function. While an N-terminal fragment of hADA3 could inhibit p53 activity, the full length actually enhances p53-mediated transcription by directly associating with p53 and attracting components of the histone acetyltransferase family. p33ING2 was identified in a homology search for proteins related to p33ING1, a previously identified p53 associated protein (Garkavtsev et al., 1998). Ectopic expression of p33ING2 induced growth arrest, apoptosis, and acetylation of p53 on lysine 382 (although no other lysines were tested), while not affecting phosphorylation. Notably, p33ING2 strongly induced a reporter construct containing the Bax promoter in the presence of p53, while only a modest twofold induction was observed for the p21WAF1 promoter reporter. These data further suggest that acetylation may have a differential effect on p53 transcriptional activity depending on the gene transactivated.

3.2.3. Deacetylation

In addition to controls over the timing of acetylation of p53 at the C-terminus, removal of those acetyl groups also appears tightly regulated (the specific acetyl group is described in the following sentence). Three reports have described the role of the NAD-dependent deacetylase Sir2a in deacetylation of p53 (Langley et al., 2002; Luo et al., 2001; Vaziri et al., 2001). This effect was specific for lysine 382. Expression of Sir2a significantly decreased the transcriptional activity of p53, and the opposite reaction was observed while knocking down the expression. Using a catalytically inactive mutant of Sir2a, the authors of all manuscripts find that this dominant negative inhibition potentiates apoptosis via p53. This particular deacetylase is dependent on nicotinamide adenine dinucleotide, and interestingly is not inhibited by TSA, a broad inhibitor of deacetylases. However, at this point the role played by NAD in p53 transcriptional regulation is unclear.

p53 deacetylation has also been found to be affected by TSA-sensitive deacetylases. An HDAC1 containing complex was found to affect the acetylation status of p53 (Juan et al., 2000; Luo et al., 2000). Included in one study was the finding of a protein, PID1, directly associated with p53 (Luo et al., 2000). PID1, previously called MTA2—metastasis associated protein 2 (Zhang et al., 1999), strongly represses p53 transcriptional activity and is known to be a component of the nucleosome remodeling and histone deacetylation (NuRD) complexes. It is quite intriguing that a protein formerly associated with metastasis is now known to significantly reduce the capacity of p53 to impart its tumor suppressive qualities.

The adenoviral E1B 55-kDa protein has long been known to inhibit the transcriptional functions of p53 (Yew and Berk, 1992), but the mechanism by which this occurs is unknown. Liu et al. (2000) discovered shortly after the identification of P/CAF as a p53 directed acetyltransferase that E1B 55kDa was able to specifically inhibit the acetylation of p53 by P/CAF, while leaving histone- and self-acetylation by P/CAF unaffected. It appears that this inhibition is rooted in the ability of E1B 55kDa to keep P/CAF from physically binding p53. This finding could partially explain the method of oncogenesis of adenovirus strains 5 and 12. Two other proteins already known to affect p53 in other respects, MDM2 and MDMX are also able to inhibit P/CAF and p300/CBP acetylation of p53, respectively (Jin et al., 2002; Sabbatini and McCormick, 2002).

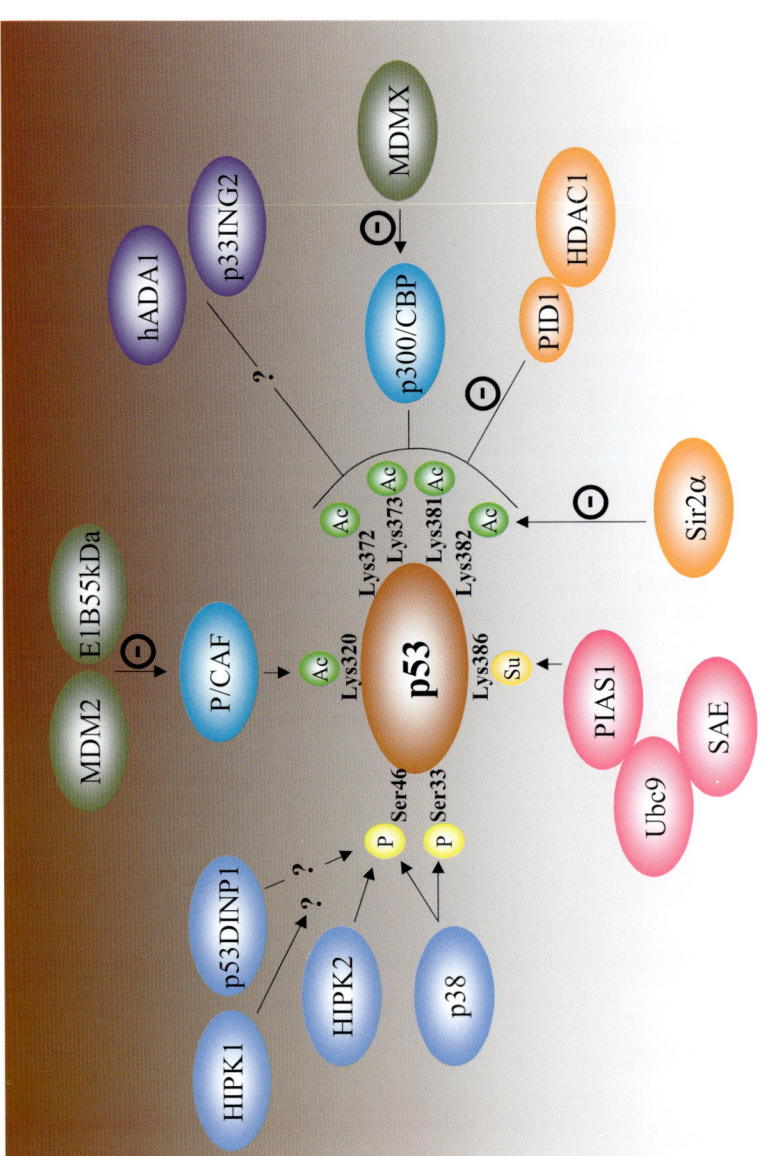

Figure 3.1. Schematic of proteins and their modification of p53 as it relates to transcriptional activity. Modifications that are indirect, or those that are not known whether they are direct, are depicted with a "?". Lines with a "-" are negative regulations, or in the case of Sir2a, deacetylation. P—Phospho group, Ac—Acetyl group, Su—SUMO group.

3.2.4. Sumoylation

An additional posttranslational modification came to light in 1996 when several reports found the covalent binding of a ubiquitin-like molecule to Ran GTPase-activating protein (Boddy et al., 1996; Kamitani et al., 1997; Mahajan et al., 1997; Matunis et al., 1996; Shen et al., 1996). This molecule, called SUMO-1 (also known as sentrin, GMP1, UBL1, and PIC1) usually binds to identical residues used for ubiquitination, however it does not target the protein for degradation as ubiquitin does. Rodriguez et al. (1999) published the finding that p53 is also sumoylated and that this modification is able to activate the transcriptional activity of p53. Contrary to other proteins, ubiquitin and SUMO1 do not compete for the same lysine (K386) that is sumoylated; therefore it is not thought that this is a mechanism to override degradation. Rather, it may cooperate with other modifications at the C-terminus that are involved in transcription control.

Sumoylation proceeds by an analogous process that is already known for ubiquitination. There are three enzymatic steps involving E1, E2, and E3 that result in SUMO1 being added to a protein (Lee et al., 1998). The SUMO activating enzyme (SAE) acts as the E1, human Ubc9 as the E2, and recently PIAS (protein inhibitor of activated STAT1) was identified as the E3 that conjugates the last step in p53 sumoylation (Kahyo et al., 2001). Mutants of PIAS were unable to catalyze the reaction, leading to the speculation that components of this pathway may be altered in cancer. Recent data involving p53 and PIAS have been conflicting, as one report indicates that sumoylation by PIAS induces transcriptional activity (Megidish et al., 2002), while another claims that it potently inhibits transcriptional activity (Schmidt and Muller, 2002). It is possible that this effect, as may be the case with acetylation, may be target gene specific.

A schematic outlining the posttranslational modifications that affect transcription specifically is outlined in Figure 3.1.

3.3. p53 BINDING PROTEINS THAT AFFECT TRANSCRIPTION

3.3.1. BRCA1

The BRCA1 breast cancer tumor suppressor was originally found to associate with p53 by two independent groups looking at the effect of BRCA1 on growth suppressing genes (Ouchi et al., 1998; Zhang et al., 1998). Direct binding between BRCA1 and p53 was found on the N-terminus of BRCA1 and the C-terminus of p53, although a third group has found binding of p53 on BRCA1 at the C-terminus (Chai et al., 1999). BRCA1 is able to stimulate p53 transcriptional activity in transient transfection experiments using reporter constructs. One of the possible mechanisms by which this occurs is based on the association of BRCA1 with the SWI/SNF-related chromatin remodeling protein, BRG1 (Bochar et al., 2000). A dominant negative form of BRG1 abrogated the ability of BRCA1 to stimulate p53-mediated transcription from either exogenously added reporter constructs or by looking at endogenous p53 target expression such as Gadd45. Additional support for the SWI/SNF complex involvement in p53 transcription has recently been reported with both BRG1 and hSNF5 binding (Lee et al., 2002).

Interestingly, BRCA1 has also been shown to be involved in stabilization of the p53 protein in both transient transfection experiments as well as mouse models (Somasundaram et al., 1999; Xu et al., 2001). However, several papers have seen no reproducible induction of apoptosis while hyperexpressing BRCA1 (Aprelikova et al., 1999, 2001; MacLachlan et al., 2000a, b, 2002; Somasundaram et al., 1997; Takimoto et al., 2002a; Zheng et al., 2000). Therefore, is it possible that BRCA1 may elicit a specific transcriptional response from p53? One publication has shown that indeed, while BRCA1 was able to stabilize p53 when overexpressed, only a subset of p53 target genes were subsequently changed in expression (MacLachlan et al., 2002). In agreement with the phenotypic response of most cells to BRCA1 expression, none of the apoptosis inducing targets of p53 were induced by p53 stabilized by BRCA1. It will be interesting to determine the posttranslationally modified status of p53 when stabilized by BRCA1 compared to known apoptosis inducing agents that also stabilize p53 such as etoposide and adriamycin.

3.3.2. p300/CBP

The interaction between p300/CBP and p53 was the first suggestion that p53 possessed a coactivator involved in transcription (Avantaggiati et al., 1997; Lill et al., 1997; Sang et al., 1997; Somasundaram and el-Deiry, 1997). Direct binding between the proteins was shown as well as potentiation of the transcriptional response elicited by both. p300/CBP was also found to be involved in p53-mediated inhibition of transcription via AP1 DNA binding sequences (Avantaggiati et al., 1997). Perhaps the most enlightening part of this barrage of studies was that half of them discovered that this was the root of adenovirus E1A mediated inhibition of p53 transcriptional regulation (Sang et al., 1997; Somasundaram and el-Deiry, 1997). E1A protein that is not able to bind p300 no longer had an effect on p53 transcription, even while retaining the pRb binding region, thereby separating the two pathways controlled by E1A. This finding uncovers the mechanism by which E1A is able to knock out the p53 half of cellular growth control.

A protein that is part of the p300 complex has recently been identified that contributes to the enhancement of p53 transcriptional activity (Demonacos et al., 2001). Strap, a protein that is composed of an unusual structure encoded almost entirely by six tetratricopeptide (TPR) motifs, facilitates p300 coactivation of p53. Interestingly, Strap is induced upon cellular stress and also interferes with Mdm2-mediated degradation of p53. There may exist a feedback loop, as previous data has implicated Mdm2 in inhibition of p53-coactivator interaction.

The human homologue of yeast Rad23 (hHR23A) has also been found to affect the association of p53 with p300/CBP (Zhu et al., 2001). hHR23A is able to bind to the C/H1 domains of p300/CBP and subsequently interfere with the interaction with p53. Naturally, p53 transcriptional activity is significantly diminished in cells expressing hHR23A.

3.3.3. p63/p73

Although included here under the header of interacting proteins, binding between wild-type p53 and the family members p63 and p73 has not been seen in vivo (Davison et al., 1999; Kaghad et al., 1997; Yang et al., 1998). Some studies however have shown that wild-type p53 is able to associate with p63, and that certain p53 mutants abolish this binding (Ratovitski et al., 2001). Mutant p53 on the other hand has been seen to directly associate with p63 and p73, and as a result negatively regulate their transcriptional activity (Di Como et al., 1999; Gaiddon et al., 2001; Strano et al., 2000). Regardless of the gray area in whether or not direct associations between wild-type p53 and p63/p73 exist, a striking result from Tyler Jack's lab pointed to an essential involvement of p63/p73 in p53-mediated transcription (Flores et al., 2002). In the study, single and double knockout p63/p73 E1A transformed mouse embryo fibroblasts were used to determine the extent of apoptosis when the cells were stressed. Surprisingly, even though endogenous wild-type p53 was stabilized after doxorubicin treatment of p63–/–;p73–/– E1A MEFs, apoptosis was absent compared to wild-type MEFs. In agreement with this was the p53 target genes that were expressed (growth arrest, but not apoptosis genes) and p53 DNA binding (no p53 bound to apoptosis promoters, yet detectable binding to promoters of growth arrest genes). A key element of this phenomenon to be determined will be what the posttranslational status is of stabilized p53 in wild-type versus p63/p73 knockout MEFs. While it is still unclear how these three wild-type proteins affect each other's activity, full induction of apoptosis-inducing genes requires all three, as the absence of any one significantly reduces, or in the case of p53 completely abolishes induction of these target genes.

3.3.4. Other Associated Proteins

AMF1 was a protein that was previously known to associate with p300 and the human papillomavirus E2 protein, and that this interaction is necessary for E2-mediated transcription (Breiding et al., 1997; Peng et al., 2000). Binding to p53 was found to involve the DNA binding region of p53, and coexpression of p53 and AMF1 greatly increased the expression from p21 promoter and synthetic p53 binding site reporter constructs (Peng et al., 2001). It is possible that AMF1 contributes to p53 activity by recruiting p300.

53BP2 was one of the first proteins identified in the now ubiquitous yeast-two hybrid system as a p53 binding protein (Iwabuchi et al., 1994). Although first found to inhibit association of p53 with a consensus DNA binding site, a longer clone of 53BP2 termed Bbp was later found to potentiate p53 activity (Iwabuchi et al., 1998). The mechanism, however, is unclear, as 53BP2/Bbp exclusively exists in the cytoplasm, even after stress induction. As it turns out, both 53BP2 and Bbp are fragments of a protein recently named ASPP2, and a homologue of 53BP2, ASPP1 was also described (Samuels-Lev et al., 2001). Both proteins stimulate the activity of p53 on apoptosis promoters specifically, thereby enhancing the apoptosis pathway of p53. A repression of ASPP expression was seen in many breast tumors that express wild-type p53, indicating an important step in tumor suppression by p53.

Mouse embryonic stem cells that are deleted for the Ets1 transcription factor have been found to possess a significantly compromised p53-mediated UV response pathway (Sampath et al., 2001; Xu et al., 2002). While the levels at which p53 was stabilized by UV irradiation were unaffected in the knockout versus wild-type cells, the induction of p53-responsive genes such as cyclin G and the ability of p53 to bind to the promoters of these genes was considerably reduced. Although the biology behind this is not yet determined, it appears that Ets1 is in the pathway to activation of p53 transcriptional activity. Interestingly, the repression of the presenillin-1 gene by p53 is dependent on the presence of Ets1 binding sites in the promoter (Pastorcic and Das, 2000), so the link between p53 and Ets1 may be very tight and extend to cotranscription of common targets.

The JMY protein is a factor originally isolated in a screen for interacting proteins of p300 (Shikama et al., 1999). JMY is able to increase the transcriptional activity of p53 directed against the Bax promoter. Interestingly, splice variants that exist of the JMY protein, specifically those that delete the C-terminal proline-rich region, are incapable of having any effect on p53 activity.

Viral proteins have as one of their main targets p53, as p53's gatekeeper function must be taken over in order to wrest control of the cell. Adenovirus E1B 55kDa protein, as described above, is involved in inhibition of acetylation of p53 (Liu et al., 2000). The simian virus-40 large T-antigen, one of the first proteins found to bind p53 (Schmieg and Simmons, 1984), is able to mask the DNA binding region of p53, inactivating it as a transcriptional regulator, although recent evidence has pointed to additional mechanisms of inhibition imposed by large T (Bargonetti et al., 1992; Sheppard et al., 1999). The LANA protein expressed from the Kaposi's sarcoma-associated herpesvirus is known to be an integral player in the development of Kaposi's sarcomas. LANA was found to perform this task in part by binding to and inactivating transcriptional activity of p53 (Friborg et al., 1999).

One would assume that a requirement of the classic tyrosine kinase c-Abl in p53 stimulation of DNA binding would involve phosphorylation. Interestingly, while binding of c-Abl to the C-terminus of p53 enhances binding of DNA and tetramerization of the protein, one mutant in the kinase domain is equally able to do so (Nie et al., 2000). The two proteins bind in response to DNA damage, and overexpression of c-Abl is able to cause G1 phase growth arrest in wild-type p53 expressing cells.

While p53 repression of some genes is known to involve HDACs, the interaction between these two proteins is bridged by the well-known transcriptional corepressor mSin3a (Murphy et al., 1999). The interaction with p53 and promoter regions of genes that are negatively controlled by p53 is always found in concert with mSin3a, explaining at least one mechanism by which p53 is able to negatively regulate the expression of certain genes.

An additional histone acetyltransferase associated protein has been found to associate with p53. The ATM-related TRRAP protein is common between the known mammalian HAT complexes SAGA, TFTC, STAGA, and Tip60 (Brown et al., 2001). Using the mdm2 gene as an example, TRRAP was able to be recruited to the mdm2 promoter by direct binding to p53 and enhance histone acetylation of the mdm2 genomic locus (Ard et al., 2002). TRRAP is apparently required for p53 transcription of mdm2, as knockdown of TRRAP by antisense diminished mdm2 gene activation.

There are several other proteins that bind and affect p53 transcription either positively (BML) (Garkavtsev et al., 2001) or negatively (S100B, ATF3, MTS1) (Grigorian et al., 2001; Lin et al., 2001; Yan et al., 2002), that are convincing in their activity and will most likely bore out additional implications in further studies.

A schematic outlining the p53 binding proteins that affect transcription specifically is outlined in Figure 3.2.

3.4. BINDING OF p53 TO REGULATORY REGIONS

In 1992, a paper that searched for the consensus promoter-binding site for p53 began to shed light on the mechanisms of transcriptional regulation imposed by p53 (el-Deiry et al., 1992). Using an assay to fish out high affinity binding oligos from a pool of random sequences, the authors established that p53 bound best to the consensus sequence consisting of two half-sites with the sequence 5'-Pu-Pu-Pu-C-(A/T)-(T/A)-G-Py-Py-Py-3' separated by 0 to 13 bases. Since then, literally hundreds of publications have identified some permutation of this sequence in regulatory regions of genes whose transcription is controlled by p53. With the majority of tumor-derived mutations in p53 taking place within the DNA binding region, it is clear that association with genomic DNA is paramount in its tumor suppressive capabilities. In recent years however, controversy has brewed with respect to certain issues including (1) whether this site also mediates transcriptional repression, (2) what additional sequence context is required for recognition, and (3) what the kinetics of binding to this site in vivo are.

3.4.1. Repression

In some cases of p53-mediated repression of gene transcription, consensus binding sites have been found in the regulatory regions of the affected gene promoter. This has included the Bcl-2, a-fetoprotein, survivin, and Tau-T genes (Han et al., 2002; Hoffman et al., 2002; Lee et al., 1999; Miyashita et al., 1994). In some of these cases, the mechanism has been proven such that p53 will displace a more potent activator bound to the regulatory region, resulting in lower overall transcription levels. However, many other p53-mediated repression events are known to occur in genes that

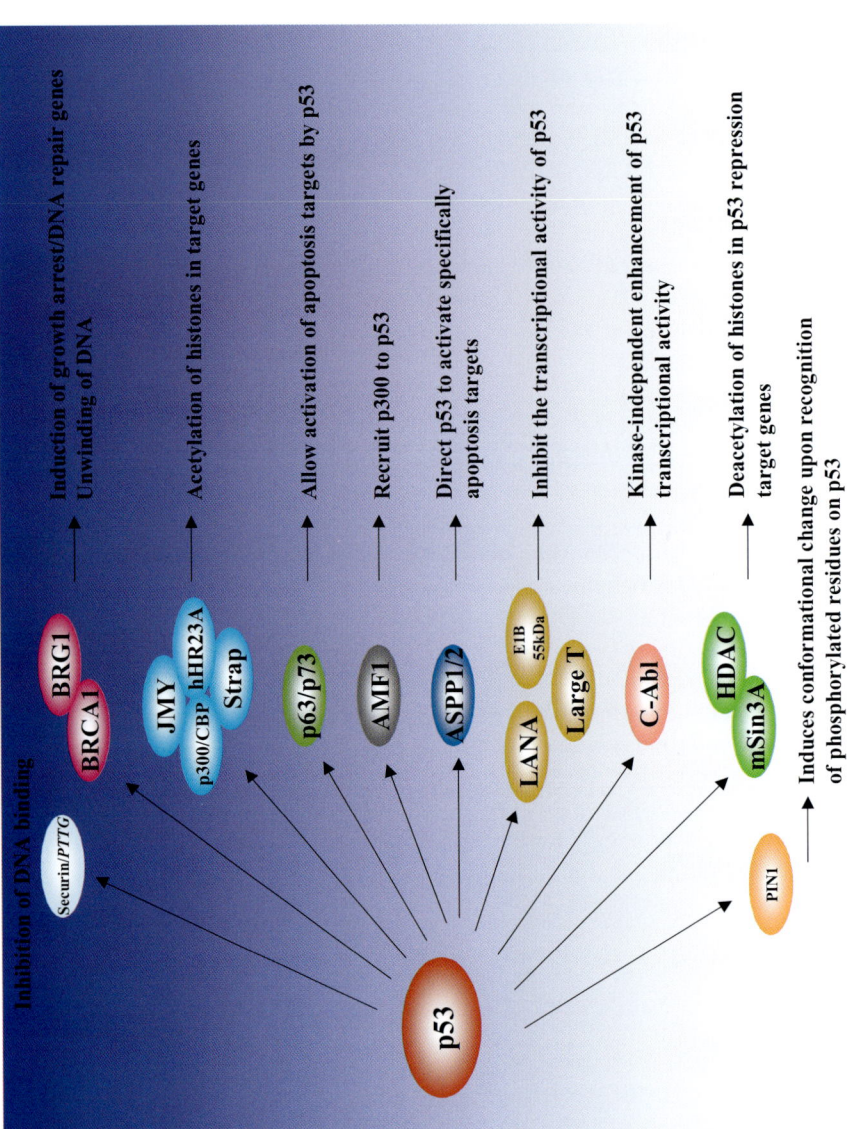

Figure 3.2. Schematic of proteins that associate with p53 to affect transcriptional activity. Proteins that directly associate with p53 are shown on the left, and any proteins that are also part of the complex, but do not directly bind p53 are to the right of the direct binder. The downstream effect is shown on the right.

do not have a high scoring consensus site upstream of a transcriptional start site, and many of these cases have witnessed p53 repression, active all the way up to the minimal promoter regions. Such targets in which this occurs includes BRCA1, Map4, stathmin, and MDR1, and may utilize the mSin3a pathway described above (Arizti et al., 2000; Chin et al., 1992; MacLachlan et al., 2000a; Murphy et al., 1996, 1999). While many reports contend that there exists no consistency between the regions of these genes that is required for repression, and have therefore suggested that it may involve indirect repression of basal transcriptional machinery, others claim to have delineated a transcriptional repression DNA element. In the case of survivin, while overlap of the consensus site with an E2F binding element may place it in the category of activator displacement, removal of the three nucleotide spacer between half-sites converts p53 to an activator of survivin, implicating a role for those nucleotides in repression (Hoffman et al., 2002). Another report, utilizing the MDR1 gene as a model, has found that when one p53 consensus quarter site of each half-site was inverted (i.e., PuPuPuC(A/T)PyPyPyG(T/A)), it acted as a repressing element, but activation occurred if the quarter site was restored to the consensus (Johnson et al., 2001). Further research will be needed to determine if either of these occurrences is global enough to be considered true p53 consensus repressing elements.

3.4.2. Sequence Context of Binding Sites

It was estimated after the identification of the p53 consensus site that approximately 200–300 genes within the human genome contain p53-tagged sites (Tokino et al., 1994). This same approximation was reached in a recent publication utilizing a computer algorithm and genomic databases (Hoh et al., 2002). This method also assessed a score to each as per reliability of the sequence actually being involved in p53-mediated transcriptional activation. However, consensus sequence alone is not sufficient, as other studies have recently pointed to an involvement in the stem–loop structure of the DNA as a critical measure of p53 binding ability (Gohler et al., 2002). Further complicating the issue is that previously identified p53 regulatory regions are now coming into question. The site identified for the PIG3 gene was established on in vitro data using reporter constructs, in which the site actually needed to be placed in tandem repeats in order to observe activation by p53 (Contente et al., 2002). Perhaps not surprisingly, the PIG3 gene is in reality activated by a polymorphic repeat of the sequence 5′-T-G-Py-C-C-3′. The number of repeats varying from 10 to 17 is proportional to the level of PIG3 gene activation. It will be interesting to determine the susceptibility to cancer relative to the number of p53 binding sites in PIG3 and those promoters that have similar situations.

3.4.3. Physical Binding

A technique that has developed quite nicely in the last two years has been chromatin immunoprecipitation (Kuo and Allis, 1999; Nal et al., 2001). This variation of looking at protein–DNA binding possesses a great advantage over the standard electrophoretic mobility shift assay in that it is able to identify association in vivo. A recent study by Szak et al. (2001) was able to delineate the binding between p53 and various promoters in a temporal fashion after adriamycin treatment. Interestingly, the association between p53 and the p21WAF1, mdm2 and PIG3 promoters occurred with separate kinetics (although it is of note that the PIG3 site studied here was deemed dispensable by the PIG3 publication described in the paragraph above).

This would indicate some differential requirement for binding to each site. As described above, some would suggest this difference lies in the posttranslational modification of p53. However, two reports on this topic debunk the hypothesis that modifications on p53 protein have any effect on DNA binding. First, Espinosa and Emerson (2001), using a p21WAF1 promoter fragment wrapped around histone to simulate chromatin, established that p53 acetylation at the C-terminus does not change affinity to DNA. All prior effects seen with respect to DNA binding by p53 were deemed artifactual by this paper due to the use of small oligonucleotides in shift assays. In the context of chromatin, p53 binds to consensus sites with the same affinity regardless of modification, especially at the C-terminus. This finding was confirmed in another study, using a combination of chromatin immunoprecipitation and quantitative PCR, where the conclusion was made that DNA binding by p53 following stabilization of the protein by a DNA damaging agent is proportional to the amount of protein present in the cell (Kaeser and Iggo, 2002). Therefore, the "latent" model of p53 activation—where p53 protein exists in DNA binding "active" and "inactive" forms—does not hold up.

These data taken together indicate that further work is required to determine what is the rate-limiting step to p53 activation/repression of transcription. Some data have suggested that the posttranslational modification steps allow the recruitment of coactivators. Whether this is indeed the rate-limiting step will need more information.

3.5. TRANSCRIPTIONAL TARGETS OF p53

The last step of the pathway of p53 as a transcriptional regulator is the amplification of mRNA from the gene to which p53 has bound to the promoter region of. As of early 2004, there existed more than 100 genes for which there is an independent publication describing the specific activation or repression of transcription. Many of these have been subjects of reviews on their own, and therefore, only those recently identified will be addressed here (please see el-Deiry (1998)) for review on classic p53 target genes). Table 3.1 outlines the majority of genes that have been found to be regulated by p53. Due to space limitations, those that have been solely described in bulk from array and SAGE screens are not shown here. Below is a general classification of the pathways in which the p53-regulated genes exist, and an elaboration on some of the more recently described targets.

Table 3.1. p53 target genes

Pathway	Target gene	Regulation	Binding site found?	Reference
Angiogenesis	BAI1	Induced	Yes	(Nishimori, et al. 1997)
Angiogenesis	EC-NOS	Repressed	Yes	(Mortensen, et al. 1999)
Angiogenesis	GD-AiF	Induced	No	(Van Meir, et al. 1994)
Angiogenesis	Tsp1	Induced	No	(Dameron, et al. 1994)
Antioxidant	Glutathione Peroxidase	Induced	Yes	(Tan, et al. 1999)
Apoptosis	actin	Repressed	No	(Guenal, et al. 1997)
Apoptosis	Aip1	Induced	Yes	(Oda, et al. 2000)
Apoptosis	Apaf1	Induced	Yes	(Moroni, et al. 2001; Robles, et al. 2001; Rozenfeld-Granot, et al. 2002)
Apoptosis	Bax	Induced	Yes	(Miyashita and Reed 1995)
Apoptosis	Bcl-2	Repressed	Yes	(Miyashita, et al. 1994; Miyashita, et al. 1994)
Apoptosis	BRCA1	Repressed	No	(Arizti, et al. 2000; MacLachlan, et al. 2000)
Apoptosis	Caspase-1	Induced	Yes	(Gupta, et al. 2001)
Apoptosis	Caspase-6	Induced	Yes	(MacLachlan and El-Deiry 2002)
Apoptosis	Cathepsin D	Induced	Yes	(Wu, et al. 1998)
Apoptosis	DINP1	Induced	Yes	(Okamura, et al. 2001)
Apoptosis	DR4	Induced	No	(Guan, et al. 2001)
Apoptosis	TRUNDD	Induced	No	(Meng, et al. 2000; Sheikh and Fornace 2000)
Apoptosis	DR5	Induced	Yes	(Wu, et al. 1997; Takimoto and El-Deiry 2000)
Apoptosis	EF-1a	Induced	Yes	(Kato, et al. 1997)
Apoptosis	Fas	Induced	No	(Owen-Schaub, et al. 1995)
Apoptosis	Fractalkine	Induced	Yes	(Shiraishi, et al. 2000)
Apoptosis	IGF-Bp3	Induced	Yes	(Buckbinder, et al. 1995)
Apoptosis	mRTVP1	Induced	No	(Ren, et al. 2002)
Apoptosis	Noxa	Induced	Yes	(Oda, et al. 2000)
Apoptosis	p85	Induced	No	(Yin, et al. 1998)
Apoptosis	PAG608/ Wig-1	Induced	No	(Israeli, et al. 1997)
Apoptosis	PERP	Induced	Yes	(Attardi, et al. 2000)
Apoptosis	PIDD	Induced	Yes	(Lin, et al. 2000)
Apoptosis	PIG3	Induced	Yes - polymorphic	(Polyak, et al. 1994; Contente, et al. 2002)
Apoptosis	Presenillin-1	Repressed	No	(Pastorcic and Das 2000)
Apoptosis	PTEN	Induced	Yes	(Stambolic, MacPherson et al. 2001)
Apoptosis	PUMA	Induced	Yes	(Nakano and Vousden 2001; Yu, et al. 2001)

Table 3.1. p53 target genes (*cont.*)

Pathway	Target gene	Regulation	Binding site found?	Reference
Apoptosis	Scotin	Induced	Yes	(Bourdon, et al. 2002)
Apoptosis	Survivin	Repressed	Yes	(Hoffman, et al. 2002)
Apoptosis	TauT	Repressed	Yes	(Han, et al. 2002)
Apoptosis	Uridine Phos-phorylase	Repressed	Yes	(Zhang, et al. 2001)
Apoptosis	Osteopontin	Induced	Yes	(Morimoto, et al. 2002)
Apoptosis	Zac-1	Induced	Yes	(Rozenfeld-Granot, et al. 2002)
Apoptosis/ Growth	EI24/PIG8	Induced	No	(Lehar, et al. 1996; Polyak, et al. 1997)
Apoptosis/ Growth	GML	Induced	No	(Furuhata, et al. 1996)
Apoptosis/ Growth	MCG10	Induced	Yes	(Zhu and Chen 2000)
DNA repair	CHK1	Repressed	No	(Gottifredi, et al. 2001)
DNA repair	DDB2	Induced	Yes	(Hwang, et al. 1999; Tan and Chu 2002)
DNA repair	p53R2	Induced	Yes	(Tanaka, et al. 2000)
DNA repair	Gadd45	Induced	Yes	(Kastan, et al. 1992)
DNA repair	HB-EGF	Induced	No	(Fang, et al. 2001)
Drug sensitivity	MDR1	Induced/ Repressed	No	(Chin, et al. 1992; Goldsmith, et al. 1995)
Feedback	Cyclin G	Induced	Yes	(Okamoto and Beach 1994)
Feedback	Mdm2	Induced	Yes	(Barak, et al. 1993)
Feedback	p14ARF	Repressed	No	(Robertson and Jones 1998)
Feedback	PTPA	Repressed	No	(Janssens, et al. 2000)
Growth	Alpha-Fetoprotein	Repressed	Yes	(Lee, et al. 1999)
Growth	14-3-3s	Induced	Yes	(Hermeking, et al. 1997)
Growth	B99	Induced	Yes	(Utrera, et al. 1998)
Growth	b-Catenin	Repressed	No	(Sadot, et al. 2001)
Growth	cdc2	Repressed	No	(Yun, et al. 1999)
Growth	cdk4	Repressed	Translational	(Ewen, et al. 1995; Miller, et al. 2000)
Growth	c-fos	Repressed/ Induced	Yes	(Ginsberg, et al. 1991; Kley, et al. 1992; Elkeles, et al. 1999)
Growth	cMet	Induced	Yes	(Seol, et al. 1999)
Growth	c-myc	Repressed	No	(Moberg, et al. 1992)
Growth	Cyclin B1	Repressed	No	(Innocente, et al. 1999)
Growth	DNA Topo II	Repressed	No	(Wang, et al. 1997)
Growth	IL2	Repressed	No	(Pesch, et al. 1996)
Growth	IL4	Repressed	No	(Pesch, et al. 1996)
Growth	IL6	Repressed	No	(Santhanam, et al. 1991)
Growth	Insulin receptor	Repressed	No	(Webster, et al. 1996)

Table 3.1. p53 target genes (*cont.*)

Pathway	Target gene	Regulation	Binding site found?	Reference
Growth	MAP4	Repressed	No	(Murphy, et al. 1996)
Growth	p21Waf1	Induced	Yes	(el-Deiry, et al. 1993)
Growth	PC3/TIS21/BTG2	Induced	No	(Rouault, et al. 1996; Guardavaccaro, et al. 2000)
Growth	Pold1	Repressed	No	(Li and 2001)
Growth	PTGFb	Induced	Yes	(Tan, et al. 2000)
Growth	Reprimo	Induced	Yes	(Ohki, et al. 2000)
Growth	RTP/rit42	Induced	No	(Kurdistani, et al. 1998)
Growth	Stathmin	Repressed	No	(Murphy, et al. 1999)
Growth	Wee1	Repressed	No	(Leach, et al. 1998)
Growth	Wip1	Induced	No	(Fiscella, et al. 1997)
Growth	P2XM	Induced	Yes	(Urano, et al. 1997)
Invasion and metastasis	KAI1	Induced	Yes	(Mashimo, et al. 1998)
Invasion and metastasis	Maspin	Induced	Yes	(Zou, et al. 2000)
Invasion and metastasis	MMP1	Repressed	No	(Sun, et al. 1999)
Invasion and metastasis	MMP13	Repressed	No	(Sun, et al. 2000)
Invasion and metastasis	PAI1	Induced	Yes	(Kunz, et al. 1995)
?	HIC-1	Induced	Yes	(Wales, et al. 1995)
?	Hsp70	Repressed	No	(Agoff, et al. 1993)
?	TP53TGI	Induced	No	(Takei, et al. 1998)

3.5.1. Apoptosis

By far, the largest category of genes regulated by p53 is those that induce cell death. The sheer number of targets that are involved in this pathway imply that there may be many different ways for a cell to die, and that these conditions may change with tissue type and apoptosis stimulus.

The PIDD gene was isolated from an RNA differential display technique using DP16.1 erythroleukemic cell lines with a temperature sensitive p53 (Lin et al., 2000). The gene bears homology to the FADD, DAPK, and RAIDD death domains. From homology and subcellular localization, it appears that PIDD may be a signaling protein on the intracellular side of a death receptor.

A group of recent papers has suggested two targets of p53 that, while not activating apoptosis on their own, lowers the threshold for apoptosis, thereby chemosensitizing the cells. Apaf1 is part of a holoenzyme that associates with caspase-9 that awaits cytochrome c to allow activation of the caspase cascade (Li et al., 1997;

Srinivasula et al., 1998; Zou et al., 1997). Both Apaf1 and caspase-9 are required for p53 dependent apoptosis (Soengas et al., 1999). Apaf1 was found by three independent groups to be a transcriptional target of p53 (Moroni et al., 2001; Robles et al., 2001; Rozenfeld-Granot et al., 2002). In one case, it was found that the transcriptional coactivator Zac-1, which is another possible transcriptional target of p53, specifically participates in the control of Apaf-1 transcription (Rozenfeld-Granot et al., 2002). Another caspase, caspase-6, also was found to be a target of p53, the activation of which was necessary for efficient apoptosis to occur in the presence of p53 overexpression and treatment with chemotherapy (MacLachlan and el-Deiry, 2002). Cotreatment of cells with a p53 expressing adenovirus as well as the DNA damaging agent Adriamycin dramatically increased the proteolytic function of caspase-6. Recently, caspase-1 was also found to be induced by p53; however, the exact role of this caspase in apoptosis is unknown (Gupta et al., 2001).

p53AIP1 was found in a yeast enhancer trap as a target gene of p53 (Oda et al., 2000b). Ectopic expression is able to cause apoptosis by dissipation of mitochondrial electrochemical gradient. Also, the target is apparently specifically activated by a serine-46 phosphorylated p53. A protein that appears to be involved in this process as well as acting as a p53 target gene is p53DINP1 (Okamura et al., 2001). Induction of p53DINP1 requires wild-type p53, and elimination of the protein by means of antisense significantly reduces phosphorylation of p53 at ser-46 as well as apoptotic gene induction by p53. Further underlining the specificity of ser-46 to apoptosis, knockdown of p53DINP1 had no effect on growth arrest targets of p53.

The PTEN tumor suppressor is known to regulate PI3 kinase products, as well as protein kinase B/Akt (Haas-Kogan et al., 1998; Maehama and Dixon, 1998; Stambolic et al., 1998). Through these effects, PTEN is able to regulate cell survival. p53 is able to activate the expression of PTEN through an intronic binding site (Stambolic et al., 2001). In cells that are mutant for PTEN, the apoptosis inducing capability of p53 is diminished.

The PUMA and Noxa genes are part of a growing family of proteins that contain BH3 domains (Nakano and Vousden, 2001; Oda et al., 2000a; Yu et al., 2001). Included in this family is a classic p53 target, Bax. All these genes are strongly induced by p53 and lead to a rapid apoptotic response. The mechanism of apoptosis induction via these proteins appears to work though cytochrome c release from the mitochondria.

The BID protein is also a member of the BH3 family of proteins that promote apoptosis. p53 was found to induce expression of BID both in vitro and in vivo, specifically in the splenic pulp and colonic epithelia. It has been suggested that regulation of BID by p53 promotes a decrease in the cell death threshold as well as chemosensitivity of cells in culture (Sax et al., 2002).

Further acting in an apoptotic pathway, p53 is able to repress such genes as survivin and Bcl-2 (Hoffman et al., 2002; Miyashita et al., 1994). Clones overexpressing survivin, and therefore not able to be repressed by p53, were not able to undergo cell death, regardless of efficient stabilization of p53.

3.5.2. Feedback, DNA Repair and Growth Control

The activation of the cyclin G gene by p53 has for some time been left in the uncharacterized target category due to the lack of functional information on cyclin G (Okamoto and Beach, 1994). A recent report identified protein phosphatase 2A as a binding protein of cyclin G (Okamoto et al., 2002). In addition, it was found that PP2A dephosphorylates mdm2, leading to mdm2's targeted destruction of p53. This is yet another example of p53 acting in a feedback loop of its own protein stability, which includes the repression of p14ARF (Robertson and Jones, 1998) and the activation of mdm2 (Barak et al., 1993).

BRCA1, as described in Section 3.3.4, represses the ability of p53 to activate apoptosis targets (MacLachlan et al., 2002). A feedback loop that allows p53 to get around this is the ability of p53 to repress transcription of BRCA1 (Arizti et al., 2000; MacLachlan et al., 2000a). The repression of BRCA1 seems to correlate with the onset of apoptosis target gene activation.

The DDB2 gene is part of a group of nucleotide excision repair proteins that is mutant or lost in XPE patients (Chu and Chang, 1988). This gene was found to be a target of p53, as well as one that is enhanced by association with BRCA1 (Takimoto et al., 2002a; Tan and Chu, 2002). Other DNA repair targets of p53 include p53R2, a ribonucleotide reductase involved in replenishing nucleotide pools after DNA damage (Tanaka et al., 2000). It is thought that this p53 to p53R2 pathway is analogous to the Rad53 → Dun1 → RNR2, 3 pathway in yeast. Gadd45, a classic target of p53, was recently knocked out in mice and found to be involved in stability of the genome (Hollander et al., 1999). Interestingly, a recent finding has suggested that gadd45 is only induced by p53 in the presence of DNA damage (Xiao et al., 2000).

Reprimo, a cytoplasmic and highly glycoslyated protein, was identified as a target of p53 in a differential display screening approach (Ohki et al., 2000). Overexpression of the protein caused cells to arrest in G2 phase of the cell cycle, indicating that Reprimo may act in a similar pathway of another G2 phase p53 target, 14-3-3σ (Hermeking et al., 1997). Another gene controlled by p53 that is thought to affect G2 phase progression is the Snk/Plk1 kinase. This protein was found to be regulated by p53, and in its absence causes mitotic catastrophe (Burns et al., 2003). Snk/Plk1 may contribute to a G2 phase checkpoint, wherein the case of damaged chromosomes during mitosis, cells will not proceed through mitosis until the damage has been repaired. There is a clear need for p53 in preventing genomic damage, and induction of this gene may be the means by which p53 controls this part of the cell cycle.

A key piece of information for all these p53 transcriptional targets will be if they all act at once, or if specific targets are activated under certain kinds of stress in particular tissues. What would be fascinating to determine, for example in induction of apoptosis by p53, is the mechanisms by which p53 recognizes and activates individual promoters of apoptosis inducing genes, while ignoring others.

3.6. THERAPEUTICS

Of course, the end goal of the massive amount of research on the transcriptional function of p53 is to find a therapy to reactivate this function in cancer cells that have lost normal p53 activity. While clinical trials are going forward with such directions as a p53- expressing adenovirus, other studies have focused on "fixing" the mutant p53 that exists within a cancer cell. Two small molecules, CP-31398 and PRIMA, have been published recently that opens up the possibility of restoring control of transcriptional regulatory functions to mutant p53 (Bykov et al., 2002; Foster et al., 1999).

In a screen for small molecules, which searched for compounds that allowed p53 to retain wild-type conformation at high temperatures, CP-31398 was identified as a protein that could also allow transcription of the p21WAF1 promoter in cells transfected with a mutant p53. This compound also forced the regression of tumors derived from A375 melanoma and DLD-1 colon carcinoma cells in mice. Recent reports however have shown that CP-31398 also stabilizes and activates wild-type p53 indicating potential toxicity (Takimoto et al., 2002b), as well as reports that the concentration of drug does not remain at high-enough levels cellularly to justify human clinical trials (Foster et al., 1999). Nevertheless, these studies led the way in looking for drugs that could reactivate p53.

PRIMA-1, a compound with a much different molecular structure than CP-31398, came out from a screen for compounds that inhibited growth of Saos2 cells with an inducible mutant p53 in a mutant p53-dependent manner. In a similar assay that was used for CP-31398, PRIMA-1 also inhibits the disappearance of the PAb1620 epitope on p53 after heating. In addition to activating an apoptotic response dependent on mutant p53, PRIMA-1 also restores transcriptional activity to mutant p53 and reduces tumor volume grown by cells that express mutant p53. A significant difference between the two molecules is that PRIMA-1 is able to convert an existing mutant p53 to a wild-type conformation, while CP-31398 needs to bind to a newly synthesized mutant p53 and hold it in a wild-type conformation before folding into a mutant one. Future experiments on the efficacy of PRIMA-1 as an actual pharmacological agent will provide important information on its use as a therapeutic.

3.7. CONCLUSIONS

The investigation into the functions of p53 as a transcriptional regulator have considerably increased our understanding of the mechanisms lost in cancers that possess mutant p53. In the past few years especially, we have discovered that the pathway to tumor suppression by p53 is more intricate and complicated than originally thought. Clearly, as proven by the discovery of potential therapies that restore the transcriptional functions of p53, this information is critical for the development of anticancer agents.

REFERENCES

Aprelikova, O., Pace, A. J., Fang, B., Koller, B. H., and Liu, E. T. (2001). BRCA1 is a selective co-activator of 14-3-3 sigma gene transcription in mouse embryonic stem cells. *J Biol Chem* 276:25647–25650.

Aprelikova, O. N., Fang, B. S., Meissner, E. G., Cotter, S., Campbell, M., Kuthiala, A., Bessho, M., Jensen, R. A., and Liu, E. T. (1999). BRCA1-associated growth arrest is RB-dependent. *Proc Natl Acad Sci USA* 96:11866–11871.

Ard, P. G., Chatterjee, C., Kunjibettu, S., Adside, L. R., Gralinski, L. E., and McMahon, S. B. (2002). Transcriptional regulation of the mdm2 oncogene by p53 requires TRRAP acetyltransferase complexes. *Mol Cell Biol* 22:5650–5661.

Arizti, P., Fang, L., Park, I., Yin, Y., Solomon, E., Ouchi, T., Aaronson, S. A., and Lee, S. W. (2000). Tumor suppressor p53 is required to modulate BRCA1 expression. *Mol Cell Biol* 20:7450–7459.

Avantaggiati, M. L., Ogryzko, V., Gardner, K., Giordano, A., Levine, A. S., and Kelly, K. (1997). Recruitment of p300/CBP in p53-dependent signal pathways. *Cell* 89:1175–1184.

Barak, Y., Juven, T., Haffner, R., and Oren, M. (1993). mdm2 expression is induced by wild type p53 activity. *EMBO J* 12:461–468.

Bargonetti, J., Reynisdottir, I., Friedman, P. N., and Prives, C. (1992). Site-specific binding of wild-type p53 to cellular DNA is inhibited by SV40 T antigen and mutant p53. *Genes Dev* 6:1886–1898.

Barlev, N. A., Liu, L., Chehab, N. H., Mansfield, K., Harris, K. G., Halazonetis, T. D., and Berger, S. L. (2001). Acetylation of p53 activates transcription through recruitment of coactivators/histone acetyltransferases. *Mol Cell* 8:1243–1254.

Bennett, M., Macdonald, K., Chan, S. W., Luzio, J. P., Simari, R., and Weissberg, P. (1998). Cell surface trafficking of Fas: a rapid mechanism of p53-mediated apoptosis. *Science* 282:290–293.

Berger, S. L. (2002). Histone modifications in transcriptional regulation. *Curr Opin Genet Dev* 12:142–148.

Bochar, D. A., Wang, L., Beniya, H., Kinev, A., Xue, Y., Lane, W. S., Wang, W., Kashanchi, F., and Shiekhattar, R. (2000). BRCA1 is associated with a human SWI/SNF-related complex: linking chromatin remodeling to breast cancer. *Cell* 102:257–265.

Boddy, M. N., Howe, K., Etkin, L. D., Solomon, E., and Freemont, P. S. (1996). PIC 1, a novel ubiquitin-like protein which interacts with the PML component of a multiprotein complex that is disrupted in acute promyelocytic leukaemia. *Oncogene* 13:971–982.

Breiding, D. E., Sverdrup, F., Grossel, M. J., Moscufo, N., Boonchai, W., and Androphy, E. J. (1997). Functional interaction of a novel cellular protein with the papillomavirus E2 transactivation domain. *Mol Cell Biol* 17:7208–7219.

Brown, C. E., Howe, L., Sousa, K., Alley, S. C., Carrozza, M. J., Tan, S., and Workman, J. L. (2001). Recruitment of HAT complexes by direct activator interactions with the ATM-related Tra1 subunit. *Science* 292:2333–2337.

Brownell, J. E., Zhou, J., Ranalli, T., Kobayashi, R., Edmondson, D. G., Roth, S. Y., and Allis, C. D. (1996). Tetrahymena histone acetyltransferase A: a homolog to yeast Gcn5p linking histone acetylation to gene activation. *Cell* 84:843–851.

Bulavin, D. V., Saito, S., Hollander, M. C., Sakaguchi, K., Anderson, C. W., Appella, E., and Fornace, A. J., Jr. (1999). Phosphorylation of human p53 by p38 kinase coordinates N-terminal phosphorylation and apoptosis in response to UV radiation. *EMBO J* 18:6845–6854.

Burns, T. F., Fei, P., Scata, K. A., Dicker, D. T. and el-Deiry, W. S. (2003): Silencing of the novel p53 target gene Snk/Plk2 leads to mitotic catastrophe in paclitaxel (taxol)-exposed cells. *Mol Cell Biol* 23:5556–5571.

Bykov, V. J., Issaeva, N., Shilov, A., Hultcrantz, M., Pugacheva, E., Chumakov, P., Bergman, J., Wiman, K. G., and Selivanova, G. (2002). Restoration of the tumor suppressor function to mutant p53 by a low- molecular-weight compound. *Nat Med* 8:282–288.

Caelles, C., Helmberg, A., and Karin, M. (1994). p53-dependent apoptosis in the absence of transcriptional activation of p53-target genes. *Nature* 370:220–223.

Chai, Y. L., Cui, J., Shao, N., Shyam, E., Reddy, P., and Rao, V. N. (1999). The second BRCT domain of BRCA1 proteins interacts with p53 and stimulates transcription from the p21WAF1/CIP1 promoter. *Oncogene* 18:263–268.

Chen, X., Ko, L. J., Jayaraman, L., and Prives, C. (1996). p53 levels, functional domains, and DNA damage determine the extent of the apoptotic response of tumor cells. *Genes Dev* 10:2438–2451.

Chin, K. V., Ueda, K., Pastan, I., and Gottesman, M. M. (1992). Modulation of activity of the promoter of the human MDR1 gene by Ras and p53. Science 255:459–462.

Chu, G., and Chang, E. (1988). Xeroderma pigmentosum group E cells lack a nuclear factor that binds to damaged DNA. *Science* 242:564–567.

Contente, A., Dittmer, A., Koch, M. C., Roth, J., and Dobbelstein, M. (2002). A polymorphic microsatellite that mediates induction of PIG3 by p53. *Nat Genet* 30:315–320.

Costanzo, A., Merlo, P., Pediconi, N., Fulco, M., Sartorelli, V., Cole, P. A., Fontemaggi, G., Fanciulli, M., Schiltz, L., Blandino, G., et al. (2002). DNA damage-dependent acetylation of p73 dictates the selective activation of apoptotic target genes. *Mol Cell* 9:175–186.

Davison, T. S., Vagner, C., Kaghad, M., Ayed, A., Caput, D., and Arrowsmith, C. H. (1999). p73 and p63 are homotetramers capable of weak heterotypic interactions with each other but not with p53. *J Biol Chem* 274:18709–18714.

DeLeo, A. B., Jay, G., Appella, E., Dubois, G. C., Law, L. W., and Old, L. J. (1979). Detection of a transformation-related antigen in chemically induced sarcomas and other transformed cells of the mouse. *Proc Natl Acad Sci USA* 76:2420–2424.

Demonacos, C., Krstic-Demonacos, M., and La Thangue, N. B. (2001). A TPR motif cofactor contributes to p300 activity in the p53 response. *Mol Cell* 8:71–84.

Di Como, C. J., Gaiddon, C., and Prives, C. (1999). p73 function is inhibited by tumor-derived p53 mutants in mammalian cells. *Mol Cell Biol* 19:1438–1449.

D'Orazi, G., Cecchinelli, B., Bruno, T., Manni, I., Higashimoto, Y., Saito, S., Gostissa, M., Coen, S., Marchetti, A., Del Sal, G., et al. (2002). Homeodomain-interacting protein kinase-2 phosphorylates p53 at Ser 46 and mediates apoptosis. *Nat Cell Biol* 4:11–19.

El-Deiry, W. S. (1998). Regulation of p53 downstream genes, *Semin Cancer Biol* 8:345–357.

El-Deiry, W. S., Kern, S. E., Pietenpol, J. A., Kinzler, K. W., and Vogelstein, B. (1992). Definition of a consensus binding site for p53. *Nat Genet* 1:45–49.

El-Deiry, W. S., Tokino, T., Velculescu, V. E., Levy, D. B., Parsons, R., Trent, J. M., Lin, D., Mercer, W. E., Kinzler, K. W., and Vogelstein, B. (1993). WAF1, a potential mediator of p53 tumor suppression. *Cell* 75:817–825.

Espinosa, J. M., and Emerson, B. M. (2001). Transcriptional regulation by p53 through intrinsic DNA/chromatin binding and site-directed cofactor recruitment. *Mol Cell* 8:57–69.

Farmer, G., Bargonetti, J., Zhu, H., Friedman, P., Prywes, R., and Prives, C. (1992). Wild-type p53 activates transcription in vitro. *Nature* 358:83–86.

Flores, E. R., Tsai, K. Y., Crowley, D., Sengupta, S., Yang, A., McKeon, F., and Jacks, T. (2002). p63 and p73 are required for p53-dependent apoptosis in response to DNA damage. *Nature* 416:560–564.

Foster, B. A., Coffey, H. A., Morin, M. J., and Rastinejad, F. (1999). Pharmacological rescue of mutant p53 conformation and function, Science 286:2507–2510.

Friborg, J., Jr., Kong, W., Hottiger, M. O., and Nabel, G. J. (1999). p53 inhibition by the LANA protein of KSHV protects against cell death. *Nature* 402:889–894.

Gaiddon, C., Lokshin, M., Ahn, J., Zhang, T., and Prives, C. (2001). A subset of tumor-derived mutant forms of p53 down-regulate p63 and p73 through a direct interaction with the p53 core domain. *Mol Cell Biol* 21:1874–1887.

Garkavtsev, I., Grigorian, I. A., Ossovskaya, V. S., Chernov, M. V., Chumakov, P. M., and Gudkov, A. V. (1998). The candidate tumour suppressor p33ING1 cooperates with p53 in cell growth control. *Nature* 391:295–298.

Garkavtsev, I. V., Kley, N., Grigorian, I. A., and Gudkov, A. V. (2001). The Bloom syndrome protein interacts and cooperates with p53 in regulation of transcription and cell growth control. *Oncogene* 20:8276–8280.

Gohler, T., Reimann, M., Cherny, D., Walter, K., Warnecke, G., Kim, E., and Deppert, W. (2002). Specific interaction of p53 with target binding sites is determined by DNA conformation and is regulated by the C-terminal domain. *J Biol Chem* 8:8.

Gottifredi, V., Shieh, S., Taya, Y., and Prives, C. (2001). From the Cover: p53 accumulates but is functionally impaired when DNA synthesis is blocked. *Proc Natl Acad Sci USA* 98:1036–1041.

Grigorian, M., Andresen, S., Tulchinsky, E., Kriajevska, M., Carlberg, C., Kruse, C., Cohn, M., Ambart-sumian, N., Christensen, A., Selivanova, G., and Lukanidin, E. (2001). Tumor suppressor p53 protein is a new target for the metastasis- associated Mts1/S100A4 protein: functional consequences of their interaction. *J Biol Chem* 276:22699–22708.

Gu, W., and Roeder, R. G. (1997). Activation of p53 sequence-specific DNA binding by acetylation of the p53 C-terminal domain. Cell 90:595–606.

Gupta, S., Radha, V., Furukawa, Y., and Swarup, G. (2001). Direct transcriptional activation of human caspase-1 by tumor suppressor p53. *J Biol Chem* 276:10585–10588.

Haas-Kogan, D., Shalev, N., Wong, M., Mills, G., Yount, G., and Stokoe, D. (1998). Protein kinase B (PKB/Akt) activity is elevated in glioblastoma cells due to mutation of the tumor suppressor PTEN/MMAC. *Curr Biol* 8:1195–1198.

Han, X., Patters, A. B., and Chesney, R. W. (2002). Transcriptional repression of taurine transporter gene (TauT) by p53 in renal cells. *J Biol Chem* 5:5.

Haupt, Y., Rowan, S., Shaulian, E., Vousden, K. H., and Oren, M. (1995). Induction of apoptosis in HeLa cells by trans-activation-deficient p53. *Genes Dev* 9:2170–2183.

Hermeking, H., Lengauer, C., Polyak, K., He, T. C., Zhang, L., Thiagalingam, S., Kinzler, K. W., and Vogelstein, B. (1997). 14-3-3 sigma is a p53-regulated inhibitor of G2/M progression. *Mol Cell* 1:3–11.

Hoffman, W. H., Biade, S., Zilfou, J. T., Chen, J., and Murphy, M. (2002). Transcriptional repression of the anti-apoptotic survivin gene by wild type p53. *J Biol Chem* 277:3247–3257.

Hofmann, T. G., Moller, A., Sirma, H., Zentgraf, H., Taya, Y., Droge, W., Will, H., and Schmitz, M. L. (2002). Regulation of p53 activity by its interaction with homeodomain- interacting protein kinase-2. *Nat Cell Biol* 4:1–10.

Hoh, J., Jin, S., Parrado, T., Edington, J., Levine, A. J., and Ott, J. (2002). The p53MH algorithm and its application in detecting p53-responsive genes. *Proc Natl Acad Sci USA* 99:8467–8472.

Hollander, M. C., Sheikh, M. S., Bulavin, D. V., Lundgren, K., Augeri-Henmueller, L., Shehee, R., Moli-naro, T. A., Kim, K. E., Tolosa, E., Ashwell, J. D., et al. (1999). Genomic instability in Gadd45a-deficient mice. *Nat Genet* 23:176–184.

Ito, A., Lai, C. H., Zhao, X., Saito, S., Hamilton, M. H., Appella, E., and Yao, T. P. (2001). p300/CBP-mediated p53 acetylation is commonly induced by p53-activating agents and inhibited by MDM2. *EMBO J* 20:1331–1340.

Iwabuchi, K., Bartel, P. L., Li, B., Marraccino, R., and Fields, S. (1994). Two cellular proteins that bind to wild-type but not mutant p53. *Proc Natl Acad Sci USA* 91:6098–6102.

Iwabuchi, K., Li, B., Massa, H. F., Trask, B. J., Date, T., and Fields, S. (1998). Stimulation of p53-mediated transcriptional activation by the p53- binding proteins, 53BP1 and 53BP2. *J Biol Chem* 273:26061–26068.

Jenkins, J. R., Rudge, K., and Currie, G. A. (1984). Cellular immortalization by a cDNA clone encoding the transformation- associated phosphoprotein p53. *Nature* 312:651–654.

Jin, Y., Zeng, S. X., Dai, M.-S., Yang, X.-J., and Lu, H. (2002). MDM2 Inhibits PCAF-mediated p53 acetylation. *J Biol Chem.* 279:20035–20043.

Johnson, R. A., Ince, T. A., and Scotto, K. W. (2001). Transcriptional repression by p53 through direct binding to a novel DNA element. *J Biol Chem* 276:27716–27720.

Juan, L. J., Shia, W. J., Chen, M. H., Yang, W. M., Seto, E., Lin, Y. S., and Wu, C. W. (2000). Histone deacety-lases specifically down-regulate p53-dependent gene activation. *J Biol Chem* 275:20436–20443.

Kaeser, M. D., and Iggo, R. D. (2002). Chromatin immunoprecipitation analysis fails to support the latency model for regulation of p53 DNA binding activity in vivo. *Proc Natl Acad Sci USA* 99: 95–100.

Kaghad, M., Bonnet, H., Yang, A., Creancier, L., Biscan, J. C., Valent, A., Minty, A., Chalon, P., Lelias, J. M., Dumont, X., et al. (1997). Monoallelically expressed gene related to p53 at 1p36, a region frequently deleted in neuroblastoma and other human cancers. *Cell* 90:809–819.

Kahyo, T., Nishida, T., and Yasuda, H. (2001). Involvement of PIAS1 in the sumoylation of tumor suppressor p53. *Mol Cell* 8:713–718.

Kamitani, T., Nguyen, H. P., and Yeh, E. T. (1997). Preferential modification of nuclear proteins by a novel ubiquitin-like molecule. *J Biol Chem* 272:14001–14004.

Kim, E. J., Park, J. S., and Um, S. J. (2002). Identification and characterization of HIPK2 interacting with p73 and modulating functions of the p53 family in vivo. *J Biol Chem* 29:29.

Kuo, M. H., and Allis, C. D. (1999). In vivo cross-linking and immunoprecipitation for studying dynamic protein:DNA associations in a chromatin environment. *Methods* 19:425–433.

Lane, D. P. (1984). Cell immortalization and transformation by the p53 gene. *Nature* 312:596–597.

Langley, E., Pearson, M., Faretta, M., Bauer, U. M., Frye, R. A., Minucci, S., Pelicci, P. G., and Kouzarides, T. (2002). Human SIR2 deacetylates p53 and antagonizes PML/p53-induced cellular senescence. *EMBO J* 21:2383–2396.

Lee, D., Kim, J. W., Seo, T., Hwang, S. G., Choi, E. J., and Choe, J. (2002). SWI/SNF complex interacts with tumor suppressor p53 and is necessary for the activation of p53-mediated transcription. *J Biol Chem* 277:22330–22337.

Lee, G. W., Melchior, F., Matunis, M. J., Mahajan, R., Tian, Q., and Anderson, P. (1998). Modification of Ran GTPase-activating protein by the small ubiquitin- related modifier SUMO-1 requires Ubc9, an E2-type ubiquitin-conjugating enzyme homologue. *J Biol Chem* 273:6503–6507.

Lee, K. C., Crowe, A. J., and Barton, M. C. (1999). p53-mediated repression of alpha-fetoprotein gene expression by specific DNA binding. *Mol Cell Biol* 19:1279–1288.

Li, P., Nijhawan, D., Budihardjo, I., Srinivasula, S. M., Ahmad, M., Alnemri, E. S., and Wang, X. (1997). Cytochrome c and dATP-dependent formation of Apaf-1/caspase-9 complex initiates an apoptotic protease cascade. *Cell* 91:479–489.

Lill, N. L., Grossman, S. R., Ginsberg, D., DeCaprio, J., and Livingston, D. M. (1997). Binding and modulation of p53 by p300/CBP coactivators. *Nature* 387:823–827.

Lin, J., Blake, M., Tang, C., Zimmer, D., Rustandi, R. R., Weber, D. J., and Carrier, F. (2001). Inhibition of p53 transcriptional activity by the S100B calcium-binding protein. *J Biol Chem* 276:35037–35041.

Lin, Y., Ma, W., and Benchimol, S. (2000). Pidd, a new death-domain-containing protein, is induced by p53 and promotes apoptosis. *Nat Genet* 26:122–127.

Liu, L., Scolnick, D. M., Trievel, R. C., Zhang, H. B., Marmorstein, R., Halazonetis, T. D., and Berger, S. L. (1999). p53 sites acetylated in vitro by PCAF and p300 are acetylated in vivo in response to DNA damage. *Mol Cell Biol* 19:1202–1209.

Liu, Y., Colosimo, A. L., Yang, X. J., and Liao, D. (2000). Adenovirus E1B 55-kilodalton oncoprotein inhibits p53 acetylation by PCAF. *Mol Cell Biol* 20:5540–5553.

Luo, J., Nikolaev, A. Y., Imai, S., Chen, D., Su, F., Shiloh, A., Guarente, L., and Gu, W. (2001). Negative control of p53 by Sir2alpha promotes cell survival under stress. *Cell* 107:137–148.

Luo, J., Su, F., Chen, D., Shiloh, A., and Gu, W. (2000). Deacetylation of p53 modulates its effect on cell growth and apoptosis. *Nature* 408:377–381.

MacLachlan, T. K., Dash, B. C., Dicker, D. T., and el-Deiry, W. S. (2000a). Repression of BRCA1 through a feedback loop involving p53. *J Biol Chem* 275:31869–31875.

MacLachlan, T. K., and el-Deiry, W. S. (2002). Apoptotic threshold is lowered by p53 transactivation of caspase-6. *Proc Natl Acad Sci USA* 99:9492–9497.

MacLachlan, T. K., Somasundaram, K., Sgagias, M., Shifman, Y., Muschel, R. J., Cowan, K. H., and el-Deiry, W. S. (2000b). BRCA1 effects on the cell cycle and the DNA damage response are linked to altered gene expression. *J Biol Chem* 275:2777–2785.

MacLachlan, T. K., Takimoto, R., and el-Deiry, W. S. (2002). BRCA1 directs a selective p53-dependent transcriptional response towards growth arrest and DNA repair targets. *Mol Cell Biol* 22:4280–4292.

Maehama, T., and Dixon, J. E. (1998). The tumor suppressor, PTEN/MMAC1, dephosphorylates the lipid second messenger, phosphatidylinositol 3,4,5-trisphosphate. *J Biol Chem* 273:13375–13378.

Mahajan, R., Delphin, C., Guan, T., Gerace, L., and Melchior, F. (1997). A small ubiquitin-related polypeptide involved in targeting RanGAP1 to nuclear pore complex protein RanBP2. *Cell* 88:97–107.

Matunis, M. J., Coutavas, E., and Blobel, G. (1996). A novel ubiquitin-like modification modulates the partitioning of the Ran-GTPase-activating protein RanGAP1 between the cytosol and the nuclear pore complex. *J Cell Biol* 135:1457–1470.

Megidish, T., Xu, J. H., and Xu, C. W. (2002). Activation of p53 by protein inhibitor of activated Stat1 (PIAS1). *J Biol Chem* 277:8255–8259.

Miyashita, T., Harigai, M., Hanada, M., and Reed, J. C. (1994). Identification of a p53-dependent negative response element in the bcl- 2 gene. *Cancer Res* 54:3131–3135.

Miyashita, T., and Reed, J. C. (1995). Tumor suppressor p53 is a direct transcriptional activator of the human bax gene. *Cell* 80:293–299.

Moroni, M. C., Hickman, E. S., Denchi, E. L., Caprara, G., Colli, E., Cecconi, F., Muller, H., and Helin, K. (2001). Apaf-1 is a transcriptional target for E2F and p53. *Nat Cell Biol* 3:552–558.

Murphy, M., Ahn, J., Walker, K. K., Hoffman, W. H., Evans, R. M., Levine, A. J., and George, D. L. (1999). Transcriptional repression by wild-type p53 utilizes histone deacetylases, mediated by interaction with mSin3a. *Genes Dev* 13:2490–2501.

Murphy, M., Hinman, A., and Levine, A. J. (1996). Wild-type p53 negatively regulates the expression of a microtubule—associated protein. *Genes Dev* 10:2971–2980.

Nagashima, M., Shiseki, M., Miura, K., Hagiwara, K., Linke, S. P., Pedeux, R., Wang, X. W., Yokota, J., Riabowol, K., and Harris, C. C. (2001). DNA damage-inducible gene p33ING2 negatively regulates cell proliferation through acetylation of p53. *Proc Natl Acad Sci USA* 98:9671–9676.

Nakamura, S., Roth, J. A., and Mukhopadhyay, T. (2000). Multiple lysine mutations in the C-terminal domain of p53 interfere with MDM2-dependent protein degradation and ubiquitination. *Mol Cell Biol* 20:9391–9398.

Nakano, K., and Vousden, K. H. (2001). PUMA, a novel proapoptotic gene, is induced by p53. *Mol Cell* 7:683–694.

Nal, B., Mohr, E., and Ferrier, P. (2001). Location analysis of DNA-bound proteins at the whole-genome level: untangling transcriptional regulatory networks. *Bioessays* 23:473–476.

Nie, Y., Li, H. H., Bula, C. M., and Liu, X. (2000). Stimulation of p53 DNA binding by c-Abl requires the p53 C terminus and tetramerization. *Mol Cell Biol* 20:741–748.

Oda, E., Ohki, R., Murasawa, H., Nemoto, J., Shibue, T., Yamashita, T., Tokino, T., Taniguchi, T., and Tanaka, N. (2000a). Noxa, a BH3-only member of the Bcl-2 family and candidate mediator of p53-induced apoptosis. *Science* 288:1053–1058.

Oda, K., Arakawa, H., Tanaka, T., Matsuda, K., Tanikawa, C., Mori, T., Nishimori, H., Tamai, K., Tokino, T., Nakamura, Y., and Taya, Y. (2000b). p53AIP1, a potential mediator of p53-dependent apoptosis, and its regulation by Ser-46-phosphorylated p53. *Cell* 102:849–862.

Ogryzko, V. V., Schiltz, R. L., Russanova, V., Howard, B. H., and Nakatani, Y. (1996). The transcriptional coactivators p300 and CBP are histone acetyltransferases *Cell* 87:953–959.

Ohki, R., Nemoto, J., Murasawa, H., Oda, E., Inazawa, J., Tanaka, N., and Taniguchi, T. (2000). Reprimo, a new candidate mediator of the p53-mediated cell cycle arrest at the G2 phase. *J Biol Chem* 275:22627–22630.

Okamoto, K., and Beach, D. (1994). Cyclin G is a transcriptional target of the p53 tumor suppressor protein. *EMBO J* 13:4816–4822.

Okamoto, K., Li, H., Jensen, M. R., Zhang, T., Taya, Y., Thorgeirsson, S. S., and Prives, C. (2002). Cyclin G recruits PP2A to dephosphorylate Mdm2. *Mol Cell* 9:761–771.

Okamura, S., Arakawa, H., Tanaka, T., Nakanishi, H., Ng, C. C., Taya, Y., Monden, M., and Nakamura, Y. (2001). p53DINP1, a p53-inducible gene, regulates p53-dependent apoptosis. *Mol Cell* 8:85–94.

Ouchi, T., Monteiro, A. N., August, A., Aaronson, S. A., and Hanafusa, H. (1998). BRCA1 regulates p53-dependent gene expression. *Proc Natl Acad Sci USA* 95:2302–2306.

Pastorcic, M., and Das, H. K. (2000). Regulation of transcription of the human presenilin-1 gene by ets transcription factors and the p53 protooncogene. *J Biol Chem* 275:34938–34945.

Pearson, M., Carbone, R., Sebastiani, C., Cioce, M., Fagioli, M., Saito, S., Higashimoto, Y., Appella, E., Minucci, S., Pandolfi, P. P., and Pelicci, P. G. (2000). PML regulates p53 acetylation and premature senescence induced by oncogenic Ras. *Nature* 406:207–210.

Peng, Y. C., Breiding, D. E., Sverdrup, F., Richard, J., and Androphy, E. J. (2000). AMF-1/Gps2 binds p300 and enhances its interaction with papillomavirus E2 proteins. *J Viroli* 74:5872–5879.

Peng, Y. C., Kuo, F., Breiding, D. E., Wang, Y. F., Mansur, C. P., and Androphy, E. J. (2001). AMF1 (GPS2) modulates p53 transactivation. *Mol Cell Biol* 21:5913–5924.

Prives, C., and Manley, J. L. (2001). Why is p53 acetylated?, Cell *107*, 815–8.

Ratovitski, E. A., Patturajan, M., Hibi, K., Trink, B., Yamaguchi, K., and Sidransky, D. (2001). p53 associates with and targets Delta Np63 into a protein degradation pathway. *Proc Natl Acad Sci USA* 98:1817–1822.

Robertson, K. D., and Jones, P. A. (1998). The human ARF cell cycle regulatory gene promoter is a CpG island which can be silenced by DNA methylation and down-regulated by wild-type p53. *Mol Cell Biol* 18:6457–6473.

Robles, A. I., Bemmels, N. A., Foraker, A. B., and Harris, C. C. (2001). APAF-1 is a transcriptional target of p53 in DNA damage-induced apoptosis. *Cancer Res* 61:6660–6604.

Rodriguez, M. S., Desterro, J. M., Lain, S., Midgley, C. A., Lane, D. P., and Hay, R. T. (1999). SUMO-1 modification activates the transcriptional response of p53. *EMBO J 18*, 6455–6461.

Rozenfeld-Granot, G., Krishnamurthy, J., Kannan, K., Toren, A., Amariglio, N., Givol, D., and Rechavi, G. (2002). A positive feedback mechanism in the transcriptional activation of Apaf- 1 by p53 and the coactivator Zac-1. *Oncogene* 21:1469–1476.

Sabbatini, P., and McCormick, F. (2002). MDMX Inhibits the p300/CBP-mediated acetylation of p53. *DNA Cell Biol* 21:519–525.

Sakaguchi, K., Herrera, J. E., Saito, S., Miki, T., Bustin, M., Vassilev, A., Anderson, C. W., and Appella, E. (1998). DNA damage activates p53 through a phosphorylation-acetylation cascade. *Genes Dev* 12:2831–2841.

Sampath, J., Sun, D., Kidd, V. J., Grenet, J., Gandhi, A., Shapiro, L. H., Wang, Q., Zambetti, G. P., and Schuetz, J. D. (2001). Mutant p53 cooperates with ETS and selectively up-regulates human MDR1 not MRP1. *J Biol Chem* 276:39359–39367.

Samuels-Lev, Y., O'Connor, D. J., Bergamaschi, D., Trigiante, G., Hsieh, J. K., Zhong, S., Campargue, I., Naumovski, L., Crook, T., and Lu, X. (2001). ASPP proteins specifically stimulate the apoptotic function of p53. *Mol Cell* 8:781–794.

Sang, N., Avantaggiati, M. L., and Giordano, A. (1997). Roles of p300, pocket proteins, and hTBP in E1A-mediated transcriptional regulation and inhibition of p53 transactivation activity. *J Cell Biochem* 66:277–285.

Sang, N., and Giordano, A. (1997). Extreme N terminus of E1A oncoprotein specifically associates with a new set of cellular proteins. *J Cell Physiol* 170:182–191.

Sax, J. K., Fei, P., Murphy, M. E., Bernhard, E., Korsmeyer, S. J. and el-Deiry, W. S. (2002): BID regulation by p53 contributes to chemosensitivity. *Nat Cell Biol.* 4:842–849.

Schmidt, D., and Muller, S. (2002). Members of the PIAS family act as SUMO ligases for c-Jun and p53 and repress p53 activity. *Proc Natl Acad Sci USA* 99:2872–2877.

Schmieg, F. I., and Simmons, D. T. (1984). Intracellular location and kinetics of complex formation between simian virus 40 T antigen and cellular protein p53. *J Virol* 52350–355.

Shen, Z., Pardington-Purtymun, P. E., Comeaux, J. C., Moyzis, R. K., and Chen, D. J. (1996). UBL1, a human ubiquitin-like protein associating with human RAD51/RAD52 proteins. *Genomics* 36:271–279.

Sheppard, H. M., Corneillie, S. I., Espiritu, C., Gatti, A., and Liu, X. (1999). New insights into the mechanism of inhibition of p53 by simian virus 40 large T antigen. *Mol Cell Biol* 19:2746–2753.

Shikama, N., Lee, C. W., France, S., Delavaine, L., Lyon, J., Krstic-Demonacos, M., and La Thangue, N. B. (1999). A novel cofactor for p300 that regulates the p53 response. *Mol Cell* 4:365–376.

Soengas, M. S., Alarcon, R. M., Yoshida, H., Giaccia, A. J., Hakem, R., Mak, T. W., and Lowe, S. W. (1999). Apaf-1 and caspase-9 in p53-dependent apoptosis and tumor inhibition. *Science* 284: 156–159.

Somasundaram, K., and el-Deiry, W. S. (1997). Inhibition of p53-mediated transactivation and cell cycle arrest by E1A through its p300/CBP-interacting region. *Oncogene* 14:1047–1057.

Somasundaram, K., MacLachlan, T. K., Burns, T. F., Sgagias, M., Cowan, K. H., Weber, B. L., and el-Deiry, W. S. (1999). BRCA1 signals ARF-dependent stabilization and coactivation of p53. *Oncogene* 18:6605–6614.

Somasundaram, K., Zhang, H., Zeng, Y. X., Houvras, Y., Peng, Y., Wu, G. S., Licht, J. D., Weber, B. L., and el-Deiry, W. S. (1997). Arrest of the cell cycle by the tumour-suppressor BRCA1 requires the CDK-inhibitor p21WAF1/CiP1. *Nature* 389:187–190.

Srinivasula, S. M., Ahmad, M., Fernandes-Alnemri, T., and Alnemri, E. S. (1998). Autoactivation of procaspase-9 by Apaf-1-mediated oligomerization. *Mol Cell* 1:949–957.

Stambolic, V., MacPherson, D., Sas, D., Lin, Y., Snow, B., Jang, Y., Benchimol, S., and Mak, T. W. (2001). Regulation of PTEN transcription by p53. *Mol Cell* 8:317–325.

Stambolic, V., Suzuki, A., de la Pompa, J. L., Brothers, G. M., Mirtsos, C., Sasaki, T., Ruland, J., Penninger, J. M., Siderovski, D. P., and Mak, T. W. (1998). Negative regulation of PKB/Akt-dependent cell survival by the tumor suppressor PTEN. *Cell* 95:29–39.

Strano, S., Munarriz, E., Rossi, M., Cristofanelli, B., Shaul, Y., Castagnoli, L., Levine, A. J., Sacchi, A., Cesareni, G., Oren, M., and Blandino, G. (2000). Physical and functional interaction between p53 mutants and different isoforms of p73. *J Biol Chem* 275:29503–29512.

Szak, S. T., Mays, D., and Pietenpol, J. A. (2001). Kinetics of p53 binding to promoter sites in vivo. *Mol Cell Biol* 21:3375–3386.

Takekawa, M., Adachi, M., Nakahata, A., Nakayama, I., Itoh, F., Tsukuda, H., Taya, Y., and Imai, K. (2000). p53-inducible wip1 phosphatase mediates a negative feedback regulation of p38 MAPK-p53 signaling in response to UV radiation. *EMBO J* 19:6517–6526.

Takimoto, R., MacLachlan, T. K., Dicker, D. T., Niitsu, Y., Mori, T., and el-Deiry, W. S. (2002a). BRCA1 transcriptionally regulates damaged DNA binding protein (DDB2) in the DNA repair response following UV-irradiation. *Cancer Biol Ther* 1:177–186; discussion 187–188.

Takimoto, R., Wang, W., Dicker, D. T., Rastinejad, F., Lyssikatos, J., and el-Deiry, W. S. (2002b). The mutant p53-conformation modifying drug, CP-31398, can induce apoptosis of human cancer cells and can stabilize wild-type p53 protein. *Cancer Biol Ther* 1:47–55; discussion 56–57.

Tan, T., and Chu, G. (2002). p53 Binds and activates the xeroderma pigmentosum DDB2 gene in humans but not mice. *Mol Cell Biol* 22:3247–3254.

Tanaka, H., Arakawa, H., Yamaguchi, T., Shiraishi, K., Fukuda, S., Matsui, K., Takei, Y., and Nakamura, Y. (2000). A ribonucleotase reductase gene involved in a p53-dependent cell-cycle checkpoint for DNA damage. *Nature* 404:42–49.

Tokino, T., Thiagalingam, S., el-Deiry, W. S., Waldman, T., Kinzler, K. W., and Vogelstein, B. (1994). p53 tagged sites from human genomic DNA. *Hum Mol Genet* 3:1537–1542.

Vaziri, H., Dessain, S. K., Ng Eaton, E., Imai, S. I., Frye, R. A., Pandita, T. K., Guarente, L., and Weinberg, R. A. (2001). hSIR2(SIRT1) functions as an NAD-dependent p53 deacetylase. *Cell* 107:149–159.

Vogelstein, B., Lane, D., and Levine, A. J. (2000). Surfing the p53 network. *Nature* 408:307–310.

Vousden, K. H. (2002). Activation of the p53 tumor suppressor protein. *Biochim Biophys Acta* 1602:47–59.

Wagner, A. J., Kokontis, J. M., and Hay, N. (1994). Myc-mediated apoptosis requires wild-type p53 in a manner independent of cell cycle arrest and the ability of p53 to induce p21waf1/cip1. *Genes Dev* 8:2817–2830.

Wang, T., Kobayashi, T., Takimoto, R., Denes, A. E., Snyder, E. L., el-Deiry, W. S., and Brachmann, R. K. (2001). hADA3 is required for p53 activity. *EMBO J* 20:6404–6413.

Xiao, G., Chicas, A., Olivier, M., Taya, Y., Tyagi, S., Kramer, F. R., and Bargonetti, J. (2000). A DNA damage signal is required for p53 to activate gadd45. *Cancer Res* 60:1711–1719.

Xu, D., Wilson, T. J., Chan, D., De Luca, E., Zhou, J., Hertzog, P. J., and Kola, I. (2002). Ets1 is required for p53 transcriptional activity in UV-induced apoptosis in embryonic stem cells. *EMBO J* 21:4081–4093.

Xu, X., Qiao, W., Linke, S. P., Cao, L., Li, W. M., Furth, P. A., Harris, C. C., and Deng, C. X. (2001). Genetic interactions between tumor suppressors Brca1 and p53 in apoptosis, cell cycle and tumorigenesis. *Nat Genet* 28:266–271.

Yan, C., Wang, H., and Boyd, D. D. (2002). ATF3 represses 72-kDa type IV collagenase (MMP-2) expression by antagonizing p53-dependent trans-activation of the collagenase promoter. *J Biol Chem* 277:10804–10812.

Yang, A., Kaghad, M., Wang, Y., Gillett, E., Fleming, M. D., Dotsch, V., Andrews, N. C., Caput, D., and McKeon, F. (1998). p63, a p53 homolog at 3q27-29, encodes multiple products with transactivating, death-inducing, and dominant-negative activities. *Mol Cell* 2:305–316.

Yew, P. R., and Berk, A. J. (1992). Inhibition of p53 transactivation required for transformation by adenovirus early 1B protein. *Nature* 357:82–85.

Yu, J., Zhang, L., Hwang, P. M., Kinzler, K. W., and Vogelstein, B. (2001). PUMA induces the rapid apoptosis of colorectal cancer cells. *Mol Cell* 7:673–682.

Zhang, H., Somasundaram, K., Peng, Y., Tian, H., Bi, D., Weber, B. L., and el-Deiry, W. S. (1998). BRCA1 physically associates with p53 and stimulates its transcriptional activity. *Oncogene* 16:1713–1721.

Zhang, Y., Ng, H. H., Erdjument-Bromage, H., Tempst, P., Bird, A., and Reinberg, D. (1999). Analysis of the NuRD subunits reveals a histone deacetylase core complex and a connection with DNA methylation. *Genes Dev* 13:1924–1935.

Zheng, L., Pan, H., Li, S., Flesken-Nikitin, A., Chen, P. L., Boyer, T. G., and Lee, W. H. (2000). Sequence-specific transcriptional corepressor function for BRCA1 through a novel zinc finger protein, ZBRK1. *Mol Cell* 6:757–768.

Zhu, Q., Wani, G., Wani, M. A., and Wani, A. A. (2001). Human homologue of yeast Rad23 protein A interacts with p300/cyclic AMP- responsive element binding (CREB)-binding protein to down-regulate transcriptional activity of p53. *Cancer Res* 61:64–70.

Zou, H., Henzel, W. J., Liu, X., Lutschg, A., and Wang, X. (1997). Apaf-1, a human protein homologous to C. elegans CED-4, participates in cytochrome c-dependent activation of caspase-3. *Cell* 90:405–413.

4

Transcriptional Repression by the p53 Tumor Suppressor Protein

Jack T. Zilfou and Maureen E. Murphy

SUMMARY

In addition to its well-characterized function as a sequence specific transcriptional activator, there is growing evidence that the p53 tumor suppressor protein is also a sequence-specific transcriptional repressor. The concept that a transcription factor can exist as both an activator and a repressor of transcription is not new. In fact, it is the rare transcription factor that can perform only one of these functions. The initial challenges for individuals studying the repression function of p53 have been met; that is, a set of genes whose expression is decreased following p53 induction has been identified. These include the genes encoding *alpha-fetoprotein*, *bcl-2*, *cyclin B*, *cdc2*, *cdc25*, *Map4*, *Mdr1*, *presenilin-1*, and *survivin*, as well as others. The promoters for many of these genes have been cloned, and p53 has been found to bind to sites within these promoters using assays that measure binding in vivo, such as chromatin immunoprecipitation. The p53 binding sites in these promoters have been identified, and in many cases the mechanism whereby p53 represses transcription, for example by promoter occlusion or by recruitment of histone deacetylases, has been elucidated. The current challenge is to create mutant forms of p53 that are capable of repressing

J. T. ZILFOU • Cold Spring Harbor Laboratory, PO Box 100, 1 Bungtown Road, Cold Spring Harbor, NY 11724, USA   M. E. MURPHY • Department of Pharmacology, Fox Chase Cancer Center, Philadelphia, PA 19111, USA

The p53 Tumor Suppressor Pathway and Cancer, edited by Zambetti.
Springer Science+Business Media, New York, 2005.

transcription but not activating it (or vice versa), such that the contribution of this activity to tumor suppression by p53 can be effectively delineated.

4.1. BACKGROUND

p53 is one of the most studied proteins in biology today, and possibly the most studied protein in cancer biology to date. It is a tumor suppressor gene that has the distinction of being the most frequently mutated gene in human cancer; together with other members of its growth suppressive pathway, the function of this protein is likely abrogated in the overwhelming majority of end-stage cancers. In response to detrimental stimuli such as DNA damage, hypoxia, or inappropriate cell proliferation, p53 becomes posttranslationally stabilized and activated as a sequence-specific DNA binding protein and transcription factor. p53 responds to such detrimental stimuli by transactivating various genes involved in the negative regulation of growth, and by repressing a separate class of genes that positively contribute to cellular proliferation or survival (for review see Gottlieb and Oren, 1996; Levine, 1997; Ko and Prives, 1996).

p53 can be viewed as the "conductor" of a stress-response cellular symphony, responsible for the simultaneous coordination and regulation of genes that determine the fate of the cell. In response to such environmental stress, p53 directs the cell either toward cell-cycle arrest, DNA repair and subsequent survival, or toward cell suicide, implemented by programmed cell death or apoptosis. In part, this decision is mediated by the particular stress incurred, but the magnitude of the stress, as well as the absolute level of p53 protein induced, also play a role in the decision between growth arrest and death (Chen et al., 1996). p53 carries out many of its functions via its ability to function as a sequence-specific transcriptional activator of genes with p53-bindings sites in their upstream promoter or intronic regions. This transactivation function of p53 has been studied extensively, and consequently, many of the p53 target genes, and coactivators necessary for their activation, have been identified and their pertinent mechanisms elucidated.

In addition to its well-characterized function as a transcriptional activator, p53 also has a lesser understood activity as a transcriptional repressor. Initial studies investigating this function of p53 were performed using transient overexpression assays where nonphysiological levels of p53 and promoter-driven reporter genes were introduced into cells. In this type of setting, a large number of promoters are repressed by p53, including many viral promoters as well as basal promoters containing TATA boxes (Ginsberg et al., 1991; Mack et al., 1993). However, for many of these initially identified promoters, there is no evidence that p53 represses genes driven by these promoters when they are stably integrated into chromatin (that is, stably transfected), or that p53 regulates the endogenous expression of these genes. Therefore, in transient overexpression reporter assays, transcriptional repression by p53 is likely nonphysiological and impossible to distinguish from nonspecific transcriptional "squelching". However, other bona fide transcriptional repressors (such as E2F family members, when complexed with pRB) also function as nonspecific transcriptional "squelchers" in transient assays. Therefore, the data generated from these studies provided the first clue that p53 might also function as a transcriptional repressor. These studies also led to the mapping of critical domains necessary for repression by p53 (Horikoshi et al., 1995). Finally, accumulated data from these early studies indicated that the repression function of p53 might be particularly critical for p53's ability to efficiently induce programmed cell death.

4.1.1. p53-Mediated Repression is Implicated in Apoptosis Induction

4.1.1.1. p53 has a Transactivation-Independent Mechanism for Cell Death

One of the first indications that p53 had a function other than transactivation involved in programmed cell death stemmed from the observations that p53-dependent apoptosis can proceed in cells even when new protein synthesis, and induction of p53-transactivated genes, is inhibited (Caelles et al., 1994; Wagner et al., 1994). Other researchers reported that a transactivation-deficient p53 mutant retains the ability to induce apoptosis in certain tumor-derived cell lines (Haupt et al., 1995). These data indicated early on that an activity of p53, other than transactivation, could function in apoptosis induction by this protein.

4.1.1.2. E1B 19K, WT-1, and bcl2 Inhibit Apoptosis
and p53-Mediated Repression

The first evidence for an involvement of p53-mediated repression in apoptosis stemmed from the surprising findings that certain antiapoptotic genes can inhibit the transrepression function of p53. Specifically, it was separately reported by several groups that the antiapoptotic proteins bcl-2, the adenovirus protein E1B 19K (which is homologous to bcl-2), and the Wilms Tumor suppressor gene (WT1) can all inhibit p53-dependent apoptosis. Significantly, each of these proteins was found to relieve p53-dependent transcriptional repression of gene expression, while not affecting p53-mediated transactivation (Shen and Shenk, 1994; Sabbatini et al., 1995; Maheswaran et al., 1995). While it could be argued that these studies focused on nonsequence specific repression (that is, transcriptional "squelching"), later studies showed that E1B 19K can also inhibit the repression of the endogenous *Map4* gene following p53 induction (Murphy et al., 1996).

4.1.1.3. A Synthetic Mutant of p53, with the Proline-Rich Domain Deleted, is
Impaired for Apoptosis Induction and Transcriptional Repression, but not
Transactivation

The correlation between p53-mediated transcriptional repression and apoptosis was further solidified by data obtained from a synthetic p53 mutant found to be impaired for repression and apoptosis. This synthetic mutant of p53 has a deletion of the proline rich domain (p53Δpro), which is located between amino acids 61 and 94. Walker and Levine first synthesized the p53Δpro mutant and characterized it for a variety of p53 functions (Walker and Levine, 1996). Specifically these authors showed that the entire proline-rich domain of p53 is dispensable for transactivation of the p21$^{Waf-1/Cip1}$ promoter, despite this domain's proximity to the N-terminal trans-activation domain. In contrast, the proline-rich domain was shown to be necessary for efficient growth suppression by p53 in clonogenic survival assays.

Sakamuro and colleagues demonstrated that p53Δpro mutant is incapable of inducing apoptosis in baby rat kidney cells transfected with the adenoviral E1A protein, or in colon carcinoma cells that are normally sensitive to p53-induced apoptosis (Sakamuro et al., 1997). Interestingly, p53Δpro retains the ability to transactivate the proapoptotic target gene BAX, again supporting the idea that additional p53 functions are required to induce apoptosis. Later Venot and colleagues extended these findings, and demonstrated that p53Δpro is incapable of inducing apoptosis, and is also unable to repress transcription, but is capable of inducing growth arrest and inducing the expression of the p53-induced genes p21$^{Waf-1/Cip1}$, MDM2, and BAX (Venot et al., 1998). The body of data provided by this synthetic p53 mutant supports the conclusion that p53-mediated repression may be a necessary component of p53-mediated apoptosis, and further that this activity of p53 may be mediated through the proline-rich domain.

4.1.1.4. Map4, bcl-2, Presenilin-1 and Survivin are Genes whose Endogenous Promoters are Repressed by p53; Repression of these Genes Plays a Role in Apoptosis

Studies such as those described above prompted a more in-depth investigation into the potential for p53 to repress gene transcription, and yielded the identification of dozens of genes whose endogenous expression is down-regulated following p53 induction (see Table 4.1 for examples). For many of these genes p53 has been shown to bind to their endogenous promoters in a sequence-specific fashion, and for several of them the mechanism whereby p53 represses them has begun to be elucidated. Four of these genes are of particular interest: *Map4* (microtubule-associated protein 4, which plays a role in assembly of microtubules), *presenilin-1* (which plays a role in Alzheimer's disease), *bcl-2* (inhibits the mitochondrial pathway of apoptosis), and *survivin* (an IAP, or inhibitor of apoptosis). These genes are of interest because their down-regulation was shown to play a direct role in apoptosis induction by p53. Specifically, antisense expression of survivin and presenilin-1 can induce apoptosis (Li et al., 1998; Roperch et al., 1998), and overexpression of Map4, bcl-2, and survivin has been shown to inhibit p53-dependent apoptosis (Shen and Shenk 1994; Murphy et al., 1996; Hoffman et al., 2002).

Perhaps the most significant evidence supporting the importance of repression to p53-mediated apoptosis comes from comparing cellular responses to hypoxia and γ radiation. Koumenis and colleagues reported that hypoxic stress, like γ radiation, can induce p53 levels and induce apoptosis in oncogenically transformed cells

Table 4.1. Reliably identified p53-repressed genes

G1 Arrest	**Reference(s)**
BRCA1	MacLachlan et al. (2000)
DNA polymerase δ	Li and Lee (2001)
Topoisomerase II α	Wang et al. (1997); Nip and Hiebert (2000)
G2/M Arrest	
Cdc2	Badie et al. (2000)
Cdc25c	Krause et al. (2001)
Chk1	Damia et al. (2001)
Cyclin B	Krause et al. (2000)
Stathmin (oncoprotein 18)	Ahn et al. (1999); Johnson et al. (2000)
Programmed cell death	
Presenilin-1	Roperch et al. (1998)
Bcl-2	Miyashita et al. (1994a, b); Budhram-Mahadeo (1999); Wu et al. (2001)
Map4	Murphy et al. (1996)
Survivin	Hoffman et al. (2002)
Mdr1	Johnson et al. (2001)
Differentiation/development	
Alpha-fetoprotein	Lee et al. (1999); Ogden et al. (2001)

(Koumenis et al., 2001). However, in contrast to radiation, p53 does not function as a transactivator during hypoxic stress. Rather, p53 appears to only function as a transcriptional repressor when it induces apoptosis in response to hypoxia. These data point to a strong cause and effect relationship between apoptosis and transcriptional repression of specific genes by p53.

4.1.2. General Mechanisms of Transcriptional Repression

It is not unusual for potent transcriptional activators to also function as transcriptional repressors in other contexts, often when bound to different DNA consensus elements, or when complexed to corepressors complexes instead of coactivators. At least four mechanisms by which DNA-bound repressors can negatively regulate transcription have been described in the literature (for review see Johnson, 1995). A transcriptional repressor can bind to specific DNA sequences: (1) and exclude a transcriptional activator from binding to an overlapping site (Type I); (2) near a DNA-bound transcriptional activator and "mask" or "quench" its activating surface thereby preventing it from stimulating the general transcriptional machinery itself (Type II); (3) and interact with the general transcriptional machinery directly and prevent transcription of the target gene (Type III). Along the lines of Type III repression, in some cases the transcription factor binds to distinct binding sites on the DNA; these altered sites may allosterically modify the DNA-bound protein in such a way that it can then interact only with corepressors, instead of coactivators (see for example Scully et al., 2000). In this fourth type of repression (Type IV), the corepressors recruit chromatin remodeling enzymes, such as histone deacetylases, which restructure the chromatin to

Mechanisms of transcriptional repression by p53

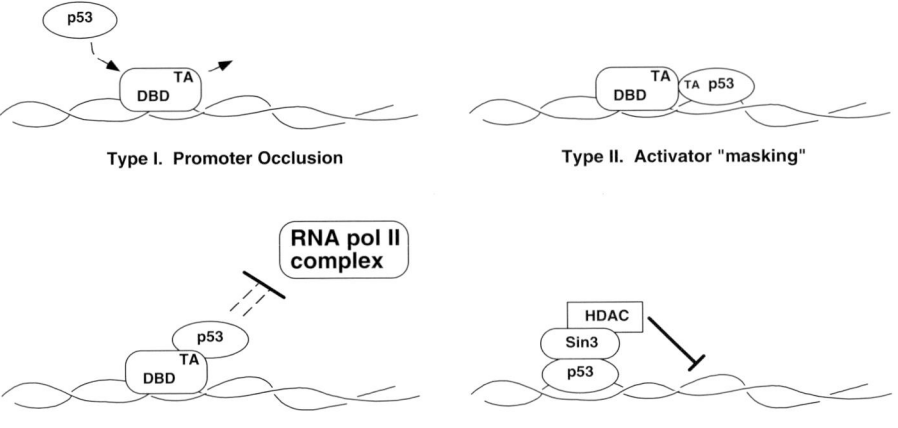

Figure 4.1. Models of transcriptional repression of gene expression by the p53 tumor suppressor protein. DBD = DNA binding domain; TA = transactivation domain. Specific examples are referenced in the text.

a configuration that is not permissive for transcription. Figure 4.1 details a depiction of all four repression mechanisms (Types I–IV). The available data indicate that p53 can utilize all four of these mechanisms to repress its target genes; examples of each of these mechanisms, and the relevant repressed genes, are discussed below.

4.2. p53 REPRESSED GENES

A large number of genes that are transcriptionally repressed following p53 induction have been identified to date; these include the genes for *alpha-fetoprotein, bcl-2, BRCA1, cdc2, cdc25c, chk1, cyclin B1, Map4, DNA polymerase delta, presenilin-1, siah1, stathmin,* and *topoisimerase IIα,* as well as others (see Table 4.1 for relevant references). For many of these genes, the mechanism whereby p53 negatively regulates them is only now becoming clear. Moreover, for the vast majority of these genes, where they fit in different pathways of p53 function is clear; Table 4.1 details where the repression of these genes plays a role in p53 function. Recent studies combining microarray with bioinformatic analysis has revealed interesting information regarding p53-repressed genes. Specifically, a comprehensive cDNA microarray analysis has indicated that the vast majority of p53-regulated genes (over 80%) are repressed by p53, as opposed to p53 activated (Mirza et al., 2003). Interestingly, most of these candidate p53-repressed genes contained a canonical p53-binding site in their promoter or regulatory regions; in 8 of 10 cases, interaction of p53 with this site was confirmed in vivo by the method of chromatin immunoprecipitation. These data support the premise, outlined below, that the majority of p53-repressed genes contain a consensus p53-binding site, or some variation thereof, in their regulatory regions. This premise is supported by recent findings of a canonical p53 consensus element in the promoter for the Cdc25C gene. This element is required for direct binding and repression of this promoter by p53, but when placed alone upstream of a minimal promoter it functions instead as an activating element (St. Clair et al., 2004). Future studies should therefore be aimed at determining the sequence and functional determinants that distinguish p53-activated from p53-repressed genes.

Several p53-repressed genes are listed in Table 4.1. For the majority of these genes, the following criteria (essential for their classification as p53-repressed genes) have been met: first, they have been shown to repress at the level of transcription following p53 induction, for example through use of the technique of nuclear run-ons. Second, these promoters have been shown to repress heterologous reporter genes when stably integrated into chromatin (as opposed to transient assays). Third, p53 has been demonstrated to interact with the regulatory regions of these genes in a sequence-specific manner; in many cases this binding is via a canonical p53-binding site, or some variation thereof. Notably, in some cases this binding has been demonstrated to occur in vivo, for example by the technique of chromatin immunoprecipitations. Finally, mutation or deletion of the p53 binding site in these repressed promoters has been shown to eliminate the ability of p53 to repress these target genes. Four examples of the specific mechanisms whereby p53 represses these target genes, and the biological consequences of this repression, are detailed below.

4.2.1. bcl-2

Bcl-2 is an oncogene whose protein product localizes to the outer mitochondrial membrane, the nuclear envelope, and parts of the endoplasmic reticulum, and acts to suppress apoptosis (Desagher and Martinou, 2000). Reed and colleagues provided insight on the mechanism of p53-mediated apoptosis by demonstrating that p53 can down-regulate bcl-2 expression in the murine leukemia cell line M1 (Miyashita et al., 1994a). Specifically, expression of a temperature-sensitive p53 mutant in the p53-deficient M1 cell line was shown to result in a decrease of bcl-2 mRNA and protein levels upon shift to the permissive temperature. By studying the effects of p53 on reporter gene constructs containing various regions of the human *bcl-2* gene, Reed and colleagues went on to identify a p53-negative response element (PNRE) corresponding to a portion of what is now known as the P2 minimal promoter region (Miyashita et al., 1994b). This PNRE, which does not contain any p53-binding sites, was found to map to the −279/−85 (195 bp) region of the *bcl-2* gene. The PNRE appears to have the characteristics of a transcriptional "silencer" because it mediates p53-dependent repression in an orientation- and position-independent manner. However, the authors did not demonstrate, through DNA-protein binding assays or chromatin immunoprecipitations, that p53 was actually bound to the PNRE in vivo, so whether repression by p53 at the PNRE is direct or indirect remains to be determined.

Latchman and colleagues identified a binding site for p53 in the *bcl-2* promoter, and a potential mechanism for repression of this gene by p53 as well. Utilizing transient reporter assays, this group found that p53 strongly inhibited the activation of the *bcl-2* promoter by the transactivator Brn-3a (Budhram-Mahadeo, 1999). This group reported finding two motifs that closely resemble p53 consensus binding sites in the region between -558 and -535 of the *bcl-2* P2 promoter; these motifs are located just proximal to the Brn-3a binding site at -598 to -581. The authors also found that p53 and Brn-3a interact with each other both in vitro and in vivo, via coimmunopreciptitation and Western blotting, and this interaction was mediated by the POU domain of Brn-3a and the DNA binding domain of p53. These data suggest that p53 represses *bcl-2*, at least in part, by inhibiting activation by a neighboring transactivator (Brn3a), via a Type II mechanism.

4.2.2. Alpha-Fetoprotein

Alpha-fetoprotein (AFP) is a protein that is normally expressed at high levels in the liver of the developing fetus and is silenced after birth. Importantly, inappropriate expression of AFP in adult liver occurs in 70 to 85% of all hepatocellular carcinomas, and consequently serves as a diagnostic tumor-specific marker. There is a defined region of the AFP promoter responsible for the developmental repression of AFP. Using computer-aided scanning for transcription factor consensus binding sites within this region, Barton and colleagues found protein binding sites for the HNF-3 fetal liver activator overlapping a dimer binding site for p53 (in other words, a "half-site") (Lee et al., 1999). The authors went on to show that AFP gene expression is controlled in part by the mutually exclusive binding of p53 and HNF-3 (Lee et al., 1999; Ogden et al., 2001). HNF-3 activates AFP transcription, while p53 represses it, both through sequence-specific binding within the previously identified AFP developmental repressor domain. The authors went on to demonstrate that recombinant p53 and HNF-3 proteins could bind to their respective DNA binding element, and that this binding is not additive or cooperative, but rather is mutually exclusive. Additionally, p53 has a much higher affinity than HNF-3 for the same AFP regulatory element, so p53 induction leads to repression of this promoter by displacement of HNF-3. Significantly, mutation of the DNA binding domain in the p53 protein, or mutation of the p53 DNA binding element, abrogates p53's ability to repress AFP.

The authors proffered a combinatorial mechanism for AFP repression by p53, a passive and an active mode of transcriptional repression. A passive mechanism involves the passive exclusion of an activator from a common DNA binding site, and this is supported by the authors' DNA binding data and transient reporter assays. This "counteracts" the activation by HNF-3. Additionally, however, active interference with the transcription of AFP by p53 is supported by approximately a fivefold downregulation of basal AFP/lacZ expression in hepatoma cells. The authors suggest that p53 can do this by directly interacting with corepressor complexes at the AFP promoter region. This combinatorial mechanism, involving "passive" and "active" modes of p53 repression, fits both Type I and Type III mechanisms of repression mentioned above.

4.2.3. Mdr1

Scotto and colleagues identified a novel orientation of the p53 binding consensus element on the promoter of the gene encoding P glycoprotein, *Mdr1* (Johnson et al., 2001). Through a series of transient reporter assays and in vitro DNA binding assays, the authors found that p53 can repress transcription by directly binding to a novel head-to-tail (HT) site within the *Mdr1* promoter (in other words, the "quarter sites" are not inverted head-to-head repeats, but rather are head-to-tail repeats). Further, the authors showed that a mutation that disrupted p53 binding to the HT site abrogated p53-mediated repression of the Mdr1 promoter. Intriguingly, when the authors replaced the HT site with a head-to-head site, p53 was converted from a transcriptional repressor to an activator, as assessed by transient reporter assays. This provocative result suggests that mere recruitment of p53 to a promoter may not be sufficient for repression; rather, the orientation of the binding element may allosterically modify p53 to determine the fate of the regulated gene.

4.2.4. Survivin

In a separate mode of repression by p53, our group demonstrated that the antiapoptotic gene *survivin*, which encodes a caspase inhibitor, was repressed by p53 (Hoffman et al., 2002). As was the case for the previous three p53-repressed genes, p53 was found to bind the *survivin* promoter to a variant of the canonical p53 consensus element. The standard p53 consensus element, which consists of two copies of the inverted repeat 5′ Pu Pu Pu C A/T T/A G Py Py Py 3′ (where Pu = purine and Py = pyrimidine) separated by a spacer of 0–13 nucleotides, was identified in immunobinding studies by el Deiry and coworkers (elDeiry et al., 1992). Interestingly, the overwhelming majority of p53-induced genes have spacers of 0–1 nucleotide. Further it was shown by Tokino and coworkers that increasing this spacer to four nucleotides still allowed p53 to bind this element, but abolished its ability to function as an enhancer (Tokino et al., 1994). The binding site for p53 on the *survivin* promoter has a spacer of three nucleotides, and deletion of these three nucleotides converted this site into a transactivation element (Hoffman et al., 2002). p53 was shown to bind to the *survivin* promoter in vivo using the technique of chromatin immunoprecipitation; this site overlapped with a binding site for E2F family members. This binding was shown to result in deacetylation of the histones surrounding this promoter. Therefore, p53 represses *survivin* via a mechanism most consistent with Type I repression (promoter occlusion). However the available data also indicate that allosteric regulation of p53 by the unusual structure of the *survivin* binding site might also work to specifically recruit chromatin remodeling enzymes to this promoter (Type IV repression) (Hoffman et al., 2002).

4.3. MECHANISMS OF REPRESSION BY p53:
p53–HDAC COMPLEXES

The reliable identification of p53-repressed genes facilitated the identification of protein complexes accessory to repression by p53. Along these lines, data from our group indicated that the histone deacetylase inhibitor Trichostatin A (TSA) could inhibit p53-mediated repression of *stathmin* and *Map4* (Murphy et al., 1999). Further, repression of *survivin* and *Map4* was shown to be accompanied by deacetylation of histone H3 within the promoter region (Hoffman et al., 2002), consistent with the activity of histone deacetylases (HDACs). HDACs remove acetyl groups from the lysine residues of promoter-associated histones (and other proteins as well), thereby increasing their affinity for DNA and compacting chromatin to create an environment unfavorable for transcription. To date, many transcriptional repressors have been found to utilize HDACs in order to repress the expression of target genes. Immunoprecipitation of HDAC 1/2 consistently reveals association with p53, and vice versa (Murphy et al., 1999); however, this interaction is not direct. Rather, it is mediated by the ubiquitous corepressor protein mSin3a (Sin3). Notably, both p53 and Sin3 can be found bound to the *Map4* promoter in vivo, while there was no evidence that Sin3 bound to the p53-induced *Mdm2* promoter (Murphy et al., 1999). The contribution of the p53–Sin3–HDAC complex to p53-dependent transcriptional repression was strengthened when it was demonstrated that deletion of the Sin3-binding domain of p53, from amino acids 61–75, significantly abrogated the ability of p53 to repress the *survivin* gene in transient assays. In contrast, this mutant was able to transactivate the Mdm2 promoter indistinguishably from wild-type p53 (Zilfou et al., 2001; Hoffman et al., 2002).

Sin3 binds to the proline-rich domain of p53 (Zilfou et al., 2001), thus validating early reports indicating that this domain was required for transcriptional repression, and solidifying the contribution of p53-dependent repression to apoptosis induction. Interestingly, p53-dependent apoptosis in response to hypoxic stress was found to occur in the absence of transactivation of p53-target genes; however, p53-dependent repression of gene expression, and p53–Sin3–HDAC complex formation, still occurs in hypoxic cells (Koumenis et al., 2001). Likewise, the HDAC inhibitor TSA can inhibit p53-dependent apoptosis (Murphy et al., 1999), as well as hypoxia-dependent apoptosis (Koumenis et al., 2001). In conclusion, the data indicating that p53 represses several genes with defined roles in apoptosis, coupled with data correlating this activity with apoptosis induction, provide compelling argument that further study of this activity will be crucial to our understanding of the ability of p53 to function in tumor suppression.

Following the discovery of the p53–Sin3–HDAC transcriptional repression complex, two other HDAC-containing complexes were identified that contain p53. These are the p53–NuRD complex, which contains HDAC-1, and the p53–Sirt1 complex, in which Sirt1 is an HDAC (Luo et al., 2000, 2001; Vaziri et al., 2001). The HDACs in these complexes have been shown to deacetylate p53, and hence to decrease p53's ability to transactivate gene expression. However, from these studies it does not appear that either the p53–NuRD or p53–Sirt1 complexes play a role in transcriptional repression (Luo et al., 2000, 2001; Vaziri et al., 2001). Instead, these complexes appear to deacetylate p53 and inhibit transcriptional activation. Whether or not Sin3–HDAC2 likewise alters the acetylation status of p53 (perhaps on novel lysine residues) is unknown. Likewise, the parameters that regulate the association of p53 with each of these complexes remain to be determined.

4.4. FUTURE CONSIDERATIONS

In 1991 Oren and colleagues published the first paper describing the repression of the *c-fos* promoter by p53 in transient reporter assays (Ginsberg et al., 1991). Later, Laimins and coworkers demonstrated that this repression activity mapped to the TATA box of reporter genes, and that TATA-less, initiator-driven promoters were not repressed (Mack et al., 1993). Since that time dozens of genes have been reported to be repressed by p53. However, too few of these studies have been performed as stringently as needed. For example, too few studies have verified that repression is at the level of transcription, or made attempts to cull out p53-dependent transcriptional repression from its downstream pathways (that is, cell cycle arrest and cell death). With the introduction of chromatin immunoprecipitations and in vivo footprinting to studies of promoter regulation, the definition of a true consensus negative response element for p53 will be facilitated. Other questions remain paramount. For example, it remains unclear how p53 distinguishes between activated and repressed promoters, and specifically how p53 recruits corepressors to repressed promoters and coactivators to induced promoters. How particular posttranslational modifications, such as acetylation, phosphorylation, or sumoylation, might regulate the decision between transactivation and transrepression also needs to be addressed. The possibility that other corepressor or protein complexes play a role in repression by p53 needs to be determined. Along these lines, recent data from the Barton laboratory indicate that the corepressor SnoN complexes with p53 and is required for the repression of the α-fetoprotein gene (Wilkinson et al., 2005). And finally, though it is clear that both transactivation and repression of gene expression play roles in apoptosis induction by p53, the relative contributions of each activity to cell death have yet to be delineated. Such delineation awaits the creation of mutant forms of p53 that can discriminate between these activities (for example, mutants that can repress but not activate and vice versa). There have been studies on a mutant form of p53 called 22/23 (amino acids 22 and 23 are changed from Leu/Trp to Gln/Ser). This mutant fails to transactivate gene expression, and in the context of a "knock-in" mouse, it also completely fails to suppress tumor development (Jimenez et al., 2000). However, these studies do not shed light on the contribution of repression, because it is known that the 22/23 mutant of p53 also fails to repress gene expression, including *survivin* and *Map4* (Murphy et al., 1996; Chao et al., 2000). A synthetic mutant of p53, containing a single point mutation if possible, must be created that fails to repress gene expression but maintains transactivation capability; the ability of this mutant to induce apoptosis must then be assessed.

REFERENCES

Ahn, J., Murphy, M., Kratowicz, S., Wang, A., Levine, A. J., and George, D. L. (1999). Down-regulation of the stathmin/Op18 and FKBP25 genes following p53 induction. *Oncogene* 18:5954–5958.

Badie, C., Itzhaki, J. E., Sullivan, M. J., Carpenter, A. J., and Porter, A. C. (2000). Repression of CDK1 and other genes with CDE and CHR promoter elements during DNA damage induced G2/M arrest in human cells. *Mol Cell Biol* 20:2358–2366.

Budhram-Mahadeo, V., Morris, P. J., Smith, M. D., Midgley, C. A., Boxer, L. M., and Latchman, D. S. (1999). p53 suppresses the activation of the Bcl-2 promoter by the Brn-3a POU family transcription factor. *J Biol Chem* 274:15237–15244.

Caelles, C., Helmberg, A., and Karin, M. (1994). p53-dependent apoptosis in the absence of transcriptional activation of p53-target genes. *Nature* 370:220–223.

Chen, X., Ko, L. J., Jayaraman, L., and Prives, C. (1996). p53 levels, functional domains, and DNA damage determine the extent of the apoptotic response of tumor cells. *Genes Dev* 10:2438–2451.

Chao, C., Saito, S., Kang, J., Anderson, C. W., Appella, E., and Xu Y. (2000). p53 transcriptional activity is essential for p53-dependent apoptosis following DNA damage. *EMBO J.* 19:4967–4975.

Damia, G., Sanchez, Y., Erba, E., and Broggini, M. (2001). DNA damage induces p53-dependent down-regulation of hCHK1. *J Biol Chem* 276:10641–10645.

Desagher, S., and Martinou, J. C. (2000). Mitochondria as the central control point of apoptosis. *Trends Cell Biol* 10:369-77.

El-Deiry, W. S., Kern, S. E., Pietenpol, J. A., Kinzler, K. W., and Vogelstein, B. (1992). Definition of a consensus binding site for p53. *Nat Genet* 1:45–49.

Ginsberg, D., Mechta, F., Yaniv, M., and Oren, M. (1991). Wild-type p53 can down-modulate the activity of various promoters. *Proc Natl Acad Sci USA* 88:9979–9983.

Gottlieb, T. M., and Oren, M. (1996). p53 in growth control and neoplasia. *Biochim Biophys Acta* 1287:77–102

Haupt, Y., Rowan, S., Shaulian, E., Vousden, K. H., and Oren, M. (1995). Induction of apoptosis in HeLa cells by trans-activation-deficient p53. *Genes Dev* 9:2170–2183.

Hoffman, W. H., Biade, S., Zilfou, J. T., Chen, J., and Murphy, M. (2002). Transcriptional repression of the anti-apoptotic survivin gene by wild type p53. *J Biol Chem* **277**:3247–3257.

Horikoshi, N., Usheva, A., Chen, J., Levine, A. J., Weinmann, R., and Shenk, T. (1995). Two domains of p53 interact with the TATA-binding protein, and the adenovirus 13S E1A protein disrupts the association, relieving p53-mediated transcriptional repression. *Mol Cell Bio* 15:227–234.

Jimenez, G. S., Nister, M., Stommel, J. M., Beeche, M., Barcarse, E. A., Zhang, X. Q., O'Gorman, S., and Wahl, G. M. (2000). A transactivation-deficient mouse model provides insights into Trp53 regulation and function. *Nat Genet* 26:37–43.

Johnson, A. D. (1995). The price of repression. *Cell* 81:655–658.

Johnson, J. I., Aurelio, O. N., Kwaja, Z., Jorgensen, G. E., Pellegata, N. S., Plattner, R., Stanbridge, E. J., and Cajot, J. F. (2000). p53-mediated negative regulation of stathmin/Op18 expression is associated with G2/M cell cycle arrest. *Int J Cancer* 88:685–691.

Johnson, R., Ince, T., and Scotto, K. (2001). Transcriptional repression by p53 through direct binding to a novel DNA element. *J Biol Chem* 276:27716–27720.

Ko L. J., and Prives C. (1996). p53: puzzle and paradigm. *Genes Dev* 10:1054–1072.

Koumenis, C., Alarcon, R., Hammond, E., Sutphin, P., Hoffman, W., Murphy, M., Derr, J., Taya, Y., Lowe, S.W., Kastan, M., and Giaccia, A. (2001). Regulation of p53 by hypoxia: dissociation of transcriptional repression and apoptosis from p53-dependent transactivation. *Mol Cell Biol* 21:1297–1310.

Krause, K., Haugwitz U., Wasner M., Wiedmann M., Mossner J., and Engeland K. (2001). Expression of the cell cycle phosphatase cdc25C is down-regulated by the tumor suppressor protein p53 but not by p73. *Biochem Biophys Res Commun* 284:743–750.

Krause, K., Wasner, M., Reinhard, W., Haugwitz, U., Dohna, C. L., Mossner, J., and Engeland, K. (2000). The tumor suppressor protein p53 can repress transcription of cyclin B. *Nuc Acids Res* 28:4410–4418.

Lee, K. C., Crowe, A. J., and Barton, M. C. (1999). p53-mediated repression of alpha-fetoprotein gene expression by specific DNA binding. *Mol Cell Biol* 19:1279–1288.

Levine, A. J. (1997). p53, the cellular gatekeeper for growth and division. *Cell* 88:323–331.

Li, F., Ambrosini, G., Chu, E. Y., Plescia, J., Tognin, S., Marchisio, P. C., Altieri, D. C. (1998). Control of apoptosis and mitotic spindle checkpoint by survivin. *Nature* 396:580–584.

Li, B., and Lee, M. Y. (2001). Transcriptional regulation of the human DNA polymerase delta catalytic subunit POLD1 by p53 tumor suppressor and Sp1. *J Biol Chem* 276:29729–29739.

Luo, J., Su F., Chen, D., Shiloh, A., Gu, W. (2000). Deacetylation of p53 modulates its effect on cell growth and apoptosis. *Nature* 408:377–381.

Luo, J., Nikolaev, A.Y., Imai, S., Chen, D., Su, F., Shiloh, A., Guarente, L., Gu, W. (2001). Negative control of p53 by Sir2alpha promotes cell survival under stress. *Cell* 107:137–148.

Mack, D.H., Vartikar, J., Pipas, J., and Laimins, L. (1993). Specific repression of TATA-mediated but not initiator-mediated transcription by wild type p53. *Nature* 363:281–283.

MacLachlan, T. K., Dash, B. C., Dicker, D. T., and el-Deiry, W. S. (2000). Repression of BRCA1 through a feedback loop involving p53. *J Biol Chem* 275:31869–31875.

Maheswaran, S., Englert, C., Bennett, P., Heinrich, G., Haber, D. A. (1995). The WT1 gene product stabilizes p53 and inhibits p53-mediated apoptosis. *Genes Dev* 9:2143–2156.

Mirza, A., Wu, Q., Wang, L., McClanahan, T., Bishop, W. R., Gheyas, F., Ding, W., Hutchins, B., Hockenberry, T., Kirschmeier, P., Greene, J. R., and Liu, S. (2003). Global transcriptional program of p53 target genes during the process of apoptosis and cell cycle progression. *Oncogene* 22:3645–3654.

Miyashita, T., Harigai, M., Hanada, M., and Reed, J. C. (1994a). Identification of a p53-dependent negative response element in the bcl- 2 gene. *Cancer Res* 54:3131–3135.

Miyashita, T., Krajewski, S., Krajewska, M., Wang, H. G., Lin, H. K., Liebermann, D. A., Hoffman, B., and Reed, J. C. (1994b). Tumor suppressor p53 is a regulator of bcl-2 and bax gene expression in vitro and in vivo. *Oncogene* 9:1799–805.

Murphy, M., Hinman, A., and Levine, A. J. (1996). Wild-type p53 negatively regulates the expression of a microtubule-associated protein. *Genes Dev* 10:2971–2980.

Murphy, M., Ahn, J., Walker, K. K., Hoffman, W. H., Evans, R. M., Levine, A. J., and George, D. L. (1999). Transcriptional repression by wild-type p53 utilizes histone deacetylases, mediated by interaction with mSin3a. *Genes Dev* 13:2490–2501.

Nip, J., and Hiebert, S. W. (2000) Topoisomerase IIalpha mediates E2F-1-induced chemosensitivity and is a target for p53-mediated transcriptional repression. *Cell Biochem Biophys* 33:199–207.

Ogden, S. K., Lee, K. C., Wernke-Dollries, K., Stratton, S. A., Aronow, B., and Barton, M. C. (2001). p53 targets chromatin structure alteration to repress alpha-fetoprotein gene expression. *J Biol Chem* 276:42057–42062.

Roperch, J. P., Alvaro, V., Prieur, S., Tuynder, M., Nemani, M., Lethrosne, F., Piouffre, L., Gendron, M. C., Israeli, D., Dausset, J., Oren, M., Amson, R., Telerman, A. (1998). Inhibition of presenilin1 is promoted by p53 and p21WAF1 and results in apoptosis and tumor suppression. *Nat Med* 4:835–838.

Sabbatini, P., Chiou, S. K., Rao, L., and White, E. (1995). Modulation of p53-mediated transcriptional repression and apoptosis by the adenovirus E1B 19K protein. *Mol Cell Biol* 15:1060–1070.

Sakamuro, D., Sabbatini, P., White, E., and Prendergast, G. C. (1997). The polyproline region of p53 is required to activate apoptosis but not growth arrest. *Oncogene* 15:887–898.

Scully, K. M., Jaconson, E. M., Jepsen, K., Lunyak, V., Viadiu, H., Carriere, C., Rose, D. W., Hooshmand, F., Aggarwal, A., and Rosenfeld, M. G. (2000). Allosteric effects of Pit-1 DNA sites on long-term repression in cell type speciation. *Science* 290:1127–1130.

Shen, Y., and Shenk, T. (1994). Relief of p53-mediated transcriptional repression by the adenovirus E1B 19-kDa protein or the cellular Bcl-2 protein. *Proc Natl Acad Sci USA* 91:8940–8944.

St. Clair, S., Giono, L., Varmeh-Ziaie, S., Resnick-Silverman, L., Liu, W., Padi, A., Dastidar, J., DaCosta, A., Mattia, M., and Manfredi, J. J. (2004). DNA damage-induced downregulation of Cdc25C is mediated by p53 via two independent mechanisms: one involves direct binding to the cdc25C promoter. *Mol Cell* 16:725–736.

Tokino, T., Thiagalingam, S., el-Deiry, W. S., Waldman, T., Kinzler, K. W., and Vogelstein, B. (1994). p53 tagged sites from human genomic DNA. *Hum Mol Genet* 3:1537–1542.

Vaziri, H., Dessain, S. K., Ng Eaton, E., Imai, S. I., Frye, R. A., Pandita, T. K., Guarente, L., Weinberg, R. A. (2001). hSIR2(SIRT1) functions as an NAD-dependent p53 deacetylase. *Cell* 107:149–159.

Venot, C., Maratrat, M., Dureuil, C., Conseiller, E., Bracco, L., and Debussche, L. (1998). The requirement for the p53 proline-rich functional domain for mediation of apoptosis is correlated with specific PIG3 gene transactivation and with transcriptional repression. *EMBO J* 17:4668–4679.

Walker, K. K., and Levine, A. J. (1996). Identification of a novel p53 functional domain that is necessary for efficient growth suppression. *Proc Natl Acad Sci USA* 93:15335–15340.

Wagner, A. J., Kokontis, J. M., Hay, N. (1994). Myc-mediated apoptosis requires wild-type p53 in a manner independent of cell cycle arrest and the ability of p53 to induce p21waf1/cip1. *Genes Dev* 8:2817–2830.

Wang, Q., Zambetti, G. P., Suttle D. P. (1997). Inhibition of DNA topoisomerase II alpha gene expression by the p53 tumor suppressor. *Mol Cell Biol* 17:389–397.

Wilkinson, D. S., Ogden, S. K., Stratton, S. A., Piechan, J. L., Nguyen, T. T., Smulian, G. A., Barton, M. C. (2005). A direct intersection between p53 and transforming growth factor β pathways targets chromatin modification and transcription repression of the α fetoprotein gene. *Mol Cell Biol* 25:1200–1212.

Wu, Y., Mehew, J. W., Heckman C. A., Arcinas, M., and Boxer L. M. (2001). Negative regulation of bcl-2 expression by p53 in hematopoietic cells. *Oncogene* 20:240–251.

Zilfou, J. T., Hoffman, W. H., Sank, M., George, D. L., Murphy, M. (2001) The corepressor mSin3a interacts with the proline-rich domain of p53 and protects p53 from proteasome-mediated degradation. *Mol Cell Biol.* 21:3974–3985.

5

Posttranslational Modifications of p53: Upstream Signaling Pathways

Carl W. Anderson and Ettore Appella

SUMMARY

The p53 tumor suppressor is a tetrameric transcription factor that is posttranslational modified at >20 different sites by phosphorylation, acetylation, or sumoylation in response to various cellular stress conditions. Specific posttranslational modifications, or groups of modifications, that result from the activation of different stress-induced signaling pathways are thought to modulate p53 activity to regulate cell fate by inducing cell cycle arrest, apoptosis, or cellular senescence. Here we review recent progress in characterizing the upstream signaling pathways whose activation in response to various genotoxic and nongenotoxic stresses result in p53 posttranslational modifications.

5.1. INTRODUCTION

Maintenance of genome integrity is critical to the well being of multicellular organisms. Elaborate mechanisms to monitor genome integrity have evolved to respond

C. W. ANDERSON • Biology Department, Brookhaven National Laboratory, Upton, NY 11973, USA.
E. APPELLA • Laboratory of Cell Biology, National Cancer Institute, National Institutes of Health, Bethesda, MD 20892, USA.

The p53 Tumor Suppressor Pathway and Cancer, edited by Zambetti.
Springer Science+Business Media, New York, 2005.

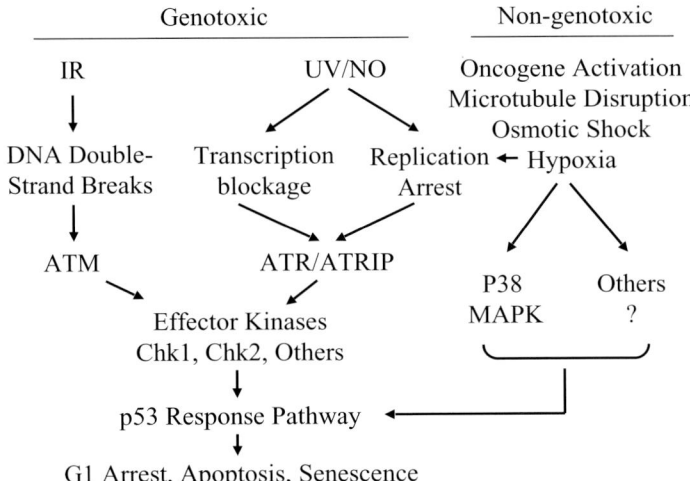

Figure 5.1. Signaling pathways for the activation of p53. The p53 tumor suppressor is stabilized and activated as a transcription factor in response to several signaling pathways that are initiated in response to genotoxic damage or nongenotoxic cellular stresses. Depicted schematically are the major genotoxic pathways that respond to DNA double-strand breaks through the activation of ATM, to bulky lesions in DNA that block transcription or DNA replication and signal through ATR, and nongenotoxic stress pathways that generally do not involve ATM or ATR but signal through p38 MAPK and other signaling systems. ATM and ATR directly phosphorylate several DNA damage associated proteins including BRCA1, 53BP1, H2AX, and p53 as well as several effector protein kinases, such as Chk1 and Chk2. The response to extreme hypoxia is exceptional in that the resulting collapsed replication forks are believed to activate ATR, resulting in the phosphorylation of p53 at Ser15 but not its subsequent acetylation at Lys382 (Hammond et al., 2002).

to a variety of environmental and cellular stresses that can disrupt the genome either directly by causing DNA damage or indirectly through disruption of normal cellular processes that involve DNA. Critical to the process of maintaining genome integrity in higher organisms is the p53 tumor suppressor, which serves to integrate signals from various DNA integrity and environmental stress-sensing signaling pathways (Fig. 5.1) (Wahl and Carr, 2001; Vousden and Lu, 2002; Oren, 2003). Human p53 is a 393 amino acid polypeptide that functions as a homotetrameric transcription factor to control cell cycle progression, cellular senescence, the induction of apoptosis, and DNA repair. Genomic approaches have shown that human p53 induces or inhibits the expression of more than 150 genes including *CDKN1A* (*p21*, *WAF1*, *CIP1*), *GADD45*, *MDM2*, *IGFBP3*, and *BAX* (Sax and el-Deiry, 2003). The arrest of cells in G_1 near the border of S phase is accomplished primarily through transcriptional induction of the cyclin kinase inhibitor $p21^{Waf1}$, and cell cycle arrest is thought to allow time for the repair of DNA damage or recovery from other cellular insults. p53 also modulates DNA repair processes either directly or through the induction of repair genes (Smith and Seo, 2002; Cline and Hanawalt, 2003). The induction of cellular senescence in response to oncogene activation also involves p53-mediated accumulation of $p21^{Waf1}$, but the role of p53 in mediating senescence is not fully understood (Itahana et al., 2001). p53-mediated apoptosis involves the induction of a number of genes that may control the release of cytochrome c from mitochondria and the activation of caspases (Vousden and Lu, 2002). Recently, it has been shown that p53 can itself directly interact with the mitochondrial membrane leading to cytochrome c release and the initiation of apoptosis (Mihara et al., 2003).

p53 normally is a short-lived protein that is rapidly degraded through ubiquitin mediated pathways and therefore is present at low levels in unstressed mammalian cells. In response to both genotoxic and nongenotoxic stresses, it becomes stabilized and accumulates in the nucleus where it binds to specific DNA sequences (Wahl and Carr, 2001; Vousden and Lu, 2002; Sax and el-Deiry, 2003) and also interacts directly with a number of other cellular and viral proteins (Fig. 5.2). Competition between repair proteins and damage sensors, as well as cell type-specific thresholds for initiating apoptosis may in part determine cellular fate. Stabilization of the p53 protein and regulation of its interaction with DNA and other proteins is regulated by posttranslational modifications, primarily phosphorylations and acetylations. Before reviewing the major stress-induced signaling pathways that lead to these modifications, we first briefly review the structure of human p53 and its posttranslational modifications.

5.2. STRUCTURE OF HUMAN p53

The structure of the intact, 393 amino acid p53 protein (Fig. 5.2) has proved difficult to study as the overall size of the tetrameric, p53-DNA complex, combined with its intrinsic flexibility, so far has prevented determination of its structure at high resolution (Kaku et al., 2001; Kaeser and Iggo, 2002). Only about 60% of the molecule is folded into compact domains, with the remainder forming flexible linkers or tails. These disordered regions contain most of the sites of posttranslational modification and are the loci for interactions with the many proteins with which p53 associates (Fig. 5.2). The N-terminal region (amino acids 1–101) is unstructured in solution, but residues 17–28 form an α-helix upon binding to Mdm2 (Kussie et al., 1996). Residues 1–42 are required for transactivation activity and interact with the transcription factors TFIID, TFIIH, several TAFs, the histone acetyltransferases CBP/p300, and possibly PCAF. Residues 11–26 are reported to function as a secondary nuclear export signal (Zhang and Xiong, 2001), while residues 63–97 comprise a proline-rich SH3 domain required for interaction with the Sin3 corepressor (Zilfou et al., 2001) and other proteins required for the induction of apoptosis. The structure of the DNA binding domain (DBD), residues 102–292, in complex with DNA, has been determined by X-ray crystallography (Cho et al., 1994) and NMR analysis (Rippin et al., 2002); the structure of the tetramerization domain (aa 325–356) also has been determined by both X-ray and NMR techniques (Clore et al., 1995; Jeffrey et al., 1995). A nuclear export signal that is masked in tetramers is located within the tetramerization domain, and the major nuclear localization signal is located with residues 312–324. The C-terminal 30 amino acids confer a structure-specific DNA binding capability to p53

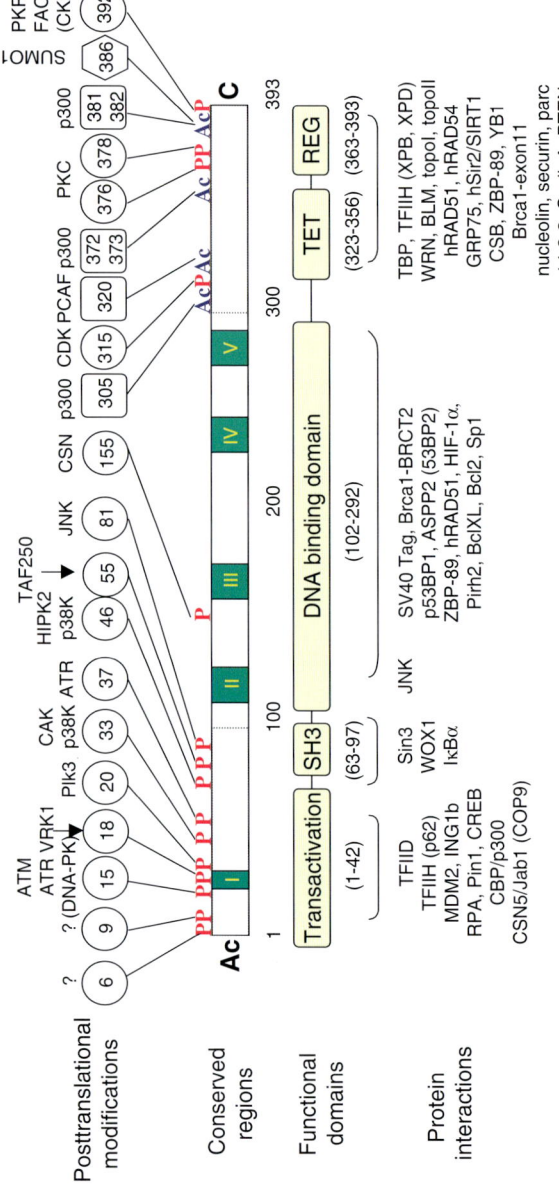

Figure 5.2. Protein domains, posttranslational modification sites, and proteins that interact with human p53. The 393 amino acid, human p53 polypeptide is represented schematically (box) with the five most highly conserved regions marked (I–V); postulated function regions and domains also are indicated. Residues ~1–42 comprise the transactivation domain; residues ~63–97 constitute a Src homology 3-like (SH3) domain that overlaps a poorly conserved proline and alanine rich segment (33–80); residues ~102–292 contain the central, sequence-specific, DNA binding core region: residues ~300–323 contain a nuclear export signal; residues 363–393 (REG) negatively regulate DNA binding by the central core to consensus recognition sites in oligonucleotides and interact in a sequence-independent manner with single- and double-stranded nucleic acids. Interaction regions for selected proteins are indicated below the polypeptide, and posttranslational modification sites (P, phosphorylation; Ac, acetylation) are indicated above the peptide together with enzymes that can accomplish the modifications in vitro. Lys386 may be modified by conjugation with SUMO1, a ubiquitin-like peptide. References are found in the text and recent reviews [e.g. Ko and Prives (1996); Anderson and Appella (2003); Craig et al. (2003).]

(Palecek et al., 1997; Mazur et al., 1999), and mutual interference between sequence-specific and structure-specific DNA binding has led to proposed regulatory roles for the C-terminal domain (Hupp and Lane, 1994; Anderson et al., 1997). However, recent structural and functional studies (Ayed et al., 2001; Espinosa and Emerson, 2001; Klein et al., 2001) raised questions regarding the mechanism of p53 activation and the role of the C-terminal domain in regulating p53 activity (Ahn and Prives, 2001; Kim and Deppart, 2003).

5.3. p53 POSTTRANSLATIONAL MODIFICATIONS

Exposure of normal cells to either genotoxic agents or nongenotoxic stresses results in the phosphorylation of p53 at approximately 15 serines or threonines in both the N- and C-terminus and acetylation at about a half-dozen lysines in the C-terminus of the p53 polypeptide (Fig. 5.2) (Appella and Anderson, 2001; Anderson and Appella, 2003). At the N-terminus, human p53 becomes phosphorylated at serines 6, 9, 15, 20, 33, 37, 46 and threonines 18, 55, and 81. Serines 33, 37, 46, and 392 are more efficiently phosphorylated after exposure to UV or adriamycin (ADR), an anticancer agent that inhibits topoisomerase II, than to ionizing radiation (IR); in contrast, phosphorylation of Thr18 is stronger in response to IR and ADR than to UV light (Saito et al., 2003). Phosphorylation of serines 15, 20, and 37, after either IR or UV light, increases the stability of p53 (Shieh et al., 1997; Chehab et al., 1999). At the C-terminus, phosphorylation at Ser315 is induced by IR, UV, or ADR, while phosphorylation at Ser392 is induced by UV light or ADR, but not by IR (Kapoor and Lozano, 1998; Lu et al., 1998). Serines 376 and 378 in the C-terminal region are reported to be constitutively phosphorylated, and treatment with IR led to the dephosphorylation of serine 376 (Waterman et al., 1998). Phosphorylation of serines 315 and 392 affects the oligomerization state of p53 (Sakaguchi et al., 1997) and its ability to bind DNA in a sequence-specific manner, at least in vitro (Hupp et al., 1992; Wang and Prives, 1995; Hao et al., 1996). Thr155 and Thr150 or Ser149, in the central, site-specific, DNA-binding domain, recently were reported to be phosphorylated by the COP9 signalosome (CSN)-associated kinase (Bech-Otschir et al., 2001); so far, these are the only sites in the central domain that have been reported to be posttranslationally modified. In fission yeast, the COP9 signalosome is required for the activation of ribonucleotide reductase (Nielsen, 2003); in mammals, it also may participate in regulating p53 degradation.

Acetylation of the p53 C-terminus is mediated through a DNA damage initiated, phosphorylated-dependent signaling cascade by the histone acetyltransferases and transcriptional coactivators p300, CBP, and PCAF (Gu and Roeder, 1997; Sakaguchi et al., 1998; Prives and Manley, 2001). The interaction of p53 with p300/CBP was shown to be enhanced by phosphorylation of p53 at serine 15 (Lambert et al., 1998; Dumaz and Meek, 1999); in turn, CBP/p300 acetylates several C-terminal lysines including 372, 373, 381, and 382. Recently, lysine 305 was also shown to be acetylated in response to IR, UV, H_2O_2, and actinomycin D (Wang et al., 2003). Peptide competition experiments suggest that phosphorylation of Thr18 and Ser20 may also enhance the recruitment of CBP/p300 to p53 (Dornan and Hupp, 2001); however, Saito et al. (2002) found that acetylation of Lys382 was decreased by mutations that changed Ser6, 9, 15, or Thr18, but not Ser20 or more distal sites, to alanine, presumably by inhibiting phosphorylation at these sites. The acetylated C-terminal lysines are also targets for ubiquitination; thus, acetylation may directly contribute to p53 stabilization (Nakamura et al., 2002). Lysine 386 is reported to be sumoylated, although only at low levels (Melchior and Hengst, 2002).

The availability of modification-specific antibodies has allowed a detailed characterization of the phosphorylation and acetylation of p53 in cultured human cells following exposures to genotoxic agents, including IR, UV, adriamycin (Saito et al., 2003), or nitrogen oxide (NO) (Hofseth et al., 2003), as well as to nongenotoxic agents, including the presence of activated oncogenes (e.g. *Ras*) (Bulavin et al., 2002b), microtubule disruptors (taxol, nocodazole), nucleoside synthesis inhibitors (PALA) (Saito et al., 2003), hypoxia (Hammond et al., 2002), and osmotic stress (Kishi et al., 2001). Use of these antibodies, most of which are now commercially available, coupled with cell lines defective in one or more signaling enzymes or the use of highly specific chemical inhibitors, has begun to elucidate the pathways that lead to specific p53 modifications. Such studies have also revealed some unexpected relationships.

In response to DNA double-strand breaks (DSBs), one of the earliest modifications to p53 that can be detected is phosphorylation of serine 15 (Siliciano et al., 1997). Although serine 15 was first identified as a site phosphorylated in vitro by the DNA-dependent protein kinase, DNA-PK (Lees-Miller et al., 1992), later it was shown that DNA-PK was not required to phosphorylate this site in vivo, nor was DNA-PK needed for the physiological responses to DNA damage that depend on p53 (Jimenez et al., 1999). DNA-PK is a member of a small family of large protein kinases that more closely resemble phosphatidylinositol-3 kinases (PI3K) in their kinase domains than the majority of serine/threonine kinases, and DNA-PK was found to preferentially phosphorylate serines or threonines that were followed by glutamine, the so-called SQ/TQ motif (Anderson and Lees-Miller, 1992). In mammalian cells, the PI3K-like kinase family includes four additional members, ATM (ataxia telangiectasia (A-T) mutated), ATR (A-T and RAD3 related), FRAP (FK506 binding protein12-rapamycin associated protein kinase), and SMG1 (also called ATX), a recently described protein kinase involved in nonsense-mediated mRNA decay (Denning et al., 2001; Yamashita et al., 2001). Each of these kinases also recognizes SQ/TQ motifs in protein substrates, and each phosphorylates serine 15 of p53 (or in a p53-related peptide) in vitro (Abraham, 2001). FRAP is involved in the regulation of translation initiation in response to nutrients and growth factors, but its activity also increases at late times after exposure of cells to UV light, where it transmits a signal for the production of immunosuppressive cytokines (Yarosh et al., 2000). Whether SMG1/ATX, FRAP, or DNA-PK are ever important for phosphorylating p53 or regulating its activity in vivo is unclear; however, ATM and ATR are both believed to directly phosphorylate p53 on Ser15 in vivo in response to DNA damage (Banin et al., 1998; Canman et al., 1998; Tibbetts et al., 1999). Cell lines that lacked ATM or that overexpressed a dominant-negative allele of ATR are deficient in p53 phosphorylation at Ser15 in vivo and are defective in the activation of DNA damage-induced cell cycle checkpoints (Abraham, 2001, 2003; Shiloh, 2003).

Studies using phospho-specific antibodies and cell lines deficient in ATM revealed that phosphorylation of p53 at Ser9, Thr18, Ser20, and Ser46 are dependent on the ATM kinase (Saito et al., 2002). These sites, all of which are phosphorylated in response to DNA damage in vivo, do not correspond to the SQ/TQ motif and are not believed to be phosphorylated by ATM or ATR directly. Rather, phosphorylation of these sites is believed to depend on effector kinases that are activated in response to ATM or ATR, or that require phosphorylation of Ser15 for recognition of p53. Two protein kinases capable of phosphorylating Ser46, p38 MAPK (Bulavin et al., 1999) and HIPK2 (D'Orazi et al., 2002; Hofmann et al., 2002), both of which are activated after exposure of cells to UV light, have been described; however, neither has been shown to be ATM dependent. Serines 6 and 9 became strongly phosphorylated in response to both IR- and UV-induced DNA damage, which indicates that Ser9 could be phosphorylated by CK1 or a CK1-like kinase in response to phosphorylation of Ser6 (Higashimoto et al., 2000). In vitro CK1 phosphorylates serines and threonines two residues distal to a phosphorylated serine or threonine. However, in response to IR, phosphorylation of Ser9 appears to be independent of phosphorylation at Ser6; thus, phosphorylation of Ser9 appears to be dependent upon activation of an unknown protein kinase that is activated by ATM. Alternatively, recognition of p53 by this kinase requires phosphorylation of p53 at Ser15.

Recently, using mutant p53s in transient transfection experiments in which individual serines were changed to alanines, Saito et al. (2003) demonstrated additional N-terminal p53 phosphorylation sites interdependencies. As had been shown previously (Bulavin et al., 1999), changing Ser33 to alanine blocked phosphorylation of Ser37, but changing Ser37 to alanine had no effect on phosphorylation at Ser33 or at other N-terminal sites. Changing Ser6 to alanine blocked phosphorylation at Ser9 and vice versa without affecting phosphorylation at the other N-terminal sites. Most strikingly, substituting alanine for Ser15 prevented IR-induced phosphorylation at Ser9, Thr18, and Ser20, while phosphorylation of Ser6, Ser33, Ser37, and Ser46 were unaffected. Similarly, changing Ser20 to alanine prevented phosphorylation of Thr18, while changing Thr18 to alanine reduced phosphorylation at Ser20 but not at the other N-terminal sites. Changing Ser37 or Ser46 to alanine had no significant effect on the phosphorylation of other sites, nor did phosphorylation of the C-terminal sites, Ser315 or Ser392, depend on any of the N-terminal phosphorylation sites or vice versa. Control experiments suggested that changing serine to alanine did not prevent recognition by phospho-specific antibodies. Thus, on the basis of single-site mutant analyses, the N-terminal p53 phosphorylation sites can be classified into four clusters: Ser6 and Ser9; Ser9, Ser15, Thr18, and Ser20; Ser33 and Ser37; and Ser46. Furthermore, phosphorylation of the Ser15 cluster (Ser9, Ser15, Thr18, and Ser20) appears to require DNA damage (Saito et al., 2003). Presently, it cannot be determined whether phosphorylation at dependent sites requires a nearby serine or the phosphorylation of that serine (or threonine); nevertheless, these results suggest that at least some site interdependencies reflect mechanisms that permit signal amplification and the integration of information from diverse signaling pathways by requiring sequential phosphorylation of sites in an ordered manner. For example, Ser9, Thr18, and Ser20 will not be phosphorylated unless Ser15 is first phosphorylated, and Ser15, Thr18, and Ser20 may all be required for efficient p53 stabilization. Furthermore, this intramolecular cascade mechanism might serve to check inappropriate p53 activation or regulate the intensity of the p53 response and would complement kinase activation cascades (Saito et al., 2002).

5.4. SIGNALING TO p53

The mechanisms by which cells detect genotoxic and nongenotoxic stresses and signal to p53 are complex and still incompletely understood. However, phosphorylation of p53 in response to DNA damage appears to be principally driven by two related signaling pathways, one mediated by ATM and the other by ATR, that are activated by different mechanisms in response to different DNA insults.

5.4.1. ATM-Dependent Signaling to p53

Although there are many different forms of DNA damage, the most dangerous among them are DNA double-strand breaks (DSBs). DNA DSBs result from exposure to external insults such as ionizing radiation and treatments with certain anticancer agents; it has been estimated, however, that even in the absence of exposure to genotoxic substances each human cell undergoes approximately eight DSBs per day from physical forces and oxidative damage generated in the course of normal cellular metabolism (Bernstein and Bernstein, 1991). While the consequences of naturally occurring DSBs were probably the evolutionary driver for development of systems that all cells have for recognizing DSBs and taking appropriate actions, treatment with ionizing radiation and radiomimetic drugs (e.g., neocarzinostatin (NCS), or toposiomerase II inhibitors [e.g., adriamycin or etoposide] are frequently used in the laboratory to study the consequences of DSBs. It must be remembered, however, that these agents have other effects. For example, IR produces far more single-stranded breaks and cluster damaged sites than simple DSBs (Sutherland et al., 2000).

In mammalian cells, cellular responses to DSBs, including phosphorylation of p53 at several sites, are heavily dependent upon the ATM protein kinase. Loss of ATM function in humans causes ataxia telangiectasia (A-T), a devastating disease characterized by progressive neurodegeneration, immunodeficiency, sterility, and a high risk of cancer (Shiloh, 2003). A-T cells are hypersensitive to killing by ionizing radiation but show normal sensitivity to UV light. While our understanding of the complex mechanism(s) by which DSBs activate ATM are incomplete, remarkable progress has recently been made. Immediately after exposure of cells to IR or radiomimetic agents, a moderate but reproducible increase in ATM kinase activity can be measured in immune complex assays (Banin et al., 1998; Canman et al., 1998). This increased activity is not accompanied by changes in ATM abundance or subcellular distribution. Purified ATM was shown to interact preferentially with the ends of double-stranded DNA fragments (Smith et al., 1999; Suzuki et al., 1999), but DNA is not required to sustain ATM activity in immune complexes; thus, the implications of this finding with respect to activation in vivo remain unclear. Nevertheless, a small fraction of the ATM molecules in cells became resistant to extraction and were detected as nuclear aggregates immediately following the induction of DSBs (Andegeko et al., 2001). Furthermore, the retained fraction of ATM colocalized with the phosphorylated form of histone H2AX (γ-H2AX) and with foci of the Nbs1 protein, suggesting that ATM associates with DSBs. DSB-induced γ-H2AX foci appear before those of most other proteins that form foci after DNA damage, and the number of γ-H2AX foci is proportional to the number of induced DSBs (Paull et al., 2000; Schultz et al., 2000; Bonner, 2003). γ-H2AX is phosphorylated at serine 139, an SQ site, by the ATM kinase in vitro, and ATM is necessary for this phosphorylation in vivo early after the induction of DSBs (Burma et al., 2001). Together, these results indicate that ATM is activated very early after DSB induction at or near the sites of DNA double-strand breaks.

A hint as to the mechanism of activation came from work in Lavin's laboratory which showed that ATM from unirradiated cells was activated in the absence of DNA after preincubation with ATP (Kozlov et al., 2003). Activation required Mn^{2+}, a required ATM cofactor, and was inhibited by wortmannin, a PI3K-specific inhibitor. Activation was reversed by phosphatase treatment, suggesting that activation involved autophosphorylation. Then, in a technical tour de force, Bakkenist and Kastan (2003) identified Ser1981, which resides in the sequence GSQS N-terminal to the kinase domain, as an IR-inducible phosphorylation site in the ATM polypeptide. Using a phospho-Ser1981-specific antibody, they then showed that a kinase-dead ATM mutant was phosphorylated in IR-treated cells that contained wild-type ATM but not in A-T cells that lack functional ATM, but this mutant was not phosphorylated in cells that expressed the related PI3Ks ATR and DNA-PK (Bakkenist and Kastan, 2003). This result strongly suggests that Ser1981 is phosphorylated as the result of self- or autophosphorylation. Ser1981 resides near the N-terminus of a FAT (*F*RAP, *A*TM, and *T*RRAP) domain, a $\sim$500 amino acid region found only in PI3K-related proteins that may serve as a structural scaffold or as a protein–protein interaction domain (Bosotti et al., 2000). Subsequent analysis of ATM protein fragments showed that the kinase domain and the FAT domain stably bound one another and that the sequences flanking Ser1981 are important for this interaction. However, mutating Ser1981 to aspartic or glutamic acid, which mimic serine phosphorylation, prevented interaction of the FAT domain with the kinase domain, suggesting that autophosphorylation results in the dissociation of a complex containing two or more inactive ATM molecules. These findings are consistent with a model in which ATM is activated in response to DSBs by autophosphorylation at Ser1981, which results in a dissociation of the ATM complex into monomers that are then capable of interacting with substrates (Fig. 5.3).

Although the above model superficially fits expectations, the astonishing finding of Bakkenist and Kastan (Bakkenist and Kastan, 2003) is that the majority of AT`

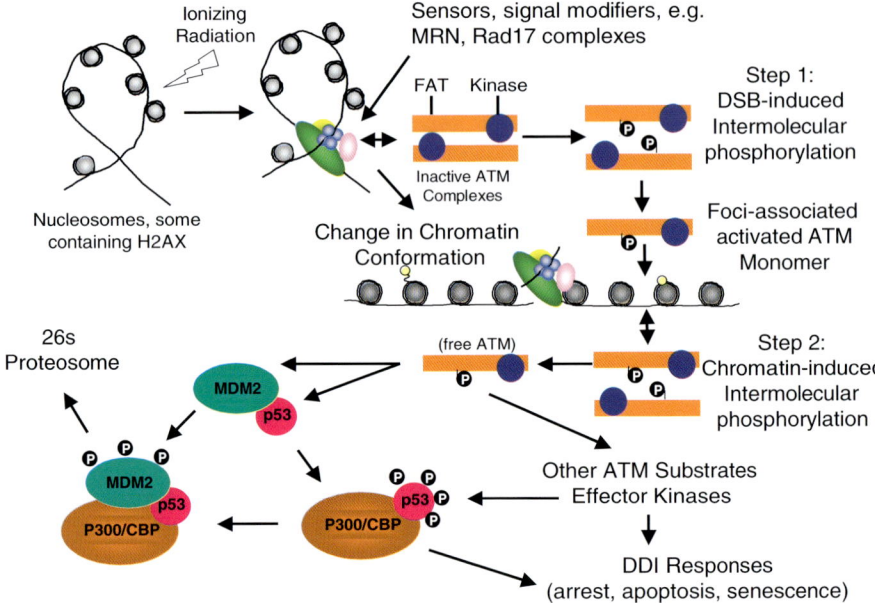

Figure 5.3. Activation of p53 in response to DNA double-strand breaks. In step 1, DNA DSBs result in the rapid activation through autophosphorylation of a fraction of a cell's ATM. This fraction becomes tightly associated with chromatin through ATM's DNA end-binding activity (Smith et al., 1999; Suzuki et al., 1999); it then phosphorylates H2AX and perhaps other substrates that assemble at the break site (Redon et al., 2002; Shiloh, 2003). H2AX is a variant of H2A with a C-terminal extension that can be directly phosphorylated (yellow circles) by ATM. It is found with RAD9, RAD1, RAD17, HUS1, and the MRN complex (Mre11, Rad50, Nbs1) in foci of DNA damage sensors and repair proteins that form at DSBs sites after DNA damage (D'Amours and Jackson, 2002; Fei and el-Deiry, 2003; Petrini and Stracker, 2003). The DSB induces a change in chromatin conformation, with which the bulk of a cell's ATM interacts to become activated, also through autophosphorylation in step 2 (Bakkenist and Kastan, 2003). Autophosphorylation at Ser1981 causes ATM to dissociate into active monomers. ATM directly phosphorylates Ser15 near the N-terminus of p53 and is required for the phosphorylation of Ser9, Ser20, Ser46, and Thr18, presumably as a consequence of ATM-dependent activation of effector protein kinases (Saito et al., 2002) and/or creation of kinase recognition sites (Saito et al., 2003). Phosphorylation of Mdm2 and p53 may promote dissociation of p53 and Mdm2, inhibit p53 degradation, and promote association of p53 with its coactivator p300/CBP. However, association of p300/CBP with the p53/Mdm2 complex may promote p53 multiubiquitination and its degradation through the 26s proteosome. ATM also phosphorylates other substrates including BRCA1, 53BP1, Mdm2, and downstream effector kinases, such as Chk2.

molecules in a cell became activated within a few minutes after exposure to IR doses that produce only a few DSBs per cell. At these low doses (0.1 Gy, which is expected to produce ~4 DSBs/cell), it is inconceivable that each ATM molecule can associate with a DSB as a requirement for activation within the time that was available. To explain this observation, Bakkenist and Kastan proposed that a DSB could reveal its presence by triggering a relatively widespread change in chromatin structure with which ATM could interact to trigger conversion of inactive ATM complexes into active monomers through autophosphorylation. Consistent with this hypothesis, the authors indeed found that treatment of cells with a histone deacetylase inhibitor induced phosphorylation on Ser1981 and resulted in the concomitant phosphorylation of p53 on Ser15. This finding suggests that activation of p53 in response to DSBs is a two-stage process (Fig. 5.3). First, a fraction of the nuclear ATM interacts with DSBs or other sensor proteins such as MRN (Mre11, Rad50, Nsb1) or the Rad17 complexes that rapidly bind to DSBs. Indeed, recent results show that the MRN complex is required for proper activation of ATM (Uziel et al., 2003). The tightly bound fraction of ATM is activated by autophosphorylation and rapidly phosphorylates H2AX and other proteins that assemble at DSB sites, recruiting additional proteins to the DSB sites. The assembled complex then triggers a change in chromatin conformation over a distance of perhaps a megabase which, in turn, provides a larger target for the interaction of additional, free ATM complexes that then autophosphorylate to become active, free monomers. The activated, free ATM molecules rapidly phosphorylate effector kinases, such as Chk2, and other substrates, e.g. p53, Mdm2, BRCA1, to accomplish control of cell cycle progression and activation of DNA repair and perhaps apoptosis. Although this model has considerable appeal, several questions remain. How does ATM sense both DNA ends and changes in chromatin structure? What is the nature of the change in chromatin structure, and how is this change distinguished from changes that accompany chromatin remodeling associated with normal transcription and DNA replication?

5.4.2. ATR-Dependent Signaling to p53

Activation of ATR, the ATM and RAD3-related kinase, is not as well understood as activation of ATM, in part because inactivation of ATR results in lethality, and only recently have genetic constructs been engineered that allow the consequences of ATR activation to be deduced at the molecular level [e.g. Cortez et al. (2001); Zou et al. (2002)]. ATR is activated after exposure of cells to UV light or alkylating agents, which produce bulky lesions in DNA, or treatment with anticancer drugs (e.g. adriamycin), hydroxyurea, or extreme hypoxia that may block transcription or replication or cause replication fork collapse (Abraham, 2001; Hammond et al., 2002; Brown and Baltimore, 2003). ATR also is activated at later times after the creation of DSBs, which probably accounts for delayed phosphorylation of p53 at Ser15 in A-T cells (Saito et al., 2002). However, it is unclear whether the DNA damage that leads to ATR activation is sensed directly or whether ATR is responding to a consequence of blocked transcription or replication, or both (Fig 5.4). As for ATM, activation of ATR is not accompanied by changes in ATR abundance or subcellular distribution. Unlike ATM, ATR isolated from cells treated with DNA damage-inducing agents does not display increased activity in kinase assays in vitro (Tibbetts et al., 2000). Furthermore, neither ATR nor the other PI3K-like kinases (DNA-PK or FRAP) have an SQ/TQ site at the N-terminus of their FAT domains equivalent to the GSQS Ser1981 autophosphorylation site in ATM (Bosotti et al., 2000), making autophosphorylation less likely as a mechanism for ATR activation in response to DNA damage.

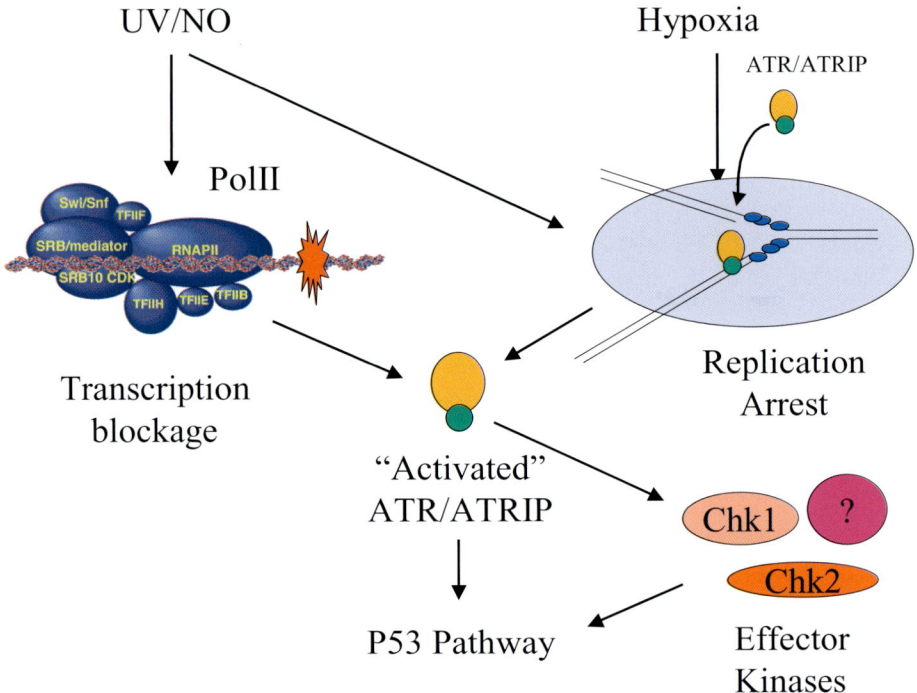

Figure 5.4. Activation of ATR in response to blockage of transcription by RNA polymerase II and the arrest of DNA replication. ATR is activated in human cells in response to UV radiation and chemicals that produced bulky lesions and oxidized DNA bases. These, in turn, may block RNA transcription by RNA polymerase II (polI) and DNA replication. In human cells, ATR exists in a stable complex with ATRIP (ATR-interacting protein) (Cortez et al., 2001). ATR is recruited to sites of DNA damage that contain single-stranded DNA segments through the interaction of ATRIP with RPA (Zou and Elledge, 2003), suggesting that RPA–ssDNA, a complex common to several DNA repair processes, may serve as a DNA damage signal for the recruitment of ATR–ATRIP. In contrast to ATM, ATR isolated from cells exposed to DNA damaging agents does not display increased kinase activity (Tibbetts et al., 2000); thus, "activation" may be achieved by the simultaneous recruitment of ATR–ATRIP and substrates to sites of DNA damage (Zou and Elledge, 2003). ATR activates the effector kinase Chk1 and is believed to phosphorylate p53 at Ser15 and Ser37. As noted above, extreme hypoxia does not cause detectable DNA damage but is believed to activate ATR by causing the collapse of DNA replication forks (Hammond et al., 2002).

In mammalian cells, ATR exists as a stable complex with ATRIP (ATR interacting protein), an 85 kDa protein that stabilizes ATR and may help regulate its activity (Cortez et al., 2001). In vitro, ATR phosphorylates ATRIP, and both proteins colocalize to intranuclear foci that may correspond to sites of DNA synthesis and repair. Recent studies by Zou and Elledge (2003) show that replication protein A (RPA), a protein complex that associates with single-stranded DNA (ssDNA) and becomes highly phosphorylated on its 34 kDa subunit following DNA damage, is required to recruit ATR–ATRIP to sites of DNA damage and to form nuclear foci. In vitro, RPA stimulated the binding of ATRIP to single-stranded DNA and the phosphorylation of Rad17 on Ser635, an in vitro and in vivo site of phosphorylation by ATR. RPA was also required for ATR-mediated activation of the Chk1 kinase in human cells. These studies suggest that ssDNA may be a common intermediate that functions as a signal for activation of ATR–ATRIP (Fig. 5.4). Single-stranded gaps are generated as an intermediate in the repair of bulky lesions by nucleotide excision repair. When replication forks encounter DNA lesions, longer stretches of ssDNA could be generated by the stalling of polymerases and/or the uncoupling of helicases and polymerases. Thus, Zou and Elledge suggest that the apparent activation of ATR may be achieved by the simultaneous enrichment of ATR–ATRIP complexes and their substrates at sites of DNA damage (Zou et al., 2002). If this is the case, it will be interesting to see whether proteins besides RPA target ATR for colocalization with substrates.

5.5. NONGENOTOXIC STRESS AND p53 EFFECTOR KINASES

ATM and ATR both phosphorylate p53 at Ser15 in vitro, and elimination of Ser15 prevents p53 phosphorylation by ATM (Banin et al., 1998; Canman et al., 1998), indicating that other kinases are responsible for phosphorylating p53 at other sites (Fig. 5.2). In response to DSBs, ATM activates the Chk2 kinase through phosphorylation of Thr68, and Thr68 is required for the full activation of Chk2 in response to IR (Ahn et al., 2000; Melchionna et al., 2000). Likewise, Chk1 is phosphorylated and activated in response to UV light in vivo in an ATR-dependent manner, and in vitro ATR phosphorylates Chk1 on serine 317 and 345 (Zhao and Piwnica-Worms, 2001). Early studies by Shieh et al. (2000) and Chehab et al. (2000) reported that Chk1 and Chk2 phosphorylated p53 at Ser20, and possibly other sites, resulting in its stabilization and activation in response to DNA damage. These results are consistent with a requirement for ATM for the phosphorylation of Ser9, Thr18, and Ser20 in response to IR (Saito et al., 2002); however, several recent studies question the role of the Chk2 effector kinase in mediating p53 phosphorylation at Ser20 as well as the role of Ser20 in stabilizing and activating p53. First, in contrast to changing Ser18 of murine p53 (the equivalent of Ser15 in human p53) to alanine (Chao et al., 2000), Wu et al. (2002) found that changing Ser23 (Ser20 in human p53) to alanine had no effect on p53 stability or activity in mouse ES cells, fibroblasts or thymocytes. Second, Takai et al. (2002) showed that mouse p53 Ser23 and human p53 Ser20 were phosphorylated equally well in cells from wild- type or Chk2 knockout mice, although p53-mediated transactivation of several target genes was abolished. The dispensability of Chk2 to phosphorylate p53 Ser20 was recently confirmed by Jallepalli et al. (2003). Third, reexamination of p53 phosphorylation in vitro by purified Chk2 indicated that p53 was a weak substrate compared to Cdc25C (Ahn et al., 2003). Furthermore, inhibition of Chk2 expression with small, interfering RNAs (siRNA) led to a marked reduction in Chk2 protein, but p53 was still stabilized and active as a transcription factor. Similar results were also seen with siRNA-mediated targeting of Chk1, suggesting that neither Chk1 nor Chk2 regulate p53 stability or activity. Together with the recently reported interdependence of p53 phosphorylation at Ser9, Ser15, Thr18, and Ser20 (Saito et al., 2003), these results indicate that the role of Ser20 in stabilizing p53 should be reevaluated.

In contrast to the Ser15 cluster, most other known phosphorylation sites in p53 (except Ser37) are phosphorylated in response to both genotoxic and nongenotoxic stresses (Saito et al., 2003). With the exception of Ser6 and Ser9, kinases that can phosphorylate most of these sites in vitro have been reported (Fig. 5.2); however, in most cases, to date there is little compelling evidence that these kinases phosphorylate p53 in vivo. Furthermore, for the most part it is not known if or how these kinases are activated in response to various forms of cellular stress.

After Ser15, a second important phosphorylation site is Ser46. Serine 46 of human p53 was shown to be phosphorylated in cells exposed to UV light (Bulavin et al., 1999). In vitro, Ser33 and Ser46 were phosphorylated by the p38 MAP kinase, and mutation of both these sites decreased p53-mediated and UV-induced apoptosis. Ser46 was also shown to be required for induction of p53AIP1, a mitochondrial localized protein whose enhanced expression leads to cell death (Oda et al., 2000). Subsequently, two laboratories showed that homeodomain-interacting protein kinase-2 (HIPK2) was activated after exposure of cells to UV light; HIPK2 also phosphorylated p53 on Ser46 in vitro (D'Orazi et al., 2002; Hofmann et al., 2002). Furthermore, HIPK2 interacts and colocalizes with p53 and CBP in PML nuclear bodies, thus facilitating p53 acetylation. As noted above, Ser46 is also phosphorylated after IR, and phosphorylation in response to IR is ATM dependent (Saito et al., 2002); however, it is unclear if either p38 MAPK or HIPK2 can be activated by ATM. Interest in the potential role of p38 MAPK in regulating p53 activity recently was stimulated by the finding that the gene (*PPM1D*) for Wip1, a p53-induced protein phosphatase that negatively regulates p38 MAPK activity (Fiscella et al., 1997; Takekawa et al., 2000), is amplified in 12 to18% of primary human breast cancers (Bulavin et al., 2002b; Li et al., 2002). Wip1 thus forms a negative feedback loop with p53 analogous to the p53–Mdm2 feedback loop. Amplification of the Wip1 gene in cancers, which would inhibit p38 MAPK-mediated activation of p53 through phosphorylation of Ser33 and Ser46, is consistent with a role for p38 MAPK in regulating p53 activity in vivo.

A large number of proteins have been shown to interact with p53, at least in vitro, and, as shown in Figure 5.2, most of these interact with the N- or C-terminal regions of p53 that are both unstructured and become highly modified in response to stress. This coincidence is unlikely to be accidental. Rather, it seems highly probable that the interaction of some of these and other proteins will be enhanced or inhibited by p53 posttranslational modifications. In turn, the complexes thus formed are likely to modulate p53 function and regulate cell fate. To date, the interaction of p53 with only a few of the proteins listed in Figure 5.2 has been shown to be modulated by phosphorylation. As described above, foremost among these are the HATs, p300/CBP. The role of phosphorylation in regulating the interaction of p53 and Mdm2 is still controversial (Schon et al., 2002; Anderson and Appella, 2003). Nevertheless, the roles for phosphorylation and acetylation in modulating interactions of proteins with p53, including protein kinases, HATs, HDACs, and their adaptors, will be a fruitful area for future research.

5.6. CONCLUSIONS

Cellular responses to both genotoxic and nongenotoxic stress are complex and involve multiple signaling pathways. This is well illustrated by the p53 tumor suppressor protein, which itself represents but one node in the cellular pathways that regulate cell function in response to both internal and external stimuli. Studies over the past 20 years have elucidated most of the posttranslational modifications to p53 that, in turn, modulate its stability and activity. The availability of reagents (antibodies) that are highly specific for p53 modified at specific sites, coupled with new genetic techniques for abrogating gene function, is facilitating elucidation of multiple, interacting pathways that posttranslationally modify p53 through phosphorylation or acetylation. Stress *signals* must first be detected through some change, the binding of a ligand to a membrane receptor or the recognition of new or unusual internal structures (e.g. DSBs or pyrimidine dimers) by *sensors* (Petrini and Stracker, 2003). Such structures may require processing by *signal modifiers* (D'Amours and Jackson, 2002), e.g. the excision of dimers leaving a region of single stranded DNA, to allow recognition by the proximal *signal transducers*, which usually are protein kinases (e.g. ATM, ATR, p38 MAPK) (Abraham, 2001; Bulavin et al., 2002a; Shiloh, 2003). Signal recognition by signal transducers may require *adaptors* (e.g. RPA) to recognize proximal processed signals (ssDNA), and *mediators* (e.g. Rad9, Mdc1) (Canman, 2003) to transmit signals to effectors (e.g. Chk1, Chk2) that ultimately modify targets, such as p53. p53 then integrates signal strength and/or signals from several sources to ultimately determine cell fate through the induction or repression of specific genes, or by direct interaction with components that mediate apoptosis. Signaling pathways are often branched and interconnected. Likewise signals, especially external environmental signals, may not be pure, thereby activating more than one signaling pathway. While substantial progress has been made in characterizing the pathways that respond to DNA damage and signal to p53, these pathways are still incompletely characterized and the actual mechanisms that detect DNA damage are only now becoming clear. Nevertheless, thanks in part to new technologies, rapid progress can be expected over the next few years.

ACKNOWLEDGMENTS

We thank Sharlyn J. Mazur and Marco Schito for constructive suggestions. We apologize to those whose publications could not be cited due to space limitations. C.W.A. was supported in part by a Laboratory Directed Research and Development Grant at the Brookhaven National Laboratory under contract with the U.S. Department of Energy.

REFERENCES

Abraham, R. T. (2001). Cell cycle checkpoint signaling through the ATM and ATR kinases. *Genes Dev* 15:2177–2196.

Abraham, R. T. (2003). Checkpoint signaling: Epigenetic events sound the DNA strand-breaks alarm to the ATM protein kinase. *Bioessays* 25:627–630.

Ahn, J., and Prives, C. (2001). The C-terminus of p53: the more you learn the less you know. *Nat Struct Biol* 8:730–732.

Ahn, J., Urist, M., and Prives, C. (2003). Questioning the role of checkpoint kinase 2 in the p53 DNA damage response. *J Biol Chem* 278:20480–20489.

Ahn, J.-Y., Schwarz, J. K., Piwnica-Worms, H., and Canman, C. E. (2000). Threonine 68 phosphorylation by ataxia telangiectasia mutated is required for efficient activation of Chk2 in response to ionizing radiation. *Cancer Res* 60:5934–5936.

Andegeko, Y., Moyal, L., Mitelman, L., Tsarfaty, I., Shiloh, Y., and Rotman, G. (2001). Nuclear retention of ATM at sites of DNA double strand breaks. *J Biol Chem* 276:38224–38230.

Anderson, C. W., and Appella, E. (2003). Signaling to the p53 tumor suppressor through pathways activated by genotoxic and non-genotoxic stresses. In R. A. Bradshaw and E. Dennis (eds),*Handbook of Cell Signaling*. Academic Press, New York, pp. 237–247.

Anderson, C. W., and Lees-Miller, S. P. (1992). The nuclear serine/threonine protein kinase DNA-PK. *Crit Rev Eukaryot Gene Expr* 2:283–314.

Anderson, M. E., Woelker, B., Reed, M., Wang, P., and Tegtmeyer, P. (1997). Reciprocal interference between the sequence-specific core and nonspecific C-terminal DNA binding domains of p53: implications for regulation. *Mol Cell Biol* 17:6255–6264.

Appella, E., and Anderson, C. W. (2001). Post-translational modifications and activation of p53 by genotoxic stresses. *Eur J Biochem* 268:2764–2772.

Ayed, A., Mulder, F. A. A., Yi, G.-S., Lu, Y., Kay, L. E., and Arrowsmith, C.H. (2001). Latent and active p53 are identical in conformation. *Nat Struct Biol* 8:756–760.

Bakkenist, C. J., and Kastan, M. B. (2003). DNA damage activates ATM through intermolecular autophosphorylation and dimer dissociation. *Nature* 421:499–506.

Banin, S., Moyal, L., Shieh, S.-Y., Taya, Y., Anderson, C.W., Chessa, L., Smorodinsky, N.I., Prives, C., Reiss, Y., Shiloh, Y., and Ziv, Y. (1998). Enhanced phosphorylation of p53 by ATM in response to DNA damage. *Science* 281:1674-1677.

Bech-Otschir, D., Kraft, R., Huang, X., Henklein, P., Kapelari, B., Pollmann, C., and Dubiel, W. (2001). COP9 signalosome-specific phosphorylation targets p53 to degradation by the ubiquitin system. *EMBO J* 20:1630–1639.

Bernstein, C., and Bernstein, H. (1991). *Aging, Sex, and DNA Repair.* Academic Press, San Diego, CA.

Bonner, W. M. (2003). Low-dose radiation: Thresholds, bystander effects, and adaptive responses. *Proc Natl Acad Sci USA* 100:4973–4975.

Bosotti, R., Isacchi, A., and Sonnhammer, E. L. L. (2000). FAT: a novel domain in PIK-related kinases. *Trends Biochem Sci* 25:225–227.

Brown, E. J., and Baltimore, D. (2003). Essential and dispensable roles of ATR in cell cycle arrest and genome maintenance. *Genes Dev* 17:615–628.

Bulavin, D. V., Amundson, S. A., and Fornace, Jr., A. J. (2002a). p38 and Chk1 kinases: different conductors for the G_2/M checkpoint symphony. *Curr Opin Genet Dev* 12:92–97.

Bulavin, D. V., Demidov, O. N., Saito, S., Kauraniemi, P., Phillips, C., Amundson, S. A., Ambrosino, C., Sauter, G., Nebreda, A. R., Anderson, C. W., Kallioniemi, A., Fornace, Jr., A. J., and Appella, E. (2002b). Amplification of *PPM1D* in human tumors abrogates p53 tumor-suppressor activity. *Nat Genet* 31:210–215.

Bulavin, D. V., Saito, S., Hollander, M. C., Sakaguchi, K., Anderson, C. W., Appella, E., and Fornace Jr., A. J. (1999). Phosphorylation of human p53 by p38 kinase coordinates N-terminal phosphorylation and apoptosis in response to UV radiation. *EMBO J* 18:6845–6854.

Burma, S., Chen, B. P., Murphy, M., Kurimasa, A., and Chen, D. J. (2001). ATM phosphorylates histone H2AX in response to DNA double-strand breaks. *J Biol Chem* 276:42462–42467.

Canman, C. E. (2003). Checkpoint mediators: relaying signals from DNA strand breaks. *Curr Biol* 13:R488–R490.

Canman, C. E., Lim, D.-S., Cimprich, K. A., Taya, Y., Tamai, K., Sakaguchi, K., Appella, E., Kastan, M. B., and Siliciano, J. D. (1998). Activation of the ATM kinase by ionizing radiation and phosphorylation of p53. *Science* 281:1677–1679.

Chao, C., Saito, S., Anderson, C. W., Appella, E., and Xu, Y. (2000). Phosphorylation of murine p53 at Ser-18 regulates the p53 responses to DNA damage. *Proc Natl Acad Sci USA* 97:11936–11941.

Chehab, N. H., Malikzay, A., Appel, M., and Halazonetis, T. D. (2000). Chk2/hCds1 functions as a DNA damage checkpoint in G_1 by stabilizing p53. *Genes Dev* 14:278–288.

Chehab, N. H., Malikzay, A., Stavridi, E. S., and Halazonetis, T. D. (1999). Phosphorylation of Ser-20 mediates stabilization of human p53 in response to DNA damage. *Proc Natl Acad Sci USA* 96:13777–13782.

Cho, Y., Gorina, S., Jeffrey, P. D., and Pavletich, N. P. (1994). Crystal structure of a p53 tumor suppressor-DNA complex: understanding tumorigenic mutations. *Science* 265:346–355.

Cline, S. D., and Hanawalt, P. C. (2003). Who's on first in the cellular response to DNA damage? *Nat Rev Mol Cell Biol* 4:361–373.

Clore, G. M., Ernst, J., Clubb, R., Omichinski, J. G., Kennedy, W. M. P., Sakaguchi, K., Appella, E., and Gronenborn, A. M. (1995). Refined solution structure of the oligomerization domain of the tumour suppressor p53. *Nat Struct Biol* 2:321–333.

Cortez, D., Guntuku, S., Qin, J., and Elledge, S. J. (2001). ATR and ATRIP: partners in checkpoint signaling. *Science* 294:1713–1716.

Craig, A. L., Bray, S. E., Finlan, L. E., Kernohan, N. M., and Hupp, T. R. (2003). Signaling to p53: The use of phospho-specific antibodies to probe for *in vivo* kinase activation. *Methods Mol Biol* 234:171–202.

D'Amours, D., and Jackson, S. P. (2002). The Mre11 complex: at the crossroads of DNA repair and checkpoint signalling. *Nat Rev Mol Cell Biol* 3:317–327.

Denning, G., Jamieson, L., Maquat, L. E., Thompson, E. A., and Fields, A. P. (2001). Cloning of a novel phosphatidylinositol kinase-related kinase: characterization of the human SMG-1 RNA surveillance protein. *J Biol Chem* 276:22709–22714.

D'Orazi, G., Cecchinelli, B., Bruno, T., Manni, I., Higashimoto, Y., Saito, S., Gostissa, M., Coen, S., Marchetti, A., Del Sal, G., Piaggio, G., Fanciulli, M., Appella, E., and Soddu, S. (2002). Homeodomain-interacting protein kinase-2 phosphorylates p53 at Ser 46 and mediates apoptosis. *Nat Cell Biol* 4:11–19.

Dornan, D., and Hupp, T. R. (2001). Inhibition of p53-dependent transcription by BOX-I phospho-peptide mimetics that bind to p300. *EMBO Rep* 2:139–144.

Dumaz, N., and Meek, D. W. (1999). Serine15 phosphorylation stimulates p53 transactivation but does not directly influence interaction with HDM2. *EMBO J* 18:7002–7010.

Espinosa, J. M., and Emerson, B. M. (2001). Transcriptional regulation by p53 through intrinsic DNA/chromatin binding and site-directed cofactor recruitment. *Mol Cell* 8:57–69.

Fei, P., and el-Deiry, W .S. (2003). P53 and radiation responses. *Oncogene* 22:5774–5783.

Fiscella, M., Zhang, H., Fan, S., Sakaguchi, K., Shen, S., Mercer, W. E., Vande Woude, G. F., O'Connor, P. M., and Appella, E. (1997). Wip1, a novel human protein phosphatase that is induced in response to ionizing radiation in a p53-dependent manner. *Proc Natl Acad Sci. USA* 94:6048–6053.

Gu, W., and Roeder, R. G. (1997). Activation of p53 sequence-specific DNA binding by acetylation of the p53 C-terminal domain. *Cell* 90:595–606.

Hammond, E. M., Denko, N. C., Dorie, M. J., Abraham, R. T., and Giaccia, A. J. (2002). Hypoxia links ATR and p53 through replication arrest. *Mol Cell Biol* 22:1834–1843.

Hao, M., Lowy, A.M., Kapoor, M., Deffie, A., Liu, G., and Lozano, G. (1996). Mutation of phosphoserine 389 affects p53 function in vivo. *J Biol Chem* 271:29380–29385.

Higashimoto, Y., Saito, S., Tong, X.-H., Hong, A., Sakaguchi, K., Appella, E., and Anderson, C.W. (2000). Human p53 is phosphorylated on serines 6 and 9 in response to DNA damage-inducing agents. *J Biol Chem* 275:23199–23203.

Hofmann, T. G., Möller, A., Sirma, H., Zentgraf, H., Taya, Y., Dröge, W., Will, H., and Schmitz, M. L. (2002). Regulation of p53 activity by its interaction with homeodomain-interacting protein kinase-2. *Nat Cell Biol* 4:1–10.

Hofseth, L.J., Saito, S., Hussain, S.P., Espey, M.G., Miranda, K.M., Araki, Y., Jhappan, C., Higashimoto, Y., He, P., Linke, S.P., Quezado, M.M., Zurer, I., Rotter, V., Wink, D.A., Appella, E., and Harris, C.C. (2003). Nitric oxide-induced cellular stress and p53 activation in chronic inflammation. *Proc Natl Acad Sci USA* 100:143–148.

Hupp, T. R., and Lane, D. P. (1994). Regulation of the cryptic sequence-specific DNA-binding function of p53 by protein kinases. *Cold Spring Harb Symp Quant Biol* 59:195–206.

Hupp, T. R., Meek, D. W., Midgley, C. A., and Lane, D. P. (1992). Regulation of the specific DNA binding function of p53. *Cell* 71:875–886.

Itahana, K., Dimri, G., and Campisi, J. (2001). Regulation of cellular senescence by p53. *Eur J Biochem* 268:2784–2791.

Jallepalli, P. V., Leaguer, C., Vogelstein, B., and Benz, F. (2003). The Chk2 tumor suppressor is not required for p53 responses in human cancer cells. *J Biol Chem* 278:20475–20479.

Jeffrey, P. D., Gorina, S., and Pavletich, N. P. (1995). Crystal structure of the tetramerization domain of the p53 tumor suppressor at 1.7 angstroms. *Science* 267:1498–1502.

Jimenez, G. F., Bryntesson, F., Torres-Arzayus, M. I., Priestley, A., Beeche, M., Saito, S., Sakaguchi, K., Appella, E., Jeggo, P. A., Taccioli,, G. E., Wahl, G. M., and Hubank, M. (1999). DNA-dependent protein kinase is not required for the p53-dependent response to DNA damage. *Nature* 400:81–83.

Kaeser, M.D., and Iggo, R.D. (2002). Chromatin immunoprecipitation analysis fails to support the latency model for regulation of p53 DNA binding activity *in vivo*. *Proc Natl Acad Sci USA* 99:95–100.

Kaku, S., Iwahashi, Y., Kuraishi, A., Albor, A., Yamagishi, T., Nakaike, S., and Kulesz-Martin, M. (2001). Binding to the naturally occurring double p53 binding site of the Mdm2 promoter alleviates the requirement for p53 C-terminal activation. *Nucleic Acids Res* 29:1989–1993.

Kapoor, M., and Lozano, G. (1998). Functional activation of p53 via phosphorylation following DNA damage by UV but not γ radiation. *Proc Natl Acad Sci USA* 95:2834–2837.

Kim, E., and Deppert, W. (2003). The complex interactions of p53 with target DNA: we learn as we go. *Biochem Cell Biol* 81:141–150.

Kishi, H., Nakagawa, K., Matsumoto, M., Suga, M., Ando, M., Taya, Y., and Yamaizumi, M. (2001). Osmotic shock induces G_1 arrest through p53 phosphorylation at Ser^{33} by activated $p38^{MAPK}$ without phosphorylation at Ser^{15} and Ser^{20}. *J Biol Chem* 276:39115–39122.

Klein, C., Planker, E., Diercks, T., Kessler, H., Künkele, K.-P., Lang, K., Hansen, S., and Schwaiger, M. (2001). NMR spectroscopy reveals the solution dimerization interface of p53 core domains bound to their consensus DNA. *J Biol Chem* 276:49020–49027.

Ko, L. J., and Prives, C. (1996). p53: puzzle and paradigm. *Genes Dev* 10:1054–1072.

Kozlov, S., Gueven, N., Keating, K., Ramsay, J., and Lavin, M. F. (2003). ATP activates ataxia-telangiectasia mutated (ATM) *in vitro*. Importance of autophosphorylation. *J Biol Chem* 278:9309–9317.

Kussie, P. H., Gorina, S., Marechal, V., Elenbaas, B., Moreau, J., Levine, A. J., and Pavletich, N. P. (1996). Structure of the MDM2 oncoprotein bound to the p53 tumor suppressor transactivation domain. *Science* 274:948–953.

Lambert, P. F., Kashanchi, F., Radonovich, M. F., Shiekhattar, R., and Brady, J. N. (1998). Phosphorylation of p53 serine 15 increases interaction with CBP. *J Biol Chem* 273:33048–33053.

Lees-Miller, S. P., Sakaguchi, K., Ullrich, S. J., Appella, E., and Anderson, C.W. (1992). Human DNA-activated protein kinase phosphorylates serines 15 and 37 in the amino-terminal transactivation domain of human p53. *Mol Cell Biol* 12:5041–5049.

Li, J., Yang, Y., Peng, Y., Austin, R. J., Van Eyndhoven, W. G., Nguyen, K. C. Q., Gabriele, T., McCurrach, M. E., Marks, J. R., Hoey, T., Lowe, S. W., and Powers, S. (2002). Oncogenic properties of *PPM1D* located within a breast cancer amplification epicenter at 17q23. *Nat Genet* 31:133–134.

Lu, H., Taya, Y., Ikeda, M., and Levine, A. J. (1998). Ultraviolet radiation, but not radiation or etoposide-induced DNA damage, results in the phosphorylation of the murine p53 protein at serine-389. *Proc Natl Acad Sci USA* 95:6399–6402.

Mazur, S. J., Sakaguchi, K., Appella, E., Wang, X. W., Harris, C. C., and Bohr, V. A. (1999). Preferential binding of tumor suppressor p53 to positively or negatively supercoiled DNA involves the C-terminal domain. *J Mol Biol* 292:241–249.

Melchionna, R., Chen, X.-B., Blasina, A., and McGowan, C. H. (2000). Threonine 68 is required for radiation-induced phosphorylation and activation of Cds1. *Nat Cell Biol* 2:762–765.

Melchior, F., and Hengst, L. (2002). SUMO-1 and p53. *Cell Cycle* 1:245–249.

Mihara, M., Erster, S., Zaika, A., Petrenko, O., Chittenden, T., Pancoska, P., and Moll, U. M. (2003). p53 has a direct apoptogenic role at the mitochondria. *Mol Cell* 11:577–590.

Nakamura, S., Roth, J. A., and Mukhopadhyay, T. (2002). Multiple lysine mutations in the C-terminus of p53 make it resistant to degradation mediated by MDM2 but not by human papillomavirus E6 and induce growth inhibition in MDM2-overexpressing cells. *Oncogene* 21:2605–2610.

Nielsen, O. (2003). COP9 signalosome: a provider of DNA building blocks. *Curr Biol* 13:R565–R567.

Oda, K., Arakawa, H., Tanaka, T., Matsuda, K., Tanikawa, C., Mori, T., Nishimori, H., Tamai, K., Tokino, T., Nakamura, Y., and Taya, Y. (2000). *p53AIP1*, a potential mediator of p53-dependent apoptosis, and its regulation by Ser-46-phosphorylated p53. *Cell* 102:849–862.

Oren, M. (2003). Decision making by p53: life, death and cancer. *Cell Death Differ* 10:431–442.

Palecek, E., Vlk, D., Stanková, V., Brázda, V., Vojtesek, B., Hupp, T. R., Schaper, A., and Jovin, T. M. (1997). Tumor suppressor protein p53 binds preferentially to supercoiled DNA. *Oncogene* 15:2201–2209.

Paull, T. T., Rogakou, E. P., Yamazaki, V., Kirchgessner, C. U., Gellert, M., and Bonner, W. M. (2000). A critical role for histone H2AX in recruitment of repair factors to nuclear foci after DNA damage. *Curr Biol* 10:886–895.

Petrini, J. H. J., and Stracker, T. H. (2003). The cellular response to DNA double-strand breaks: defining the sensors and mediators. *Trends Cell Biol* 13:458–462.

Prives, C., and Manley, J. L. (2001). Why is p53 acetylated? *Cell* 107:815–818.

Redon, C., Pilch, D., Rogakou, E., Sedelnikova, O., Newrock, K., and Bonner, W. (2002). Histone H2A variants H2AX and H2AZ. *Curr Opin Genet Dev* 12:162–169.

Rippin, T. M., Freund, S. M. V., Veprintsev, D. B., and Fersht, A. R. (2002). Recognition of DNA by p53 core domain and location of intermolecular contacts of cooperative binding. *J Mol Biol* 319:351–358.

Saito, S., Goodarzi, A. A., Higashimoto, Y., Noda, Y., Lees-Miller, S. P., Appella, E., and Anderson, C. W. (2002). ATM mediates phosphorylation at multiple p53 sites, including Ser[46], in response to ionizing radiation. *J Biol Chem* 277:12491–12494.

Saito, S., Yamaguchi, H., Higashimoto, Y., Chao, C., Xu, Y., Fornace, Jr., A. J., Appella, E., and Anderson, C. W. (2003). Phosphorylation site interdependence of human p53 post-translational modifications in response to stress. *J Biol Chem* 278:37536–37544.

Sakaguchi, K., Herrera, J. E., Saito, S., Miki, T., Bustin, M., Vassilev, A., Anderson, C. W., and Appella, E. (1998). DNA damage activates p53 through a phosphorylation-acetylation cascade. *Genes Dev* 12:2831–2841.

Sakaguchi, K., Sakamoto, H., Lewis, M. S., Anderson, C. W., Erickson, J. W., Appella, E., and Xie, D. (1997). Phosphorylation of serine 392 stabilizes the tetramer formation of tumor suppressor protein p53. *Biochemistry* 36:10117–10124.

Sax, J. K., and el-Deiry, W. S. (2003). p53 downstream targets and chemosensitivity. *Cell Death Differ* 10:413–417.

Schon, O., Friedler, A., Bycroft, M., Freund, S. M. V., and Fersht, A. (2002). Molecular mechanism of the interaction between MDM2 and p53. *J Mol Biol* 323:491–501.

Schultz, L. B., Chehab, N. H., Malikzay, A., and Halazonetis, T. D. (2000). p53 binding protein 1 (53BP1) is an early participant in the cellular response to DNA double-strand breaks. *J Cell Biol* 151:1381–1390.

Shieh, S.-Y., Ahn, J., Tamai, K., Taya, Y., and Prives, C. (2000). The human homologs of checkpoint kinases Chk1 and Cds1 (Chk2) phosphorylate p53 at multiple DNA damage-inducible sites. *Genes Dev* 14:289–300.

Shieh, S.-Y., Ikeda, M., Taya, Y., and Prives, C. (1997). DNA damage-induced phosphorylation of p53 alleviates inhibition by MDM2. *Cell* 91:325–334.

Shiloh, Y. (2003). ATM and related protein kinases: safeguarding genome integrity. *Nat Rev Cancer* 3:155–168.

Siliciano, J. D., Canman, C. E., Taya, Y., Sakaguchi, K., Appella, E., and Kastan, M. B. (1997). DNA damage induces phosphorylation of the amino terminus of p53. *Genes Dev* 11:3471–3481.

Smith, G. C. M., Cary, R. B., Lakin, N. D., Hann, B. C., Teo, S.-H., Chen, D. J., and Jackson S. P. (1999). Purification and DNA binding properties of the ataxia-telangiectasia gene product ATM. *Proc Natl Acad Sci USA* 96:11134–11139.

Smith, M. L., and Seo, Y. R. (2002). p53 regulation of DNA excision repair pathways. *Mutagenesis* 17:149–156.

Sutherland, B. M., Bennett, P. V., Sidorkina, O., and Laval, J. (2000). Clustered DNA damages induced in isolated DNA and in human cells by low doses of ionizing radiation. *Proc Natl Acad Sci USA* 97:103–108.

Suzuki, K., Kodama, S., and Watanabe, M. (1999). Recruitment of ATM protein to double strand DNA irradiated with ionizing radiation. *J Biol Chem* 274:25571–25575.

Takai, H., Naka, K., Okada, Y., Watanabe, M., Harada, N., Saito, S., Anderson, C. W., Appella, E., Nakanishi, M., Suzuki, H., Nagashima, K., Sawa, H., Ikeda, K., and Motoyama, N. (2002). Chk2-deficient mice exhibit radioresistance and defective p53-mediated transcription. *EMBO J* 21:5195–5205.

Takekawa, M., Adachi, M., Nakahata, A., Nakayama, I., Itoh, F., Tsukuda, H., Taya, Y., and Imai, K. (2000). p53-inducible Wip1 phosphatase mediates a negative feedback regulation of p38 MAPK-p53 signaling in response to UV radiation. *EMBO J* 19:6517–6526.

Tibbetts, R. S., Brumbaugh, K. M., Williams, J. M., Sarkaria, J. N., Cliby, W. A., Shieh, S.-Y., Taya, Y., Prives, C., and Abraham, R. T. (1999). A role for ATR in the DNA damage-induced phosphorylation of p53. *Genes Dev* 13:152–157.

Tibbetts, R. S., Cortez, D., Brumbaugh, K. M., Scully, R., Livingston, D., Elledge, S. J., and Abraham, R. T. (2000). Functional interactions between BRCA1 and the checkpoint kinase ATR during genotoxic stress. *Genes Dev* 14:2989–3002.

Uziel, T., Lerenthal, Y., Moyal, L., Andegeko, Y., Mittelman, L., and Shiloh, Y. (2003). Requirement of the MRN complex for ATM activation by DNA damage. *EMBO J* 22:5612–5621.

Vousden, K. H., and Lu, X. (2002). Live or let die: the cell's response to p53. *Nat Rev Cancer* 2:594–604.

Wahl, G.M., and Carr, A.M. (2001). The evolution of diverse biological responses to DNA damage: insights from yeast and p53. *Nat Cell Biol* 3:E277–E286.

Wang, Y., and Prives, C. (1995). Increased and altered DNA binding of human p53 by S and G2/M but not G1 cyclin-dependent kinases. *Nature* 376:88–91.

Wang, Y.-H., Tsay, Y.-G., Tan, B.C.-M., Lo, W.-Y., and Lee, S.-C. (2003). Identification and characterization of a novel p300-mediated p53 acetylation site, lysine 305. *J Biol Chem* 278:25568–25576.

Waterman, M. J. F., Stavridi, E. S., Waterman, J. L. F., and Halazonetis, T. D. (1998). ATM-dependent activation of p53 involves dephosphorylation and association with 14-3-3 proteins. *Nat Genet* 19:175–178.

Wu, Z., Earle, J., Saito, S., Anderson, C. W., Appella, E., and Xu, Y. (2002). Mutation of mouse p53 Ser23 and the response to DNA damage. *Mol Cell Biol* 22:2441–2449.

Yamashita, A., Ohnishi, T., Kashima, I., Taya, Y., and Ohno, S. (2001). Human SMG-1, a novel phosphatidylinositol 3-kinase-related protein kinase, associates with components of the mRNA surveillance complex and is involved in the regulation of nonsense-mediated mRNA decay. *Genes Dev* 15:2215–2228.

Yarosh, D. B., Cruz, Jr., P. D., Dougherty, I., Bizios, N., Kibitel, J., Goodtzova, K., Both, D., Goldfarb, S., Green, B., and Brown, D. (2000). FRAP DNA-dependent protein kinase mediates a late signal transduced from ultraviolet-induced DNA damage. *J Invest Dermatol* 114:1005–1010.

Zhang, Y., and Xiong, Y. (2001). A p53 amino-terminal nuclear export signal inhibited by DNA damage-induced phosphorylation. *Science* 292:1910–1915.

Zhao, H., and Piwnica-Worms, H. (2001). ATR-mediated checkpoint pathways regulate phosphorylation and activation of human Chk1. *Mol Cell Biol* 21:4129–4139.

Zilfou, J. T., Hoffman, W. H., Sank, M., George, D. L., and Murphy, M. (2001). The corepressor mSin3a interacts with the proline-rich domain of p53 and protects p53 from proteasome-mediated degradation. *Mol Cell Biol* 21:3974–3985.

Zou, L., Cortez, D., and Elledge, S. J. (2002). Regulation of ATR substrate selection by Rad17-dependent loading of Rad9 complexes onto chromatin. *Genes Dev* 16:198–208.

Zou, L., and Elledge, S. J. (2003). Sensing DNA damage through ATRIP recognition of RPA-ssDNA complexes. *Science* 300:1542–1548.

6

p53 in Human Cancer – Somatic and Inherited Mutations and Mutation-independent Mechanisms

Ute M. Moll and Nicole Concin

SUMMARY

The p53 tumor suppressor protein plays a central role in maintaining genomic integrity. It does so by occupying a nodal point in the DNA damage control pathway. When cells are subjected to ionizing radiation or other mutagenic events, p53 mediates cell cycle arrest, senescence or programmed cell death (apoptosis). Furthermore, some evidence suggests that p53 plays a role in the recognition and repair of damaged DNA. p53 is a tetrameric transcription factor but also has transcription-independent proapoptotic functions.

Conversely, disruption of the p53 response pathway strongly correlates with tumorigenesis. p53 is functionally inactivated by structural mutations, neutralization by viral products, cytoplasmic sequestration, and alterations in upstream regulators or downstream effectors in the vast majority of human cancers. p53-deficient mice

U. M. MOLL • Department of Pathology, Health Sciences Center, State University of New York at Stony Brook, Stony Brook New York 11794-8691, USA   N. CONCIN • Department of Obstetrics & Gynecology, University of Innsbruck, Anichstrasse 35, A-6020 Innsbruck, Austria

The p53 Tumor Suppressor Pathway and Cancer, edited by Zambetti.
Springer Science+Business Media, New York, 2005.

have a highly penetrant tumor phenotype with over 90% tumor incidence within nine months. In some cancers direct physical evidence exists that identify the p53 gene as a target of known environmental carcinogens such as UV light and benzo[a]pyrene in cancers of the skin and lung. When p53 loss occurs, cells do not get repaired or eliminated but rather proceed to replicate damaged DNA, which results in more random mutations, gene amplifications, chromosomal rearrangements, and aneuploidy. In some experimental models, loss of p53 confers resistance to anticancer therapy due to loss of apoptotic competence. The translational potential of these discoveries are beginning to be tested in novel p53-based therapies.

6.1. INTRODUCTION

Perhaps the most intense research effort ever mounted in the field of cancer genetics centers around the p53 gene—for good reasons, as it turns out. The p53 tumor suppressor gene plays a preeminent role in protecting cells from malignant transformation. p53 protein is an astute watchdog over the physical integrity of the cellular genome. When DNA damage occurs, p53 acts as an emergency brake on the cell cycle, directing several powerful biological responses that yield an effective damage control. The inactivation of p53 function through mutational and nonmutational mechanisms eliminates a major roadblock in tumorigenesis. Indeed, disruption of p53 activity occurs with extraordinarily high frequency in diverse types of human cancers.

6.2. BIOLOGICAL ACTIVITIES OF p53

6.2.1. p53 Maintains Genomic Stability

Genes involved in maintaining genomic stability integrate the identification of a nonpermissible genome status with the execution of cellular responses that lead to repair or elimination of a damaged cell. p53 is critical in performing this integrative function. The loss of wild-type p53 alone without any other genetic abnormalities is sufficient to permit genomic instability to occur. In experimental systems, instability is measured as cells acquiring the potential for amplification of the CAD gene (trifunctional enzyme carbamoyl-P synthetase, aspartate transcarbamylase, dihydroorotase), selected for by resistance to the purine synthesis inhibitor PALA (N-phosphonacetyl-L-aspartate). Normal primary diploid fibroblasts with two or even just one wild-type alleles of p53 arrest their growth in PALA and have an undetectable (10^{-9}) frequency of CAD amplification. In contrast, cells that had lost the second allele do not exhibit growth arrest but amplify the CAD gene with high frequency (10^{-3}–10^{-5}) (Livingstone et al., 1992; Yin et al., 1992). The converse occurs after wild-type p53 is restored in cells that contain only mutant p53 alleles. When wild-type p53 cDNA is transfected into postcrisis Li-Fraumeni fibroblasts as well as a glioblastoma cell line (both homozygous for mutant p53), those expressing high amounts of wild-type p53 protein show growth arrest under PALA challenge and suppress gene amplification (Yin et al., 1992). This data fits beautifully with the in vivo observation that primary human tumors have such an extraordinary frequency of both p53 abnormalities and aneuploidy.

6.2.2. Tumor Suppression In Vitro and In Vivo

p53 completely suppresses cell transformation in primary embryo fibroblasts transformed by potent viral and cellular oncogenes such as activated *ras* plus *myc* or E1A. The few foci that do grow out have mutated p53 cDNA due to rearrangement during the process of genomic integration (Finlay et al., 1989; Eliyahu et al., 1989). p53 transgenic mice which express tumor-derived dominant negative p53 mutant proteins have a high incidence of spontaneous lung carcinomas, lymphomas, and sarcomas and show accelerated induction of leukemia by Friend retrovirus (Lavigueur et al., 1989). This spontaneous tumor susceptibility is even more dramatic in p53 knockout mice. Mice homozygous for the p53 null allele have a spontaneous tumor incidence of over 95% by the age of nine months. The tumor spectrum comprises mainly lymphomas and soft tissue sarcomas with very few carcinomas and the mean time to clinical appearance is four months (Donehower et al., 1992). Heterozygous p53+/− animals have a tumor incidence in between that of the wild-type and homozygously-deficient animals. Tumors that develop in heterozygous animals have typically lost their remaining wild-type allele. In addition, p53 deficiency also leads to increased susceptibility to chemical and physical carcinogens in these mice (Harvey et al., 1993; Kemp et al., 1994).

6.3. p53 STRUCTURE AND FUNCTION

p53 is a master regulator of growth arrest in response to a broad spectrum of cellular stress signals such as DNA damage, oncogene deregulation hypoxia, or nucleotide depletion. In response to such insults, p53 induces cell cycle arrest, senescence, or apoptosis. p53 initiates G1 and G2 cell cycle arrest and senescence principally by transactivating the cyclin-dependent kinase inhibitor p21Waf1 and the G2 checkpoint protein 14-3-3g (el-Deiry et al., 1993; Hermeking et al., 1997).

In contrast, the mechanism of p53-mediated apoptosis remains poorly understood. The elucidation of this pathway, however, is particularly important, as there is in vivo evidence that it is primarily the activation of apoptosis by p53 rather than its arrest/senescence function that is crucial in tumor suppression. The basis of its unique apoptotic potency lies in its pleiotropic actions that include transcription-dependent and -independent functions. p53 is a critical activator of the *mitochondrial* death pathway (Johnstone et al., 2002). Antiapoptotic Bcl2, BclXL completely block p53-dependent apoptosis (Schuler and Green, 2001). p53 can mediate apoptosis by transcriptionally activating proapoptotic genes including the BH3-only family members Noxa (Oda et al., 2000a,b) and Puma (Nakano and Vousden, 2001; Yu et al.,

2001), Bax (Miyashita and Reed, 1995), p53 AIP1(Oda et al., 2000b), Apaf-1 (Moroni et al., 2001), and the cytoplasmic membrane protein PERP (Attardi et al., 2000), and by transcriptionally repressing the antiapoptotic proteins Bcl2 (Wu et al., 2001) and IAPs (survivin) (Hoffman et al., 2002). For Noxa, Puma, and PIDD, downregulation of the endogenous genes by antisense methods decreases—but does not abolish—the extent of death after a stress stimulus. It is to be noted that the induction of these p53-induced gene products exhibits variable kinetics, with some being rather delayed in their response (24 hour or longer), e.g. BAX and p53AIP1 (Nakano and Vousden, 2001; Attardi et al., 2000). Analysis of p53-regulated global gene expression patterns demonstrate that the nature of the transcriptional p53 response and the target gene profile depends on p53 levels, stress type, and cell type (Zhao et al., 2000). This strongly suggests that only individual genes will be chosen from the complex spectrum of potentially inducible genes to mediate a specific p53 response in a given physiological situation. The pleiotropic p53 function is further stressed by the fact that in genetic deletion experiments, none of the p53-induced apoptotic response genes tested so far proved to be required to execute apoptosis. Moreover, most known p53 target genes are induced to *similar* levels during p53-mediated G1 arrest *and* apoptosis (Attardi et al., 2000). This strongly suggests that they function more *generally* in transducing p53 stress signals, but that they are not the decisive death determinant in the cell's decision fork whether to arrest or to undergo cell death. p53 also has transcription-independent apoptotic activities that may amplify the transcription-dependent functions (Caelles et al., 1994; Wagner et al., 1994; Haupt et al., 1995; Chen et al., 1996). We recently discovered that a fraction of induced p53 rapidly translocates to mitochondria where it mainly locates to the organellar surface in primary and cultured cells. This occurs only during p53-dependent death but not during p53-independent death or p53-dependent arrest. Moreover, bypassing the nucleus by deliberately targeting p53 to mitochondria is sufficient to launch apoptosis in tumor cells (Marchenko et al.; 2000, Sansome et al., 2001).

6.4. MECHANISMS OF p53 INACTIVATION IN HUMAN TUMORS

Disruption of the p53 response pathway strongly correlates with tumorigenesis. Indeed, the p53 stress response is lost one way or another in virtually all cancers. This frequency parallels the almost universal deregulation of the E2F family of transcription factors in over 90% of human cancers (Phillips and Vousden, 2001). Functional inactivation of the p53 gene is the single most common event in human malignancies and occurs in at least 50% of all cancers (Fig. 6.1) (see p53 databases online at http://p53.curie.frhttp://p53.curie.fr or http://www.iarc.fr)

Mutational inactivation of p53 is the most common mechanism and occurs in a large spectrum of sporadic and familial cancers of, e.g., the breast, gastrointestinal tract, lung, brain, and soft tissues. The largest databases on any cancer-causing gene exist for p53, e.g. the Curie database currently contains 13,789 human tumors from around the world, 201 germline mutations, 55 normal individuals without cancer,

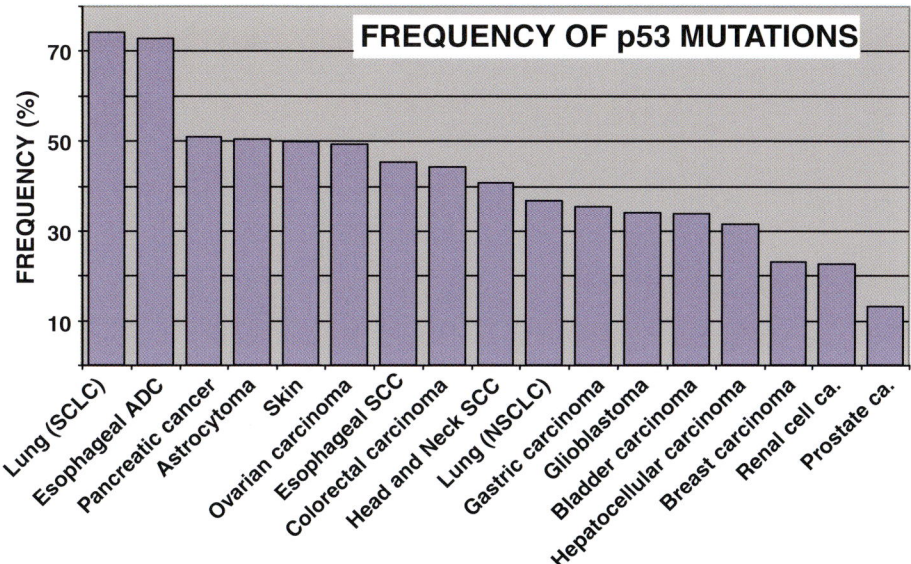

Figure 6.1. Frequency of p53 mutations in a spectrum of human cancers. Reproduced from p53 database, December 2001 issue at http://p53.curie.fr with permission from Th. Soussi.

and 890 cell lines. Deletion of one allele accompanied by a missense mutation in the central DNA-binding domain of the remaining allele is classical. Analysis of the mutational spectrum reveals that only 5% of the mutations occur in the N-terminal and C-terminal domains, while 95% fall into the specific DNA binding domain with hotspots localizing to four highly conserved domains. Most missense mutations (84%) target codons that are completely evolutionarily conserved. In addition, the nature of the mutations depends on the domains. While the rare mutations that occur in the N- and C-termini are nonsense and frameshift mutations (leading to truncated and scrambled proteins, respectively), the core domain shows predominantly missense (point) mutation (95% of the total), leading to amino acid exchanges. Crystallographic evidence of the three-dimensional structure of p53 bound to specific DNA confirmed that the hotspot missense mutations occur at those residues that either directly contact DNA or are critical for the stability of correct protein folding (contact and structural mutants) (Cho et al., 1994). Most mutations that occur in human tumors produce an abnormal protein that cannot bind to DNA, crippling its transactivation function. While most other tumor suppressor genes typically select for truncations, frameshifts, and deletions, p53 is unique among tumor suppressor genes in that mutant p53 protein is actively retained and in fact grossly overexpressed in 95% of cases due to missense mutations. Most point mutations produce a protein with a prolonged half-life that accumulates in the nucleus and becomes readily detectable by immunocytochemistry. Given the two hit requirement, a heterozygous p53 mutation becomes oncogenic via dominant negative action on the coexpressed wild-type protein. Mixed p53 tetramers have altered activity that varies for different mutants (Milner and Medcalf, 1991). In addition, some but not all p53 mutants appear to be gain-of function mutants by acquiring novel transforming activities in the absence of wild-type p53 (Blandino et al., 1999; Cadwell and Zambetti, 2001). This could be due to the fact that in cultured cells mutant p53 can act as dominant negative inhibitor of the family members p63 and p73 by forming mixed complexes (Di Como et al., 1999). Whether other mechanisms beyond family members exist is unclear.

6.4.1. p53 Mutational Profiles

As already stated, only 5% of p53 mutations occur in the N-terminal and C-terminal domains, while 95% fall into the specific DNA-binding domain, with "hotspots" localizing to codons 175, 248, and 273 in all cancers (Fig. 6.2). A few specific tumors exhibit additional hotspots, e.g. lung cancer (codon 157) and hepatocellular carcinoma (codon 249) (see below). Among the 393 codons of the human p53 gene, 222 codons are the target of 698 different events (excluding nonsense or frameshift mutations) (http://p53.curie.frhttp://p53.curie.fr). The biological activity of the hotspot mutants 175, 248, 273, and 282 has been extensively tested (Table 6.1), while most of the other mutants have not been rigorously analyzed, leaving open the possibility that some mutants retain partial activity. Analysis of the exact type of base changes reveals differences among tumor sites. The p53 mutational profile supports theoretical models of carcinogenesis and represents an excellent example of molecular epidemiology.

p53 mutation patterns can be of two kinds: endogenous, i.e. spontaneously arising base changes during replication (misincorporation of nucleotides on the

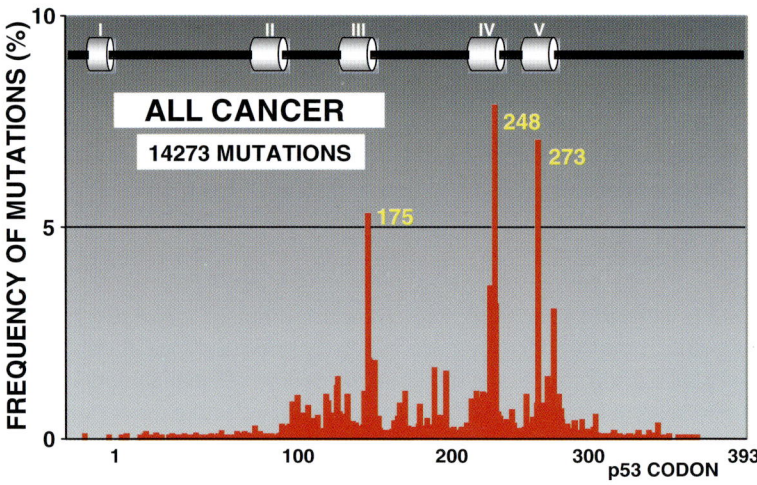

Figure 6.2. Distribution of mutations within the p53 gene in 14, 273 human cancers representing all tissue types. Reproduced from p53 database, December 2001 issue at http://p53.curie.fr with permission from Th. Soussi.

Table 6.1. The most frequent p53 mutations and their effects on protein structure. The following mutations represent about 30% of all mutations.

Codon	Residue	Mutant	Effects on protein structure
175	Arg	His	Breaks crucial H-bond bridging loops in L2 and L3
248	Arg	Gln	Breaks main contact with DNA in minor groove
273	Arg	His	Breaks main contact with DNA in major groove
248	Arg	Trp	Breaks main contact with DNA in minor groove
273	Arg	Cys	Breaks main contact with DNA in major groove
282	Arg	Trp	Destabilizes H2 helix and DNA binding in the major groove, and breaks contacts on the β-hairpin

Source: Adapted from Olivier and Hainaut, IARC p53 database (www.iarc.fr).

complementary DNA strand due to DNA polymerase errors), or exogenous, i.e. in-duced base miscoding due to carcinogen attack on the DNA. Exogenous carcinogens are often related to special classes and locations of p53 mutations, leaving a "mu-tational signature." Analyzing the mutational profile therefore provides clues to the etiology and molecular pathogenesis of tumors.

The GC→AT transitions are the most common base substitution in the p53 gene (Fig. 6.3). (Transversion is the change of a pyrimidine to purine or vice versa, and transition is the change of a pyrimidine to another pyridmidine or a purine to another purine). A well-studied *endogenous* mechanism of DNA damage is the phenomenon of deamination of cytosine and 5-methylcytosine. Cytosine and 5-methylcytosine can spontaneously deaminate to uracil and thymine, respectively, which, if not repaired will result in GC→AT transitions. These mutations occur most frequently at CpG dinucleotides (a cytosine followed by a guanine), which are frequently methylated (Ehrlich, 1990). In internal cancers, the most common single class of mutations are C→T changes at 5-methyl-CpG dinucleotides, a hallmark of spontaneous sequence drift in mammalian evolution (Cooper and Krawczak, 1990). They are generated by spontaneous deamination at 5-methylcytosine sites. In colorectal carcinoma, 50% of all tumors have p53 mutations of this type. In contrast, in patients with head and neck cancers, endogenous mutations at these CpG sites are rare.

6.4.1.1. Skin Cancer

Sunlight-induced skin cancer, i.e. basal cell and squamous cell carcinoma, is a good example of a mutation signature of UV irradiation. UV irradiation-induced mutations are mainly located at dipyrimidine sites (TT, CC, CT, or TC) and correspond to a C→T transition (Rady et al., 1992). A particular characteristic of the action of

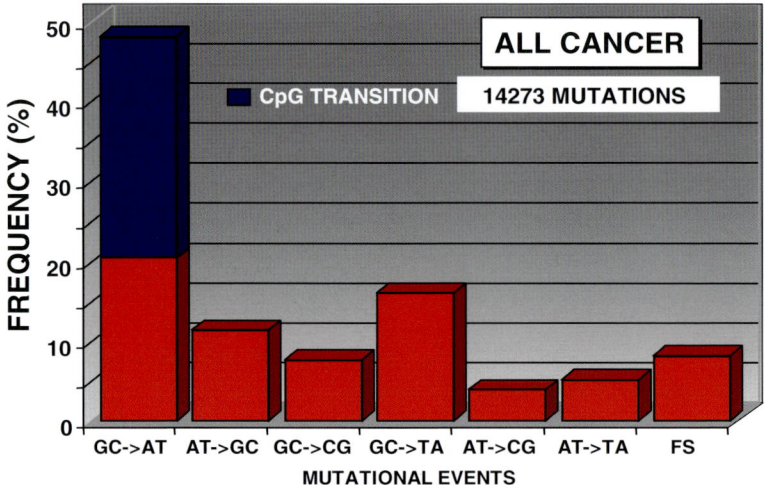

Figure 6.3. Frequency of transitions and transversions. The GC→AT transitions is the most common base substitution in the p53 gene. These mutations occur typically at CpG dinucleotides, which are frequently methylated. Reproduced from p53 database. December 2001 issue at http://p53.curie.fr with permission from Th. Soussi.

UV radiation is tandem mutations involving two adjacent dipyrimidines, CC→TT (Ziegler et al., 1993). In skin cancer, a series of mutations have been detected at pyrimidine dimers in over 90% of cases (Dumaz et al., 1994) and a 20% prevalence of p53 tandem mutations has been described (Figs. 4A and B) (Brash et al., 1991). In patients with genetic DNA repair deficiencies, such as xeroderma pigmentosum (XP), this phenotype is even more marked. All mutations found in skin cancers of these patients are located on the pyrimidine dimers and almost 60% are tandem mutations CC→TT (Dumaz et al., 1994; Sato et al., 1993). In contrast, less than one per thousand internal cancers harbor this type of p53 mutation (Hollstein et al., 1994; Dumaz et al., 1994). While in skin tumors and in internal cancer no special trends regarding the location of these UV irradiation-induced mutations are observed, in XP patients more than 95% of the mutations are located on the noncoding strand of the p53 gene, suggesting a preferential repair of the coding strand (Dumaz et al., 1994). This is consistent with a study of Tornaletti and Pfeifer (1994), showing that the repair rate of pyrimidine dimers vary highly within the p53 gene with an especially low rate in the codons that are often subjected to mutations in skin cancer.

6.4.1.2. Lung Cancer

Likewise, a direct etiological link between benzo[*a*]pyrene (in cigarette smoke) and lung cancer has been established. About 60% of human lung cancers contain p53 mutations and a majority of these are G→T transversions at hotspot codons 157, 248, and 273. A clear-cut dose–response relationship between the amount of tobacco smoked daily and the subsequent risk of lung cancer has been proven (Doll

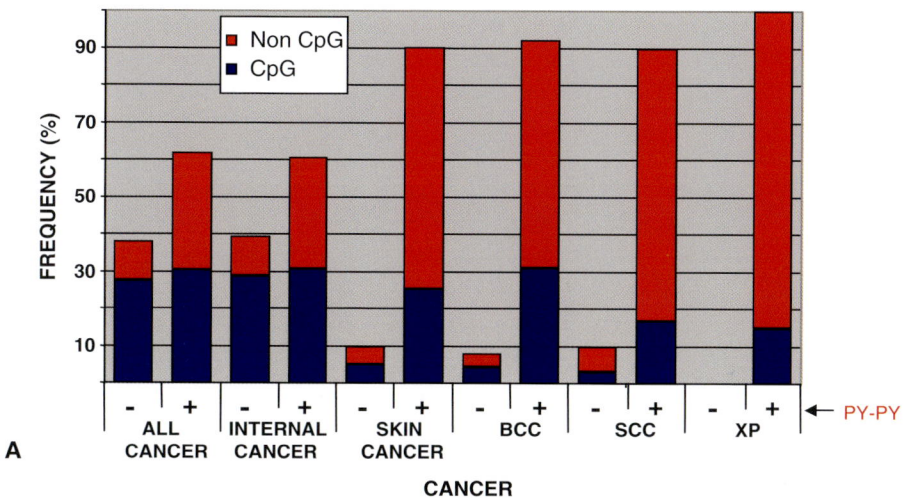

Figure 6.4. A: Mutation signature of UV irradiation in sunlight-induced cancers. In skin cancer series, mutations have been found to be located on pyrimidine dimers in over 90% of cases (reviewed by Dumaz et al., (1994)). BCC: Basal cell sarcoma; SCC: Squamous cell carcinoma; XP: Xeroderma Pigmentosum. Reproduced from p53 database/ December 2001 issue at http://p53.curie.fr with permission from Th. Soussi. **B:** Mutation signature of UV irradiation in sunlight-induced cancers. Non-XP skin cancers show 20% prevalence of tandem mutations and XP skin cancers have an even higher rate of almost 60% (reviewed by Dumaz et al., 1994). BCC: Basal cell sarcoma; SCC: Squamous cell carcinoma; XP: Xeroderma Pigmentosum. Reproduced from p53 database/ December 2001 issue at http://p53.curie.fr with permission from Th. Soussi.

and Hill, 1999). The strongest relationship to smoking is with squamous cell and small cell carcinoma. More than 98% of smokers who developed lung cancer have these histological subtypes (Saunders, 1999). p53 mutations range from 70% in small cell carcinomas to 33% in adenocarcinomas. In contrast to other cancers, the majority of these mutations are GC→TA transversions occurring at hotspot codons 157, 175, 248, and 273. Codon 157 is lung cancer specific (Figs. 5A and B) (Maher et al.,

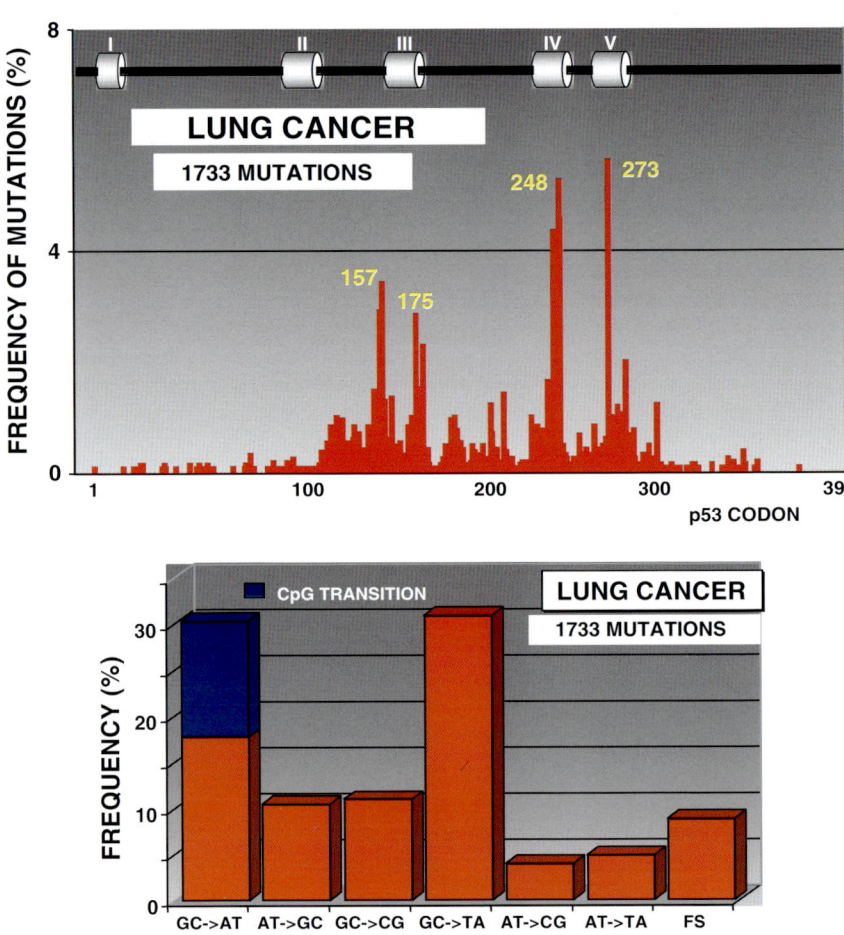

Figure 6.5. A: In lung cancer, the majority of p53 mutations are GC→TA transversions at hotspot codons 157, 175, 248, and 273. Codon 157 is lung cancer specific and the mutational signature of benzo[*a*]pyrene (Maher et al., 1990; Ruggeri et al., 1993; Denissenko et al., 1996). Reproduced from p53 database/ December 2001 issue at http://p53.curie.fr with permission from Th. Soussi. **B:** In lung cancer, the majority of p53 mutations are GC → TA transversions at hotspot codons 175, 248, 273, and 157. A derivative product of benzo[*a*]pyrene has been shown to bind predominantly to guanine and gives rise to specific GC → TA transversions (Hussain et al., 2001). Reproduced from p53 database, December 2001 issue with permission from Th. Soussi at http://p53.curie.fr

1990; Ruggeri et al., 1993; Denissenko et al., 1996). One of the derivative products of benzo[*a*]pyrene, a highly carcinogenic compound of cigarette smoke, has been shown to bind predominantly to guanine and gives rise to specific GC→TA transversions (Friedberg et al., 1995). When normal bronchial epithelial cells were exposed to benzo[*a*]pyrene for only 30 minutes in culture, targeted adduct formation occurred at the hotspot codons of p53 observed in lung cancer (Denissenko et al., 1996). These studies clearly incriminate tobacco in lung cancer; thus p53 mutations carry the "fingerprints" of this carcinogen.

6.4.1.3. Hepatocellular Carcinoma

The p53 mutational profile in hepatocellular carcinoma (HCC) is another instructive example. A strong association between infection with hepatitis B virus and liver cancer is known. In addition, aflatoxin B1, a highly carcinogenic compound produced by the fungal strain *Aspergillus flavus*, is considered to be a significant etiological factor for this cancer in South Africa and Asia (Ozturk, 1991; Bressac et al., 1991; Shimizu et al., 1999). In Mozambique and China, where food is contaminated by aflatoxin B1, a predominance of GC→TA transversions at the third base of codon 249 (Arg to Ser) has been reported (Figs. 6A and B) (Hsu et al., 1991; Bressac et al., 1991; Aguilar et al., 1994). This high mutation rate at codon 249 was not found in Transkei, a country which borders on Mozambique and has a similar rate of chronic HBV infection, but less aflatoxin B1 contamination (Ozturk, 1991). A similar situation has been reported in various parts of China, in which the prevalence of these mutations vary according to the level of aflatoxin exposure (Montesano et al., 1997). In contrast, HCC from Western populations show far fewer and heterogeneous p53 mutations (Harris and Hollstein, 1993). In Europe and the USA, which do not consume aflatoxin contaminated food, a low rate of p53 mutations is seen in hepatocelluar carcinoma and the mutations are scattered along the central region of p53 (Ozturk, 1991; Aguilar et al., 1994, Montesano et al., 1997). In vitro and in vivo studies have shown a very high sensitivity of the p53 codon 249 to the action of aflatoxin B1 (Aguilar et al., 1993).

6.4.1.4. Other Cancer Types

More than 90% of cervical carcinomas contain DNA from high-risk human papillomaviruses (HPV), mostly HPV serotypes 16 and 18 (Howley, 1991). E6 protein from high-risk HPV can induce p53 hyperdegradation and therefore functionally inactivate p53 (Scheffner et al., 1990; Howley, 1991). This pathway is suggested to be involved in tumorigenesis of cervical cancer. Indeed, several studies found HPV-positive cervical cell lines and primary carcinomas of the cervix to express wild-type p53, whereas in HPV-negative cell lines and carcinomas p53 mutations have been found (Crook et al., 1992; Scheffner et al., 1991). This model is in accordance with the fact that HPV-negative carcinomas have a worse prognosis than HPV-positive ones, as somatic mutations result in the expression of an altered p53 protein, which

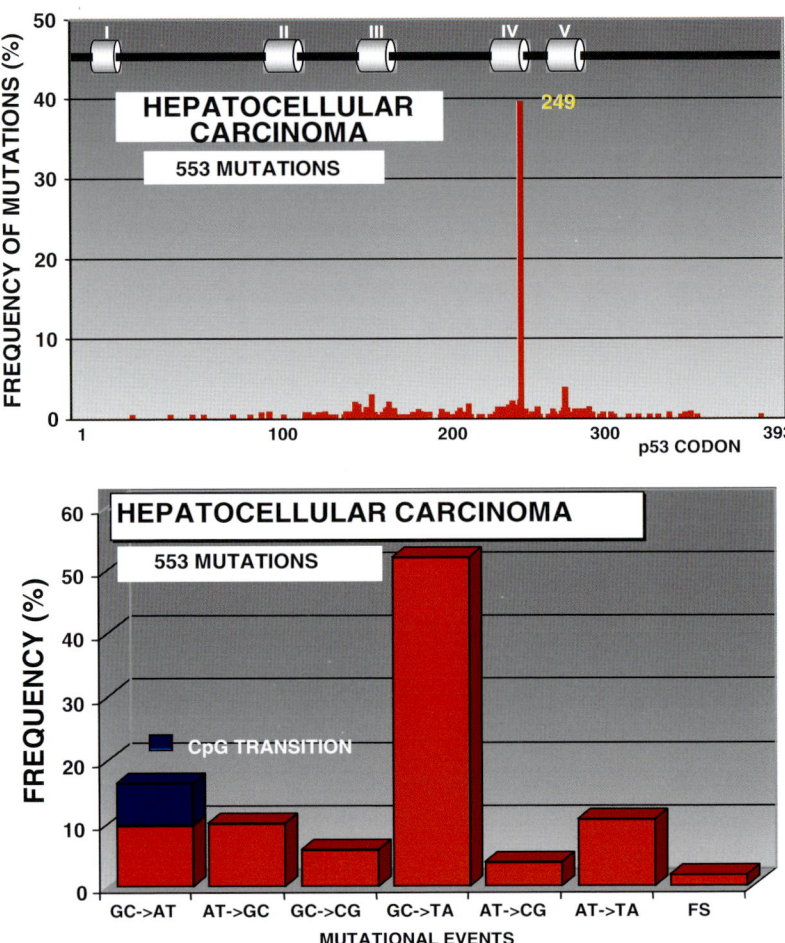

Figure 6.6. The p53 mutational profile in hepatocellular carcinoma in third world countries where food is contaminated by Aflatoxin B1 and hepatitis B virus infection is endemic. A predominance of GC→TA transversions at the third base of codon 249 (Arg to Ser) is striking. (Hsu et al., 1991; Bressac et al., 1991; Aguilar et al., 1994). Reproduced from p53 database, December 2001 issue at http://p53.curie.fr with permission from Th. Soussi.

can elicit positive transforming activity. However, doubt has risen upon this attractive model of un-inverse correlation between HPV positivity and p53 mutation in cervical cancer. Several independent studies showed that in the small percentage of cervical cancers with mutated p53, no correlation to the presence or absence of HPV infection exists (Tommasino et al., 2003).

In colon cancer p53 mutations are known to be a late event in carcinogenesis and an association with the conversion from colorectal adenoma to early carcinoma is suggested. About 40% to 50% of colorectal carcinomas harbor p53 mutations. 80%

of these mutations are GC→AT transitions, which are predominantly located at the CpG dinucletoides. In fact, the three hotspot codons in colon cancer, i.e. 175, 248, and 273, contain such dinucleotides. Additionally, various studies found codons 248 and 273 to be methylated in vivo. Taken together, these observations suggest that most of the p53 mutations found in this cancer are due to endogenous processes related to the deamination of 5-methycytosine. Based on the high frequency of mutations at CpG dinucleotides the p53 mutational profile in this cancer is characterized by a strong concentration of mutations in just a few codons.

Breast cancer is the third most common malignancy in the world, although the prevalence of breast cancer varies considerably in different geographical regions, with high-risk areas in North America and western Europe and low-risk areas in China and Japan. This suggests the importance of environmental factors in the etiology of breast cancer, emphasized through the fact of changed risk in migrant populations. Substantial diversity of p53 mutational pattern is seen in this cancer in terms of frequency and type of mutations (Hartmann et al., 1997; Blaszyk et al., 1996). In rural white US midwest women, e.g., an unusually high frequency of deletions and insertions has been described (Sommer et al., 1992; Saitoh et al., 1994; Kovach et al., 1996; Blaszyk et al., 2000). A high frequency of GC→CG and GC→TA transversions has been found in white women from Tennessee in contrast to the dominance of transitions in breast cancer in general (Caleffi et al., 1994). Among Austrian women, the pattern is characterized by a high frequency of AT→TA transversions (Hartmann et al., 1995). This difference in p53 mutational patterns in geographically and racially diverse populations may reflect the exposure to particular environmental carcinogens as well as intrinsic, endogenous mechanisms that might be active in this cancer.

In summary, p53 mutational data suggest that environmental and endogenous mutagenesis is operational in human tumor formation, and the relative importance of each depends on cancer site, exposure, and the reparability of the host tissue.

6.5. INHERITED MUTATIONS OF p53

6.5.1. Germline p53 Mutations: Li-Fraumeni Syndrome

Consistent with the dictum of a tumor suppressor gene, p53 germline mutations occur in the Li-Fraumeni cancer syndrome (Malkin et al., 1992). This is a rare familial cancer syndrome characterized by the development of a first malignancy (breast, brain, sarcomas, adrenocortical carcinoma, acute lymphocytic leukemia, and others) before the age of 30 and, if survived, a second cancer later on. The clinical definition of LFS requires: (1) an individual with a sarcoma diagnosed before the age of 45; (2) a first-degree relative with cancer before age 45; and (3) another first-degree or second-degree relative with either a sarcoma diagnosed at any age or any cancer diagnosed before age 45 (Wolf and Rotter, 1985). Its clinical inheritance is autosomal dominant (similar to familial retinoblastoma susceptibility due to inactivating pRB

Table 6.2. Inherited germline mutation of the p53 gene.

p53 codon	Amino acid change	Observed in
133	MetαThr	LFS
151 152	Frame-shift	LFS
181	ArgαCys	BC
	His	BC
245	GlyαCys	LFS
	Asp	LFS
	Ser	OS
248	ArgαTrp	LFS
	Gln	LFS
258	GluαLys	LFS
282	ArgαTrp	SMN, OS, MOS
128	GlyαVal	SMN

Source: Modified from Frebourg T, Friend SH. Cancer risks from germline p53 mutations. J Clin
Invest 1992;90:1637–1641.
OS = Osteosarcoma; MPC = multiple primary cancers; BC = breast cancer; SMN = second malignant
neoplasm; MOS = multifocal osteogenic sarcoma; LFS = Li-Fraumeni syndrome.

mutations). The heterozygous mutant p53 allele is passed on from parent to child
or generated de novo in the germ cells of a patient. Affected individuals carrying
the mutant germline p53 allele have a 50% increase in cancer risk by age 30 and
a 90% increase in risk by age 70 (Malkin et al., 1990). Transgenic mice carrying
tumor-derived hotspot p53 mutations have a high incidence of malignancies, including bone and soft tissue sarcomas, reminiscent of phenotypes of the LFS affected
individuals (Oliner et al., 1992; Lavigueur et al., 1989). Mutations in the germline
are not restricted to "hotspots" of p53 (Malkin et al., 1990; Srivastava et al., 1990).
Expanded surveys of LFS families have clearly shown that germline p53 mutations
may span most of the coding sequences (Law et al., 1991; Metzger et al., 1991;
Malkin et al., 1992) (Table 6.2), and are also found in introns (Jolly et al., 1994). The
germline of LFS individuals has one mutated p53 allele, while the second allele is
wild type. The exact contribution of a heterozygous p53 mutation to LFS cancer predisposition is unclear. Lymphoblastoid cell lines, peripheral blood lymphocytes and
fibroblasts from LFS patients with various germline p53 mutations showed defects
in their response to DNA damage, apoptosis, and cell cycle arrest (Goi et al., 1997;
Camplejohn et al., 1995; Sproston et al., 1996). However, other studies found no
cell cycle arrest defect in such LFS cells (Williams et al., 1996, 1997; Bech-Hansen
et al., 1981; Parshad et al., 1993; Lalle et al., 1995). Nevertheless, whether or not
a dominant negative inactivation of the remaining wild-type allele by the germline
mutant plays a role during LFS tumorigenesis or not, the inactivation could not be
sufficient since selection pressure against the wild-type allele remains. Tumors from
LFS patients often lose the second, normal p53 allele, fulfilling the "Two-Hit" genetic
requirement.

In the general population of cancer patients the prevalence of p53 germline
mutations is low, and even among patients with a component of LFS, the frequency

of constitutional mutations of p53 is only about 1% (Toguchida et al., 1992; Prosser et al., 1992; Borresen et al., 1992). However, when criteria for the "classic LFS" are expanded, additional patients with "LFS-like" syndromes and diverse germline mutations of p53 are captured. These include:

1. Children and young adults with second malignancies but no family history of LFS (Malkin et al., 1992).
2. Patients with sarcomas whose background includes either multiple primary cancers *or* a family history of cancer (Toguchida et al., 1992).
3. Patients with familial breast cancer (Sidransky et al., 1992). Clearly, early detection of a germline p53 mutation may identify patients and family members at high risk of the development of tumors.

6.5.2. Germline CHK2 Mutations: Li-Fraumeni Syndrome

While 50–60% of LFS families carry a p53 germline mutation in the p53 gene, the remainder of the classic LFS families does not have p53 mutations, but likely carry constitutional mutations in positive upstream regulators of p53. This indicates that the molecular basis of the syndrome may vary from family to family. CHK2 (also called CHEK2) is a factor controlling a G2 cell cycle checkpoint that is activated by ATM in response to DNA damage. CHK2 encodes the human homolog of the yeast Cds1 and Rad53 G2 kinase. Its role as a mammalian G2 stress checkpoint depends on its ability to phosphorylate downstream effectors such as p53, cdc25, and BRCA1. CHK2 directly phosphorylates p53 on Serine 20, which interferes with Mdm2 binding. Thus, CHK2 activation in response to DNA damage leads to increased p53 stability by preventing its ubiquitination (Hirao et al., 2000). Recently, heterozygous germline mutations in hCHK2 were found in several families with wild-type p53 Li-Fraumeni syndromes harboring sarcoma, breast cancer, and brain tumors (Bell et al., 1999). In 44 Finnish families with LFS and LFS-like syndrome, one disease-causing inactivating CHK2 mutation was observed in two different families (2 of 44 families; 4.5%). However, the cancer phenotype in the CHK2 families was not characteristic of LFS, suggesting variable phenotypic expression in the rare families with CHK2 mutations (Vahteristo et al., 2001). A particular protein-truncating mutation of CHK2, 1100delC in exon 10, abolishes the kinase function of CHEK2, and has been found in LFS and LFS-like families including those with breast cancer. In an unselected population of 1,035 breast cancer patients, the frequency of 1100delC was 2.0%, which was not significantly higher than the 1.4% frequency found among 1,885 population control subjects. However, a significantly elevated frequency was found among those 358 patients with a positive family history of only two affected relatives (11/358 or 3.1%) and patients with *bilateral* breast cancer were sixfold more likely to be 1100delC carriers than were patients with unilateral cancer (Vahteristo et al., 2002). Moreover, in 507 patients with familial breast cancer with no BRCA1 and BRCA2 mutations, an increased frequency of 1100delC (28/507 or 5.5 %) were found. In the knockout mouse, CHK2−/− thymocytes are resistant to DNA damage-induced apoptosis and are defective for p53 stabilization and induction of p53-dependent transcripts. Reintroduction of the CHK2 gene restored p53 function (Hirao et al., 2000). Together, these data indicate that hCHK2 is a tumor suppressor gene that when mutated predisposes the carrier to cancer. In addition, somatic CHK2 mutations, although rare, have been found in breast, vulval, lung, ovarian carcinoma, lymphoma, and osteosarcoma (Ingvarsson et al., 2002; Reddy et al., 2002; Miller et al., 2002; Tavor et al., 2001).

Additional genetic heterogeneity within the LFS complex, beyond p53 and CHK2, are still likely to be discovered.

6.6. TARGETING p53 REGULATORS

6.6.1. HDM2 Overexpression

p53 levels in unstressed cells are very low due to constant protein degradation via the ubiquitin–proteasome pathway. The principal negative regulator of p53 stability and function is HDM2, an E3 ubiquitin ligase specific for p53, which mediates p53 (and its own) degradation via the ubiquitin–26S proteasome pathway (Haupt et al., 1997; Kubbutat et al., 1997; Honda et al., 1997). p53 degradation occurs both on cytoplasmic and nuclear proteasomes (Freedman and Levine, 1998; Shirangi et al., 2002). The mouse homolog is named Mdm2. The E3 ligase activity of HDM2/mdm2 maps to a cysteine- and histidine-rich RING finger motif at the C-terminus, which ubiquitinates p53 on multiple lysine residues throughout its C-terminus (Fang et al., 2000). A basic but insufficient requirement for destruction is a direct interaction between p53 and mdm2 through their N-termini. Mdm2-binding site mutants of human p53 (14/19 and 22/23 double-point mutants) are resistant to degradation by mdm2 (Haupt et al., 1997; Kubbutat et al, 1997). The crystallographic analysis of the interacting domains has shown a tight key-lock configuration of the p53–mdm2 interface (Kussie et al., 1996). The hydrophobic side of the amphipathic p53 alpha-helix aa19-26 fits deeply into the hydrophobic cleft of mdm2.

p53 and HDM2 are linked through an autoregulatory feedback loop, in which activated p53 stimulates HDM2 transcription, while induced HDM2 in turn inhibits p53 function (Wu et al., 1993). The significance of this autoregulatory feedback loop becomes dramatically apparent in MDM2-deficient mice which die early in embryogenesis due to unchecked p53 activity (Montes de Oca Luna et al., 1995; Jones et al., 1995). Simultaneous ablation of p53 in MDM2$-/-$p53$-/-$ double null mice, however, fully rescues embryonic lethality. Thus, HDM2 is responsible both for the low steady state levels of p53 in unstressed cells and for switching off a p53 stress response after cell damage is repaired.

Under stress conditions, p53 becomes rapidly stabilized by different modes of downregulating the mdm2 pathway, all of them resulting in interference with mdm2-dependent degradation. Interestingly, different activators target the mdm2 degradation pathway by completely different mechanisms. Ionizing radiation acts through a cascade of stress kinases involving the ATM kinase signaling to the hCHK1 and hCHK2 kinases to phosphorylate p53 (Siliciano et al., 1997; Shieh et al., 2000; Chehab et al., 2000) (and also mdm2 itself; Khosravi et al., 1999) at several N-terminal serine residues. One of these p53 residues, Ser20, lies directly within the mdm2-binding domain and its phosphorylation interferes with the proper fit into the mdm2-binding pocket. The overall effect of these modifications is a reduced affinity of the p53–mdm2 complex, with mdm2 being "bumped off", thereby stabilizing p53. In contrast, UV radiation and hypoxia reduce the levels of mdm2 transcripts (Wu and Levine, 1997; Alarcon et al., 1999), hence reducing degradation. Moreover, UV damage blocks ubiquitination and instead favors sumoylation of p53 on Lys 386 (SUMO is a ubiquitin-like protein, but involved in p53 activation) (Rodriguez et al., 1999).

Overexpression of the mdm2 gene in cultured cells increases their tumorigenic potential and overcomes the growth-suppressive activity of p53 (Finlay, 1993). Consistent with p53 inactivation by HDM2 (the human mdm2), the HDM2 gene is frequently amplified or overexpressed in the absence of p53 mutations in most types of sarcomas, but also in leukemias/lymphomas, melanoma, and glioblastoma. The earliest report linking HDM2 deregulation in human tumors to an escape from p53-regulated growth control found HDM2 gene amplification in over a third of 71 sarcomas, including bone and soft tissue sarcomas (Oliner et al., 1992; Leach et al, 1993). Further evidence was that chromosome 12q13-14, the chromosomal position amplified in many sarcomas, contained the HDM2 locus. (It is to be noted that this chromosome locus also harbors CDK4 and GLI, both of which can be coamplified in human sarcomas) (Khatib et al., 1993). In 83 cases of osteosarcomas, p53 mutations were found in 26.5% and HDM2 amplification in 6.6% of tumors. The overall frequency of mutually exclusive alterations of the p53–mdm2 pathway was in 34% of cases (Lonardo et al., 1997). HDM2 gene amplification might be associated with tumor progression and metastasis, since it was detected only in metastatic or locally recurrent osteosarcomas but not in primary osteosarcomas (Ladanyi et al., 1993). In a Swedish series of 94 mesenchymal tumors, 20 tumors showed HDM2 amplification between 3–20-fold (malignant fibrous histiocytomas, pleomorphic liposarcomas, atypical lipomas, and typical lipomas). Interestingly, HDM2 amplification correlated with the presence of marker ring chromosomes, which, as shown by this study, often harbor the chromosome 12-derived HDM2 locus (Nilbert et al., 1994). Here, the human gene mimics the mouse mdm2 gene, originally found in ring chromosomes of transformed fibroblasts (*m*urine *d*ouble *m*inute-2 gene).

The notion that complementary mechanisms are used for inactivating the same growth-suppressing pathway is also played out in glioblastoma multiforme, the most malignant brain tumor. p53 mutations occur in over two-thirds of secondary glioblastomas (those that progress from low-grade or anaplastic astrocytomas), but rarely in de novo (primary) glioblastomas. Conversely, HDM2 overexpression is frequent in de novo glioblastoma (52% in a series of 29 tumors), but rare in secondary glioblastoma (11% in a series of 27 tumors). In this tumor, HDM2 overexpression occurs on a transcriptional level, without gene amplification (Biernat et al., 1997).

Leukemias are another group where transcriptional deregulation rather than gene amplification plays a significant role in HDM2 overexpression. HDM2 mRNAs, frequently as various isoforms, were overexpressed in 53% of 64 leukemias; and in some cases levels were comparable to sarcomas harboring 50-fold HDM2 gene amplification. In contrast, no gene amplification was found in 48 cases of leukemia (Bueso-Ramos et al., 1993).

HDM2 protein overexpression, but not gene amplification, is also associated with tumorigenesis of cutaneous melanoma, one of the few tumor types lacking p53 mutations. Among 172 cases representing different degrees of melanocyte transformation (dysplastic nevi, melanomas in situ, invasive primaries, and metastatic lesions), HDM2 overexpression correlated with malignant progression (6% of 16 dysplastic nevi, 27% of 11 melanomas in situ, and 56% of 145 invasive primary and metastatic melanoma (Polsky et al., 2001). A final example of a nonmesenchymal tumor with HDM2 overexpression is lung tumors. Mdm2 was overexpressed in 31% of 192 lung tumors of all histologic types by immunostaining. By western analysis, overexpression of at least one isoform was seen in 50% of 28 of these lung tumors. Interestingly, an inverse relationship between p14(ARF) loss and HDM2 overexpression was present, supporting the notion that p14ARF and HDM2 act in a common pathway to regulate p53 (Eymin et al., 2002).

6.6.2. Overexpression of Dominant Negative p73

Full-length p73 (TAp73) has significant structural and functional homology to p53. However, tumor-associated upregulation of full-length wild-type p73 in multiple primary tumor types (Moll et al., 2001) and genetic data from both human tumors (lack of TP73 mutations, deletions, or silencing) and p73-deficient mice exclude a classical tumor suppressor role. However, data is emerging that points to a striking similarity in the relationships between HDM2 and p53 on the one hand, and TP73 and p53 on the other with respect to their role in human tumorigenesis by counteracting the negative growth control of p53.

First, TP73 can produce three different N-terminally truncated isoforms, collectively called Delta TAp73. This occurs either by aberrant splicing of the first one or two translated exons of full-length transcripts (called Ex2Del or Ex2/3Del p73), or by using an alternative internal P2 promoter in intron 3. The latter transcript, called DeltaNp73, has been shown to function as a potent dominant negative inhibitor of wtp53 and full-length TAp73 in tissue culture (Zaika et al., 2002; Grob et al., 2001). In the developing mouse brain, DeltaNp73 is the predominant form associated with in vivo counteraction of p53-driven developmental apoptosis of excess neurons (Pozniak et al., 2000). Second, parallel to the HDM2–p53 relationship, DeltaNp73 is connected to TAp73 and p53 by an autoregulatory negative feedback loop (Grob et al., 2001; Nakagawa et al., 2002): DeltaNp73 is a target gene of TAp73 and wtp53 through a p53-responsive element located on the DeltaN promoter (Grob et al., 2001). In transfected cultured cells, expression of DeltaNp73 inhibits the function of p53 and TAp73, but also shuts off its own expression (Grob et al., 2001; Nakagawa et al., 2002). Third, studies show that some primary human tumors exhibit overexpression of dominant negative N-terminally deleted isoforms lacking the transactivation domain compared to normal tissue, often together with overexpression of TAp73. This is the case in 81% of 37 matched normal/tumor pairs of mainly gynecological cancers, which exhibited tumor-specific upregulation of DeltaNp73 and/or Ex2Del p73 (Zaika et al., 2002). Moreover, TAp73 was upregulated in 49% of 37 tumors. Among the 31 tumor pairs with upregulation of any one or all of these three p73 transcripts, 71% exhibited preferential upregulation of DeltaNp73 (19 tumors) or Ex2Del p73. More-over, 67% of these 22 tumors exhibited *exclusive* upregulation of DeltaNp73 and/or Ex2Del. Excellent correlation was seen between transcript and protein expression data p73 (Zaika et al., 2002). Similar tumor-specific upregulation of N-terminally truncated p73 isoforms was seen in six of six hepatocellular carcinomas (Stiewe et al., 2002). Moreover, DeltaNp73 focuses on inhibitory complex with wild-type p53 in cultured cells and primary tumors, indicating mixed complexes as one mechanism of interference without p53 function (Nakagawa et al., 2002; Zaika et al., 2002).

Moreover, when DeltaNp73 transcript levels were measured in 52 unmatched breast cancers and compared to 8 normal breast tissues, 31% of breast cancers overex-pressed DeltaNp73 levels between 6- and 44-fold and an additional 10 tumors showed DeltaNp73 upregulation between 2- and 6-fold. Among the 16 cancers with a 6- to 44-fold increase of DeltaNp73, 12 cancers again showed preferential upregulation of DeltaNp73 over TAp73 (Zaika et al., 2002). Significant expression of DeltaTAp73 isoforms is also present in human cancer cell lines from various tumor types (Fillippovich et al., 2001; Grob et al., 2001; Sayan et al., 2001; Stiewe et al., 2002).

Taken together, DeltaNp73 likely mediates a novel inactivation mechanism of p53 and TAp73 via a dominant-negative family network in vivo. Deregulated expres-sion of DeltaNp73 can bestow oncogenic activity upon the TP73 gene by functionally inactivating the suppressor action of p53 and TAp73. This trait may be selected in hu-man cancers. A thorough analysis of expression patterns of DeltaTAp73 isoforms in a large spectrum of primary tumors will need to be done to substantiate this principle of p53 inactivation.

6.6.3. Overexpression of Wip1/ PPM1D, a p53 Ser46 Phosphatase, in Breast Cancer

In response to ionizing irradiation, p53 is extensively phosphorylated at the N-terminus, which correlates with functional activation. In particular, the phospho-rylation of p53 on Ser 46 correlates with the ability of p53 to induce an apoptotic response via induction of the apoptotic target gene *p53AIP1*, but not the mediator of cell-cycle arrest *p21* (Oda et al., 2000b). Control of Ser46 phosphorylation might therefore be crucial in directing a p53 response toward apoptosis rather than arrest. Both the homeo-domain-interacting protein kinase 2 (HIPK2) and the p38 mitogen-activated protein kinase (MAPK) can mediate Ser46 phosphorylation in response to

UV irradiation (D'Orazi et al., 2002; Hofmann et al., 2002; Takekawa et al., 2000). In contrast, Ser46 phosphorylation in response to ionizing radiation requires both ATM (Saito et al., 2002) and the p53-inducible gene *p53DINP1* (Okamura et al., 2001). The activity of MAPK is regulated by the phosphatase Wip1, which dephosphorylates p53 on Ser46 and inhibits the apoptotic function of p53, for example after p38MAPK signaling. To guarantee the fine-tuning of the system, Wip1 itself is a p53 target gene, providing again a negative feed back loop (Takekawa et al., 2000).

Overexpression of Wip1/PPM1D confers oncogenic phenotypes on cells in culture such as attenuation of apoptosis induced by serum starvation and transformation of primary cells in cooperation with RAS (Bulavin et al., 2002; Li et al., 2002). Wip1/PPM1D is located in a hotspot of genomic amplification in breast cancer (at 17q23). Wip 1 is amplified in human breast cancer cell lines and in 11% of primary human breast cancers, most of which harbor wild-type p53 (Bulavin et al., 2002; Li et al., 2002). These findings suggest that inactivation of the p38 MAPK through PPM1D overexpression contributes to at least some human cancers by suppressing p53 activation.

6.6.4. Loss of an Upstream Activator: p16INK4a/p14ARF

Cooperation between RB and p53 in tumor suppression is strong and many tumor types exhibit mutations in both RB and p53. Also, *Rb+/− p53−/−* mice show a broader tumor spectrum at a younger age than mice that are either *Rb+/−* or *p53−/−* alone. Successfully transformed RB−/− cells often exhibit p53 inactivation (Symonds et al., 1994; Weinberg, 1995). The *INKa/ARF* locus occupies the regulatory nexus of the Rb and p53 growth control pathways, because it encodes two distinct tumor suppressor products: the p16*INK4a* cyclin-dependent kinase inhibitor that functions upstream of RB; and p19*ARF*, which sequesters MDM2, thus unleashing p53 activity. Functional inactivation of both pathways appears to be an almost ubiquitous requirement for all successful cancer cells because it robs the cell of two major defense mechanisms against transformation: cell cycle arrest and apoptosis.

The *INK4a/ARF* locus utilizes alternative first exons via two separately regulated promoters, but common downstream exons. Exon 1 alpha encodes part of p16*INK4a* and exon 1 beta reads an alternative reading frame, called p14*ARF*. Mice carrying a targeted deletion of the *INK4a/ARF* locus that eliminates both p16 and p19 (the mouse homolog of p14) develop spontaneous tumors at an early age and are highly sensitive to carcinogenic treatments (Serrano et al., 1996). On the other hand, mice lacking only p19ARF but expressing functional p16INK4a also develop tumors early in life (Kamijo et al., 1997). In fact, their cancer phenotype is strikingly similar to the phenotype generated by codeletion of both *Ink4a* gene products. Conversely, loss of p16INK4a with retention of p19ARF predisposes mice to spontaneous and carcinogen-induced tumorigenesis, albeit late in life. This "weaker" effect, at least in mice, is mirrored in the fact that p16-null MEFs exhibit a rate of immortalization below the ones observed for MEFs null for INK4a/ARF, pARF, or p53 (Sharpless et al., 2001).

p14ARF stabilizes p53 by counteracting the destabilizing effect of HDM2 both by binding and inhibiting the E3 ligase activity of HDM2 and by sequestering HDM2 into the nucleolus (Weber et al., 1999). In vitro and animal experiments gave strong evidence for a pathway which places p14ARF as an upstream positive regulator of p53: (a) p53 is required for p19ARF-induced G1 arrest (Kamijo et al., 1997), (b) the mutual exclusiveness of either p53 mutations or p19ARF loss in a myc-driven mouse lymphoma model (Schmitt et al., 1999), (c) in cell culture, p19*ARF* potently suppresses oncogenic transformation either by Myc and RAS, or E1a and RAS and this suppression requires intact p53 function (Pomerantz et al., 1998). However, the human tumor genetic/expression data does not support a clear-cut linear hierarchy between p14ARF and p53, at least not in the few studies that addressed this point.

Examples of p14*ARF*-specific deletions in human tumors, although rarer than codeletions or p16*INK4a* deletions, exist (see Table 6.3 with examples of big human tumor studies). These include an exon 1 beta specific insertion that affects only p19*ARF* expression, without impacting upon p16INK4a protein expression, in a patient with multiple primary melanomas (Dobrowolski et al., 2002). Also, the critical region deleted in glioblastomas maps to the region *between* the *INK4B* and *INK4A* exonic sequences, where exon 1 beta lies (Larsen, 1996). However, p14ARF-only deletions are not predominant in brain tumors, since in another study of 105 gliomas, deletion of p14ARF was always associated with codeletion of p16INK4A and increased in frequency upon progression from low to high grade gliomas (Labuhn et al., 2001). Most significantly, p19*ARF*, rather than p16*INK4a* or p15*INK4b*, is the crucial target in T-cell acute lymphocytic leukemia (ALL) that exhibit 9p21 lesions (Gardie et al., 1998).

Germline p16*INK4a*-specific mutations are associated with cancer susceptibility in familial melanoma kindreds. Somatic p16*INK4a*-specific mutations are frequently found in sporadic melanoma, pancreatic adenocarcinoma, lung and bladder carcinoma (Kamb, 1995). However, the majority of tumor-specific *INK4a/ARF* alterations are exon 2-region deletions in human (Kamb, 1995) and mouse tumors (Chin et al., 1997) and therefore affect *both* the p16*INK4a* and p19*ARF* ORFs (see Table 6.3). The frequency of INK4a/ARF inactivation, irrespective of patient age and tumor type comes close to that of p53 (Hainaut et al., 1997). This is a direct reflection of the paramount importance of the RB/p53 suppressor double nexus in human cancers. Parenthetically, relative to other cyclin dependent kinase inhibitor genes, the predominance of the INK4a/ARF locus in tumor suppression is remarkable.

6.6.5. p53 and PML

The PML gene, involved in the t(15;17) chromosomal translocation of acute promyelocytic leukemia (APL), encodes a protein which localizes to the PML-nuclear body (NB), a huge subnuclear macromolecular structure of >30 different proteins. PML controls apoptosis, cell proliferation, and senescence in part in a p53-dependent fashion. PML acts as a scaffolding molecule, recruiting both p53 and the

Table 6.3. Examples of human tumors with alterations in the INK4a/p16 ARF/OCW

Tumor	Cases	p14(ARF)	p16 (INK4a)	p53 status	Ref
Hepatocellular carcinoma	71	15% inactivation 7% homozygous deletion of entire INK4A-ARF locus 55% p16 methylation but unmethylated p14 promoter	66% loss of mRNA transcription 39 cases p16 methylation with an unmethylated p14(ARF) promotor	42% p53 mutations, all but 1 case with wtp14	Tannapfel et al., Oncogene 2001
Non-small cell lung cancer	38	14/22 tumors with p16 inactivation and ip14ARF inactivation: 2 mutations exon 2, 12 homozygous deletions	58% inactivated: 31% homzygous deletion, 21% p16 promoter hypermethylation, 5% point mutation in exon 2. loss of p16 protein in all cases	47% p53 mutations 9/18 harbored p14ARF inactivation, no correlation between p14ARF and p53 mutations	Sanchez-Cespedes et al., Oncogene 1999
Lung cancer	64	8% (4/52) of lung tumors showed p14 methylation Association with tobacco smoking: 42% (21/50) of NSCLC from ever smokers exhibited p16 methylation with a significant higher risk of p16 methylation in former smokers compared to current smokers (odds ratio 5.1, 95% confidence interval 1.3-2.2)	34% (22/64) of lung tumors showed p16 methylation. p16 methylation occurred only in non-small cell lung cancer (NSCLC) (41%, 22/54) with the highest frequency in large cell carcinomas (71%, 5/7)		Jarmalaite et al., Int J Cancer 2003

Cancer	n				Reference
Head and neck cancer (HNSCC)	160	p16 methylation *in parallel* with p53 mutation or p14 methylation occurred more frequently in former smokers than in current smokers (44% versus 14%, $P = 0.035$)			Gruttgen et al., J. Pathol. 2001
		15% loss of p14, 4% selective loss of p14 only; no homozygous deletion although 60% allelic imbalance at 9p21, p14 protein readily detetable, 11% of tumors lost both proteins	54% loss of p16 mRNA and protein expression 37% selective loss of p16 only		
	100	22% sequence alterations in exons 1 alpha, and 2 mainly microdeletions, insertions, and single nucleotide substitutions			Poi et al., Mol Carcinog. 2001
Neuroblastoma	40	No loss of p14 (ARF) expression	No alteration in the p16 gene, expression of p16 mRNA and protein even in the unfavorable stages	p53 mutations very infrequent	Omura-Minamisawa et al., Clin Cancer Res. 2001
Melanoma	?	In general: 20–40% of melanoma kindreds associated with germline mutations of INK4a/ARF, most commonly impaired p16, while mutations uniquely targeting p14 are rare, >40% of INK4a/ARF alterations in exon 2 affect both p16 and p14 This study: p14 functionally impaired in melanoma kindreds with INK4a/ARF mutations: 3/7 tested INK4a/ARF mutations diminished the ability of p14 to activate p53 pathway		p53 mutations very infrequent	Rizos et al., J. Biol Chem. 2001
Melanoma	1	Case report: 16 base pair/ exon 1 beta germline insertion, specifically altering p14ARF but not p16 in an individual with multiple primary melanomas (sporadic)		This mutation failed to stabilize p53 and to arrest growth of a p53 expressing melanoma cell line	Rizos et al., Oncogene 2001

(continued)

Table 6.3. (Continued)

Tumor	Cases	p14(ARF)	p16 (INK4a)	p53 status	Ref
Melanoma	32	Inverse correlation between progression of melanoma and lack of p14ARF protein expression by immunostaining: Positive p14 protein staining in 11/14 benign nevi, in 3/12 melanomas, in 0/6 melanoma metastases −> p14 inactivation important in the development of melanomas			Dobrowolski *et al.*, Arch Dermatol Res. 2002
Melanoma	45 sporadic	71% LOH at one or more of the 6 examined polymorphic microsatellite markers on locus 9p21, 9% homozygous deletion at the INK4a/ARF locus	Subset of melanomas with LOH at the INK4 locus have inactivating mutations of p16		Kumar *et al.*, Melanoma Res. 1994
Breast cancer	100	no mutation in exon 1beta, 24% p14 methylation, 17% overexpression, but no gene alterations in majority of overexpressing cases; 26% decreased expression, mainly due to promotor hypermethylation, LOH and homozygous deletion, LOH of INK4a/ARF locus in 21% informative tumors, 4% homozygous deletion of exon 2			Silva *et al.*, Oncogene 2001

Tumor type	N	Findings		Reference
Colon cancer	60	33% methylation in carcinomas, no methylation in adjacent normal mucosa, 1 mutation in exon 1beta, 19% deletion of p14/p16 allele, 6 of these were p16 methylated, none of them p14 methylated, 15% simultaneous p14 and p16 methylation- no statistical correlation between p14 and p16 methylation	32% methylation, no methylation in adjacent tumor free colon mucosa, no tumor specific mutations of p16 exons 1, 2 or 3	25 carcinomas p53 positive by staining p14 Methylation inversely correlated with p53 overexpression No correlation with p16 methylation status Burri et al., Lab Invest. 2001
Colon cancer	95	33% (31/95) showed aberrant p14ARF methylation 6% (6/95) showed concomitant hypermethylation of both genes	20% (19/95) showed p16INK4a hypermethylation	Dominguez et al., Mutation Research 2003
Glioma	105	41% homozygous and 7% hemizygous deletion of INK4a locus, p14/p16 expression always linked deletion of p14 was always associated with co-deletion of p16 increased frequency of 14/16 co-deletion during progression from low to high grade gliomas		Labuhn et al., Oncogene 2001
Intracranial germ cell tumors	21	71% genetic alterations in the INK4a/ARF genes: 14 homozygous deletions, 1 frameshift mutation		Iwato et al., Cancer Res. 2000

acetyltransferase CBP/p300 into NBs. Acetylation of p53 enhances its ability to bind DNA and activate transcription. PML directly interacts with the DNA binding domain of p53 and colocalizes with p53 in PML-NBs. Supporting this idea is the observation that HIPK2 (a Ser46 kinase) colocalizes with p53 in the PML-NBs (Hofmann et al., 2002; D'Orazi et al., 2002). While *p53−/−* thymocytes are completely resistant to radiation-induced apoptosis, *PML−/−* thymocytes are partially resistant. Moreover, in *PML−/−* thymocytes, resistance correlates with impaired induction of p53 target genes such as *bax* and *p21* (Guo et al., 2000). In *PML−/−* cells radiation-induced acetylation of p53 is substantially impaired, indicating that PML might regulate p53 transcriptional function by promoting its acetylation (Guo et al., 2000). However, only a specific isoform of a large spectrum of PML variants can modulate p53 transcription (Fogal et al., 2000). Nevertheless, at least in part through the PML-NB, PML modulates p53 function, in turn potentiating its tumor-suppressive activity. Thus, PML controls the targeting of p53 into the PML-NB, its acetylation and transcriptional activation. As a consequence, p53 target genes relevant for either apoptosis or senescence are transcribed (Salomoni and Pandolfi, 2002).

In APL blast cells, PML-RARα causes a mislocalization of PML into aberrant microspeckled nuclear structures and disruption of the bone fide PML-NBs. Currently, however, it is difficult to make a direct connection between the expression of chimeric PML oncoproteins in APL and the disruption of the p53 pathway. Leukemia in PML-RARα transgenic mice develops after a long latency of over one year and only in approximately 20% of the mice, which suggests that additional genetic events have to occur. This suggests a general role for PML in allowing genomic instability to accumulate (Salomoni and Pandolfi, 2002).

6.6.6. Inactivation of LKB1

Peutz–Jeghers syndrome (PJS) is a rare, hereditary intestinal polyposis syndrome associated with an increased risk of gastrointestinal and reproductive organ cancers. Most cases of PJS are associated with inactivating mutations in the tumor suppressor gene *LKB1*, which encodes a ubiquitously expressed serine/threonine kinase. Recent studies have begun to illustrate the molecular mechanisms by which LKB1 functions as an important new tumor suppressor. However, although evidence for a tumor-promoting role of LKB1 in PJS patients is solid, there is currently little evidence to support a role for LKB1 in sporadic cancers. *LKB1* is highly expressed in apoptotic cells of the small intestine. LKB1 regulates p53-dependent apoptosis and forced overexpression of LKB1 induces classical apoptosis in a p53-dependent manner. LKB1 physically associates with p53 and phosphorylates p53 at low levels, which might be required for p53 activation, at least in epithelial cells of the small intestine. Interestingly, LKB1 protein is present in both the cytoplasm and nucleus and translocates to mitochondria during apoptosis (Karuman et al., 2001). LKB1-induced cell death is dependent on the kinase activity of the enzyme. LKB1 has also been shown to control cell proliferation. It also interacts with the chromatin remodeling protein brahma-related gene-1 (*BRG1*), and with the cell-cycle regulatory proteins LKB1-interacting protein 1 (LIP1) and WAF1 (Yoo et al., 2002).

6.7. VIRAL TARGETING OF WILD-TYPE p53

Other mutation-independent mechanisms that directly target p53 are also utilized in human malignancies, albeit in select tumor types. Here, the common theme is sequestration and inactivation of wild-type p53 protein. This was first discovered in rodent models of viral oncogenesis. The viral oncoproteins SV-40 large T-antigen and Adenovirus type 5 E1B form stabilized complexes with p53, thereby functionally inactivating it (Mietz et al., 1992; Debbas and White, 1993; Yew and Berk, 1992).

HPV E6 and E7 act by uncoupling the checkpoint controls of the cell cycle and do so principally by inhibiting the normal function of p53 and pRb, respectively. In HPV-positive genitoanal cancers, the high-risk HPV-16/18 E6 oncoproteins form a p53 complex resulting in rapid p53 degradation (Scheffner et al., 1990). HPV E6, in association with the cellular protein E6-AP, is a potent p53-directed ubiquitin ligase. By in large, E6 and E7 of nononcogenic viruses do not have irreversible effects on growth properties. Other oncoviral proteins that complex and inactivate p53 include Hepatitis B X-protein associated with hepatocellular carcinoma (Feitelson et al., 1993; Wang et al., 1994; Ueda et al., 1995) Hepatitis B X-protein inhibits p53 transcriptional activity and its association with ERCC3 repair factor (Wang et al., 1994). Moreover, in primary wild-type and p53-null mouse hepatocytes, p53 is required for global genomic DNA repair and HBx expression suppresses global nucleotide excision repair in a p53-dependent manner. This suggests that in viral hepatitis, the hepatitis B virus could inhibit the p53-dependent component of DNA repair, leading over time to accumulation of genetic defects and promoting carcinogenesis (Prost et al., 1998).

Restenotic and atherosclerotic lesions often contain smooth muscle cells (SMCs) with high rates of proliferation and apoptosis. It has been postulated that human cytomegalovirus (HCMV) increases the incidence of restenosis and predisposes to atherosclerosis (Speir et al., 1994). HCMV IE-84 protein binds to and inhibits p53 transcriptional activity (Speir et al., 1994) and expression of IE2-84, but not the other major immediate-early gene product IE-72, protects SMCs from p53-mediated apoptosis. This suggests the intriguing possibility that HCMV infection predisposes to SMC accumulation and thereby contributes to restenosis and atherosclerosis, in part by p53 inactivation (Tanaka et al., 1999).

6.8. SUBCELLULAR LOCALIZATION OF p53

Wild-type p53 is localized to the cytoplasm in a subset of human tumors, blocking its ability to act as a transcription factor. Almost 100% of undifferentiated neuroblastomas (NB) and NB cell lines—but not benign differentiated ganglioneuromas—(Moll et al., 1995; Ostermeyer et al., 1996), retinoblastoma (Schlamp et al., 1997), PC12 pheochromocytoma cells (Eizenberg et al., 1996), and a subset of breast cancer (Moll et al., 1992; Stenmark-Askmalm et al. 1994; Lou et al., 1997) and colon cancer (Sun et al., 1992, 1996; Bosari et al., 1995) constitutively accumulate high levels of elevated levels of wild-type p53 protein in their cytoplasm in the absence of stress. The sequestered wild-type p53 is stabilized due to resistance to HDM2 degradation (Zaika et al., 1999). Cytoplasmic p53 accumulation is associated with a defect in p53 function in response to genotoxic stress (Moll et al., 1996; Isaacs et al., 1998; Aladjem et al., 1998). This was shown dramatically in A1-5 cells expressing high levels of temperature sensitive p53val135 (tsp53) in the nucleus. After chemical mutagenization, 22 independent A1-5 clones were selected for escape from p53 suppression (growth at permissive temperature). Most clones exhibited cytoplasmic sequestration as the mechanism by which p53 was inactivated (Gaitonde et al., 2000).

It is of note that neuroblastoma is among those very few tumor types that virtually never acquire neither p53 mutations nor INK4a/ARF mutations, genetically supporting the notion that p53 sequestration is the inactivating mechanism in this tumor. Interestingly, sequestration-induced p53 inactivation might be reversible. One mechanism of cytoplasmic sequestration of p53 depends on Glucocorticoid Receptor (GR), which form a complex with p53 in vitro and in vivo, resulting in cytoplasmic sequestration and inactivation of both p53 and GR. In NB cells, p53 and GR form a complex that can be dissociated by GR antagonists, resulting in accumulation of p53 in the nucleus, activation of p53-responsive genes, growth arrest, and apoptosis. These results suggest that molecules that efficiently disrupt GR–p53 interactions would have a therapeutic potential for the treatment of neuroblastoma (Sengupta et al., 2000). Alternatively, when CHP134 neuroblastoma cells were differentiated with retinoic acid, massive apoptosis, associated with nuclear translocation of p53, was induced (Takada et al., 2001).

Physiologically, mouse embryonic stem cells (Aladjem et al., 1998) and ductal epithelium cells of quiescent mammary gland (Kuperwasser et al., 2000) also exhibit constitutively sequestered p53 in the cytoplasm. Likewise, hormonal stimulation of quiescent ductal breast cells resulted in nuclear accumulation of p53, induction of p21, and increased apoptosis after ionizing radiation. Ionizing radiation alone failed to recruit p53 to the nucleus, and p53-dependent responses were minimal (Kuperwasser et al., 2000). Normal endothelial cells infected with cytomegalovirus are resistant to p53-mediated apoptosis. CMV infection sequesters p53 in the cytoplasm by blocking its nuclear localization signal. It is unclear whether HCMV IE-84 mediates this effect. The selective resistance to apoptosis might be important during CMV replication and may explain the oncogenic potential of CMV as well as its pathogenic role in intimal-proliferation-mediated vascular diseases and atherogenesis (Kovacs et al., 1996; Wang et al., 2001). On the other hand, normal human keratinocytes respond to UV rays by developing a fast adaptive response aimed at maintaining function and survival. Protection against UVB-induced apoptosis depends on p38-mediated phosphorylation and stabilization of p53 and is tightly associated with cytoplasmic sequestration of wild-type p53 (Chouinard et al., 2002).

6.9. INACTIVATION OF DOWNSTREAM EFFECTORS: Apaf-1

Primary and metastatic melanomas are strongly chemoresistant and are unable to execute apoptosis in response to p53 activation, yet they show a remarkable absence of p53 mutations. However, metastatic melanomas often lose Apaf-1, a downstream cell-death effector that acts with cytochrome c to activate procaspase-9 and mediate p53-dependent apoptosis. Loss of Apaf-1 expression can be due to allelic loss or promoter silencing by methylation. Restoring normal levels of Apaf-1 markedly enhances chemosensitivity of resistant melanoma cell lines and rescues their apoptotic defects. Apaf-1 loss may therefore contribute to the low frequency of p53 mutations in melanoma (Soengas et al., 2001). Apaf-1 protein deficiency also confers resistance to cytochrome c-dependent apoptosis in human leukemic cells (Jia et al., 2001). Interestingly, Apaf-1 is also a transcriptional target of p53 in DNA damage-induced apoptosis (Robles et al., 2001). However, Apaf-1 loss is probably rather tumor selective because another candidate tumor type, male germ cell tumors, which harbor recurrent deletions in the Apaf-1 locus chromosome 12q22, maintained Apaf-1 expression. This did not support the notion that Apaf-1 is the critical tumor suppressor target in this region (Bala et al., 2000).

6.10. p53 STATUS AND PROGNOSIS

The most powerful prognostic markers for most tumors are tumor size, clinical spread (stage), and histologic grade. Among the few molecular markers in use, N-Myc amplification in neuroblastomas remains the best one. On the basis of what we know about p53, there are strong biological reasons, derived from cell and animal based studies, to suggest that abnormalities of p53 are indicative of a poor prognosis in cancer. Yet, while a massive literature has compiled since 1991 analyzing the prognostic (clinical outcome for the patient) and predictive (response to treatment) value of p53 abnormalities in many different cancers, p53 has not become a useful prognostic and predictive molecular marker. The reason is that p53 failed to give consistent results among independent clinical studies. To make it clear, though, to show such correlations across this vast spectrum of human cancers is intrinsically a very difficult task. The reason for conflicting results from clinico-pathologic studies that try to link the presence of p53 mutations to poor prognosis in human tumors is many fold. On one hand, they can be traced back to clinical parameters such as variable cohort sizes, inaccurate diagnostic criteria, heterogenous treatment, and variations in statistical methods. On the other hand they are technical in nature due to variations in techniques used for p53 mutation detection; for example, p53 immunohistochemistry using abnormal overexpression as a surrogate marker for p53 mutations versus PCR based mutational SSCP screening or other methods that rely on biophysical properties of amplicons versus screening based on functional yeast assay versus the definitive nucleotide sequencing. Moreover, because 95% of mutations occur in the core domain, the vast majority of p53 mutational searches limited themselves to the DNA-binding region of exons 5–8, biasing against mutations in the regulatory regions. This view is supported by the fact that 40% of studies examined only exons 5 to 8, whereas only 14% examined the entire p53 gene. This dilemma is exemplified by two large breast cancer studies (Caleffi et al., 1994; Elledge et al., 1993). Furthermore, the question of independence of p53 as a prognostic marker is unsolved. By the very nature of p53 as a guardian against stress-induced mutations, p53 abnormalities are likely to coincide with well-known classical parameters of poor prognosis such as morphologic pleomorphism and atypia, aneuploidy, higher grade/stage, and resistance to therapy. Only multivariate analysis can resolve this issue. Currently, the available information suggests that p53 indeed does have prognostic power in non-Hodgkin's lymphoma, breast cancer, and non-small-cell lung cancer. The next step is to evaluate this potential in large, prospective multicenter case controlled studies.

6.11. DIRECTIONS FOR FUTURE RESEARCH

Intense research over the past 15 years have established p53's preeminent role as a "smoking gun" in cancer biology. Abrogation of normal p53 function is at the heart of a majority of human tumors. This realization has the potential for huge clinical payoffs. With this straight application in mind, great efforts in academia and industry are underway to restore normal p53 function of mutant or inactivated p53 proteins in tumors that directly target p53 itself. Gene therapy, oncolytic virus therapy, and reactivating small molecules are all currently being pursued. Over the past few years, there has been growing evidence that tumor cells can also cripple their p53 failsafe barrier by targeting regulators or effectors of p53 rather than p53 itself. Obviously, much remains to be discovered in this area before we have a better understanding of which are the critically important regulators to become rational therapeutic targets. In this respect, disruption of the p53–HDM2 interaction in wild-type tumors is already being tried (Moll and Zaika, 2000).

ACKNOWLEDGMENTS

The authors wish to thank Dr. Thierry Soussi for granting permission to incorporate some plots from the p53 database. This work was supported by a grant from the National Cancer Institute and from Phillip Morris Extramural Research Program to U.M.M., as well as a grant from the Schroedinger Fellowship Program from the Austrian Research Council FWF to N.C.

REFERENCES

Aguilar, F., Harris, C. C., Sun, T., Hollstein, M., and Cerutti, P. (1994). Geographic variation of p53 mutational profile in nonmalignant human liver. *Science* 264:1317–1319.
Aguilar, F., Hussain, S. P., and Cerutti, P. (1993). Aflatoxin B1 induces the transversion of G→T in codon 249 of the p53 tumor suppressor gene in human hepatocytes. *Proc Natl Acad Sci USA* 90:8586–8590.
Aladjem, M. I., Spike, B. T., Rodewald, L. W., Hope, T. J., Klemm, M., Jaenisch, R., and Wahl, G. M. (1998). ES cells do not activate p53-dependent stress responses and undergo p53-independent apoptosis in response to DNA damage. *Curr Biol* 8:145–155.
Alarcon, R., Koumenis, C., Geyer, R. K., Maki, C. G., and Giaccia, A. J. (1999). Hypoxia induces p53 accumulation through MDM2 down-regulation and inhibition of E6-mediated degradation. *Cancer Res* 59:6046–6051.

Attardi, L. D., Reczek, E. E., Cosmas, C., Demicco, E. G., McCurrach, M. E., Lowe, S. W., and Jacks, T. (2000). PERP, an apoptosis-associated target of p53, is a novel member of the PMP-22/gas3 family. *Genes Dev* 14:704–718.

Bala, S., Oliver, H., Renault, B., Montgomery, K., Dutta, S., Rao, P., Houldsworth, J., Kucherlapati, R., Wang, X., Chaganti, R. S., and Murty, V. V. (2000). Genetic analysis of the APAF1 gene in male germ cell tumors. *Genes Chromosomes Cancer* 28:258–268.

Bech-Hansen, N. T., Blattner, W. A., Sell, B. M., McKeen, E. A., Lampkin, B. C., Fraumeni, J. F., Jr., and Paterson, M. C. (1981). Transmission of in-vitro radioresistance in a cancer-prone family. *Lancet* 1:1335–1337.

Bell, D. W., Varley, J. M., Szydlo, T. E., Kang, D. H., Wahrer, D. C., Shannon, K. E., Lubratovich, M., Verselis, S. J., Isselbacher, K. J., Fraumeni, J. F., et al. (1999). Heterozygous germ line hCHK2 mutations in Li-Fraumeni syndrome. [see comments.]. *Science* 286:2528–2531.

Biernat, W., Kleihues, P., Yonekawa, Y., and Ohgaki, H. (1997). Amplification and overexpression of MDM2 in primary (de novo) glioblastomas. *J Neuropathol Exp Neurol* 56:180–185.

Blandino, G., Levine, A. J., and Oren, M. (1999). Mutant p53 gain of function: differential effects of different p53 mutants on resistance of cultured cells to chemotherapy. *Oncogene* 18:477–485.

Blaszyk, H., Hartmann, A., Cunningham, J. M., Schaid, D., Wold, L. E., Kovach, J. S., and Sommer, S. S. (2000). A prospective trial of midwest breast cancer patients: a p53 gene mutation is the most important predictor of adverse outcome. *Int J Cancer* 89:32–38.

Blaszyk, H., Hartmann, A., Tamura, Y., Saitoh, S., Cunningham, J. M., McGovern, R. M., Schroeder, J. J., Schaid, D. J., Ii, K., Monden, Y., et al. (1996). Molecular epidemiology of breast cancers in northern and southern Japan: the frequency, clustering, and patterns of p53 gene mutations differ among these two low-risk populations. *Oncogene* 13:2159–2166.

Borresen, A. L., Andersen, T. I., Garber, J., Barbier-Piraux, N., Thorlacius, S., Eyfjord, J., Ottestad, L., Smith-Sorensen, B., Hovig, E., and Malkin, D. (1992). Screening for germ line TP53 mutations in breast cancer patients. *Cancer Res* 52:3234–3236.

Bosari, S., Viale, G., Roncalli, M., Graziani, D., Borsani, G., Lee, A. K., and Coggi, G. (1995). p53 gene mutations, p53 protein accumulation and compartmentalization in colorectal adenocarcinoma. *Am J Pathol* 147:790–798.

Brash, D. E., Rudolph, J. A., Simon, J. A., Lin, A., McKenna, G. J., Baden, H. P., Halperin, A. J., and Ponten, J. (1991). A role for sunlight in skin cancer: UV-induced p53 mutations in squamous cell carcinoma. *Proc Natl Acad Sci USA* 88:10124–10128.

Bressac, B., Kew, M., Wands, J., and Ozturk, M. (1991). Selective G to T mutations of p53 gene in hepatocellular carcinoma from southern Africa. [see comments.]. *Nature* 350:429–431.

Bueso-Ramos, C. E., Yang, Y., deLeon, E., McCown, P., Stass, S. A., and Albitar, M. (1993). The human MDM-2 oncogene is overexpressed in leukemias. *Blood* 82:2617–2623.

Bulavin, D. V., Demidov, O. N., Saito, S., Kauraniemi, P., Phillips, C., Amundson, S. A., Ambrosino, C., Sauter, G., Nebreda, A. R., Anderson, C. W., et al. (2002). Amplification of PPM1D in human tumors abrogates p53 tumor-suppressor activity. *Nat Genet* 31:210–215.

Burri, N., Shaw, P., Bouzourene, H., Sordat, I., Sordat, B., Gillet, M., Schorderet, D., Bosman, F. T., and Chawbert, P. (2001). Methylation silencing and mutations of the p14ARF and p16INK4a genes in colon cancer. *Lab Invest* 2:217–229.

Cadwell, C., and Zambetti, G. P. (2001). The effects of wild-type p53 tumor suppressor activity and mutant p53 gain-of-function on cell growth. *Gene* 277:15–30.

Caelles, C., Helmberg, A., and Karin, M. (1994). p53-dependent apoptosis in the absence of transcriptional activation of p53-target genes. [see comments.]. *Nature* 370:220–223.

Caleffi, M., Teague, M. W., Jensen, R. A., Vnencak-Jones, C. L., Dupont, W. D., and Parl, F. F. (1994). p53 gene mutations and steroid receptor status in breast cancer. Clinicopathologic correlations and prognostic assessment. *Cancer* 73:2147–2156.

Camplejohn, R. S., Perry, P., Hodgson, S. V., Turner, G., Williams, A., Upton, C., MacGeoch, C., Mohammed, S., and Barnes, D. M. (1995). A possible screening test for inherited p53-related defects based on the apoptotic response of peripheral blood lymphocytes to DNA damage. *Br J Cancer* 72:654–662.

Chehab, N. H., Malikzay, A., Appel, M., and Halazonetis, T. D. (2000). Chk2/hCds1 functions as a DNA damage checkpoint in G(1) by stabilizing p53. *Genes Dev* 14:278–288.

Chen, X., Ko, L. J., Jayaraman, L., and Prives, C. (1996). p53 levels, functional domains, and DNA damage determine the extent of the apoptotic response of tumor cells. *Genes Dev* 10:2438–2451.

Chin, L., Pomerantz, J., Polsky, D., Jacobson, M., Cohen, C., Cordon-Cardo, C., Horner, J. W., 2nd, and DePinho, R. A. (1997). Cooperative effects of INK4a and ras in melanoma susceptibility in vivo. *Genes Dev* 11:2822–2834.

Cho, Y., Gorina, S., Jeffrey, P. D., and Pavletich, N. P. (1994). Crystal structure of a p53 tumor suppressor-DNA complex: understanding tumorigenic mutations. [see comments.]. *Science* 265:346–355.

Chouinard, N., Valerie, K., Rouabhia, M., and Huot, J. (2002). UVB-mediated activation of p38 mitogen-activated protein kinase enhances resistance of normal human keratinocytes to apoptosis by stabilizing cytoplasmic p53. *Biochem J* 365:133–145.

Cooper, D. N., and Krawczak, M. (1990). The mutational spectrum of single base-pair substitutions causing human genetic disease: patterns and predictions. *Hum Genet* 85:55–74.

Crook, T., Wrede, D., Tidy, J. A., Mason, W. P., Evans, D. J., and Vousden, K. H. (1992). Clonal p53 mutation in primary cervical cancer: association with human-papillomavirus-negative tumours. [see comments.]. *Lancet* 339:1070–1073.

D'Orazi, G., Cecchinelli, B., Bruno, T., Manni, I., Higashimoto, Y., Saito, S., Gostissa, M., Coen, S., Marchetti, A., Del Sal, G., et al. (2002). Homeodomain-interacting protein kinase-2 phosphorylates p53 at Ser 46 and mediates apoptosis. *Nat Cell Biol* 4:11–19.

Debbas, M., and White, E. (1993). Wild-type p53 mediates apoptosis by E1A, which is inhibited by E1B. *Genes Dev* 7:546–554.

Denissenko, M. F., Pao, A., Tang, M., and Pfeifer, G. P. (1996). Preferential formation of benzo[a]pyrene adducts at lung cancer mutational hotspots in P53. *Science* 274:430–432.

Di Como, C. J., Gaiddon, C., and Prives, C. (1999). p73 function is inhibited by tumor-derived p53 mutants in mammalian cells. *Mol Cell Biol* 19:1438–1449.

Dobrowolski, R., Hein, R., Buettner, R., and Bosserhoff, A. K. (2002). Loss of p14ARF expression in melanoma. *Arch Dermatol Res* 293:545–551.

Doll, R., and Hill, A. B. (1999). Smoking and carcinoma of the lung. Preliminary report. 1950. *Bull World Health Organ* 77:84–93.

Dominguez, G., Silva, J., Garcia, J. M., Silva, J. M., Rodriguez, R., Munoz, C., Chacon, I., Sanchez, R., Carballido, J., Colas, A., Espuna, P., and Bonilla, F. (2003). Prevalence of aberrant methylation of p14ARF over p16INK4a is some human tumors. *Mutant Res* 530:9–17.

Donehower, L. A., Harvey, M., Slagle, B. L., McArthur, M. J., Montgomery, C. A., Jr., Butel, J. S., and Bradley, A. (1992). Mice deficient for p53 are developmentally normal but susceptible to spontaneous tumours. *Nature* 356:215–221.

Dumaz, N., Stary, A., Soussi, T., Daya-Grosjean, L., and Sarasin, A. (1994). Can we predict solar ultraviolet radiation as the causal event in human tumours by analysing the mutation spectra of the p53 gene? *Mutat Res* 307:375–386.

Ehrlich, M., Zhang, X. Y., and Inamdar, N. M. (1990). Spontaneous deamination of cytosine and 5-methylcytosine residues in DNA and replacement of 5-methylcytosine residues with cytosine residues. *Mutat Res* 238:277–286.

Eizenberg, O., Faber-Elman, A., Gottlieb, E., Oren, M., Rotter, V., and Schwartz, M. (1996). p53 plays a regulatory role in differentiation and apoptosis of central nervous system-associated cells. *Mol Cell Biol* 16:5178–5185.

el-Deiry, W. S., Tokino, T., Velculescu, V. E., Levy, D. B., Parsons, R., Trent, J. M., Lin, D., Mercer, W. E., Kinzler, K. W., and Vogelstein, B. (1993). WAF1, a potential mediator of p53 tumor suppression. *Cell* 75:817–825.

Eliyahu, D., Michalovitz, D., Eliyahu, S., Pinhasi-Kimhi, O., and Oren, M. (1989). Wild-type p53 can inhibit oncogene-mediated focus formation. *Proc Natl Acad Sci USA* 86:8763–8767.

Elledge, R. M., Fuqua, S. A., Clark, G. M., Pujol, P., Allred, D. C., and McGuire, W. L. (1993). Prognostic significance of p53 gene alterations in node-negative breast cancer. *Breast Cancer Res Treat* 26:225–235.

Eymin, B., Gazzeri, S., Brambilla, C., and Brambilla, E. (2002). Mdm2 overexpression and p14(ARF) inactivation are two mutually exclusive events in primary human lung tumors. *Oncogene* 21:2750–2761.

Fang, S., Jensen, J. P., Ludwig, R. L., Vousden, K. H., and Weissman, A. M. (2000). Mdm2 is a RING finger-dependent ubiquitin protein ligase for itself and p53. *J Biol Chem* 275:8945–8951.

Feitelson, M. A., Zhu, M., Duan, L. X., and London, W. T. (1993). Hepatitis B x antigen and p53 are associated in vitro and in liver tissues from patients with primary hepatocellular carcinoma. *Oncogene* 8:1109–1117.

Fillippovich, I., Sorokina, N., Gatei, M., Haupt, Y., Hobson, K., Moallem, E., Spring, K., Mould, M., McGuckin, M. A., Lavin, M. F., and Khanna, K. K. (2001). Transactivation-deficient p73alpha (p73Deltaexon2) inhibits apoptosis and competes with p53. *Oncogene* 20:514–522.

Finlay, C. A. (1993). The mdm-2 oncogene can overcome wild-type p53 suppression of transformed cell growth. *Mol Cell Biol* 13:301–306.

Finlay, C. A., Hinds, P. W., and Levine, A. J. (1989). The p53 proto-oncogene can act as a suppressor of transformation. *Cell* 57:1083–1093.

Fogal, V., Gostissa, M., Sandy, P., Zacchi, P., Sternsdorf, T., Jensen, K., Pandolfi, P. P., Will, H., Schneider, C., and Del Sal, G. (2000). Regulation of p53 activity in nuclear bodies by a specific PML isoform. *EMBO J* 19, 6185–6195.

Freedman, D. A., and Levine, A. J. (1998). Nuclear export is required for degradation of endogenous p53 by MDM2 and human papillomavirus E6. *Mol Cell Biol* 18:7288–7293.

Friedberg, E. C., Walker, G. C., and Siede, W. (1995). DNA repair and mutagenesis. Friedberg, Walker, Siede (eds.) pp. 407–434.

Gaitonde, S. V., Riley, J. R., Qiao, D., and Martinez, J. D. (2000). Conformational phenotype of p53 is linked to nuclear translocation. *Oncogene* 19:4042–4049.

Gardie, B., Cayuela, J. M., Martini, S., and Sigaux, F. (1998). Genomic alterations of the p19ARF encoding exons in T-cell acute lymphoblastic leukemia. *Blood* 91:1016–1020.

Goi, K., Takagi, M., Iwata, S., Delia, D., Asada, M., Donghi, R., Tsunematsu, Y., Nakazawa, S., Yamamoto, H., Yokota, J., et al. (1997). DNA damage-associated dysregulation of the cell cycle and apoptosis control in cells with germ-line p53 mutation. *Cancer Res* 57:1895–1902.

Grob, T. J., Novak, U., Maisse, C., Barcaroli, D., Luthi, A. U., Pirnia, F., Hugli, B., Graber, H. U., De Laurenzi, V., Fey, M. F., et al. (2001). Human delta Np73 regulates a dominant negative feedback loop for TAp73 and p53. *Cell Death Differ* 8:1213–1223.

Gruttgen, A., Reichenzeller, M., Junger, M., Schlien, S., Affolter, A., and Bosch, F. X. (2001). Detailed gene expression analysis but not microsatellite marker analysis of 9p21 reveals differential defects in the INK4a gene locus in the majority of head and neck cancers. *J Pathol* 194(3):311–317.

Guo, A., Salomoni, P., Luo, J., Shih, A., Zhong, S., Gu, W., and Paolo Pandolfi, P. (2000). The function of PML in p53-dependent apoptosis. *Nat Cell Biol* 2:730–736.

Hainaut, P., Soussi, T., Shomer, B., Hollstein, M., Greenblatt, M., Hovig, E., Harris, C. C., and Montesano, R. (1997). Database of p53 gene somatic mutations in human tumors and cell lines: updated compilation and future prospects. *Nucleic Acids Res.* 25:151–157.

Harris, C. C., and Hollstein, M. (1993). Clinical implications of the p53 tumor-suppressor gene. [see comments.]. *New Engl J Med* 329:1318–1327.

Hartmann, A., Blaszyk, H., Kovach, J. S., and Sommer, S. S. (1997). The molecular epidemiology of p53 gene mutations in human breast cancer. *Trends Genet* 13:27–33.

Hartmann, A., Rosanelli, G., Blaszyk, H., Cunningham, J. M., McGovern, R. M., Schroeder, J. J., Schaid, D. J., Kovach, J. S., and Sommer, S. S. (1995). Novel pattern of P53 mutation in breast cancers from Austrian women. *J Clin Invest* 95:686–689 [Erratum in: *J Clin Invest* (1995) June 95(6):2991].

Harvey, M., Sands, A. T., Weiss, R. S., Hegi, M. E., Wiseman, R. W., Pantazis, P., Giovanella, B. C., Tainsky, M. A., Bradley, A., and Donehower, L. A. (1993). In vitro growth characteristics of embryo fibroblasts isolated from p53-deficient mice. *Oncogene* 8:2457–2467.

Haupt, Y., Maya, R., Kazaz, A., and Oren, M. (1997). Mdm2 promotes the rapid degradation of p53. *Nature* 387:296–299.

Haupt, Y., Rowan, S., Shaulian, E., Vousden, K. H., and Oren, M. (1995). Induction of apoptosis in HeLa cells by trans-activation-deficient p53. *Genes Dev* 9:2170–2183.

Hermeking, H., Lengauer, C., Polyak, K., He, T. C., Zhang, L., Thiagalingam, S., Kinzler, K. W., and Vogelstein, B. (1997). 14-3-3 sigma is a p53-regulated inhibitor of G2/M progression. *Mol Cell* 1:3–11.

Hirao, A., Kong, Y. Y., Matsuoka, S., Wakeham, A., Ruland, J., Yoshida, H., Liu, D., Elledge, S. J., and Mak, T. W. (2000). DNA damage-induced activation of p53 by the checkpoint kinase Chk2. [see comments.]. *Science* 287:1824–1827.

Hoffman, W. H., Biade, S., Zilfou, J. T., Chen, J., and Murphy, M. (2002). Transcriptional repression of the anti-apoptotic survivin gene by wild type p53. *J Biol Chem* 277:3247–3257.

Hofmann, T. G., Moller, A., Sirma, H., Zentgraf, H., Taya, Y., Droge, W., Will, H., and Schmitz, M. L. (2002). Regulation of p53 activity by its interaction with homeodomain-interacting protein kinase-2. *Nat Cell Biol* 4:1–10.

Hollstein, M., Rice, K., Greenblatt, M. S., Soussi, T., Fuchs, R., Sorlie, T., Hovig, E., Smith-Sorensen, B., Montesano, R., and Harris, C. C. (1994). Database of p53 gene somatic mutations in human tumors and cell lines. *Nucleic Acids Res* 22:3551–3555.

Honda, R., Tanaka, H., and Yasuda, H. (1997). Oncoprotein MDM2 is a ubiquitin ligase E3 for tumor suppressor p53. *FEBS Lett* 420:25–27.

Howley, P. M. (1991). Role of the human papillomaviruses in human cancer. *Cancer Res* 51, 5019s–5022s.

Hsu, I. C., Metcalf, R. A., Sun, T., Welsh, J. A., Wang, N. J., and Harris, C. C. (1991). Mutational hotspot in the p53 gene in human hepatocellular carcinomas. [see comments.]. *Nature* 350:427–428.

Hussain, S. P., Amstad, P., Raja, K., Sawyer, M., Hofseth, L., Shields, P. G., Hewer, A., Phillips, D. H., Ryberg, D., Haugen, A., and Harris, C. C. (2001). Mutability of p53 hotspot condons to benzo(a)pyrene diol epoxide (BPDE) and the frequency of p53 mutations in nontumorous human lung. *Cancer Res* 61(17):6350–6355.

Ingvarsson, S., Sigbjornsdottir, B. I., Huiping, C., Hafsteinsdottir, S. H., Ragnarsson, G., Barkardottir, R. B., Arason, A., Egilsson, V., and Bergthorsson, J. T. (2002). Mutation analysis of the CHK2 gene in breast carcinoma and other cancers. *Breast Cancer Res* 4:R4.

Isaacs, J. S., Hardman, R., Carman, T. A., Barrett, J. C., and Weissman, B. E. (1998). Differential subcellular p53 localization and function in N- and S-type neuroblastoma cell lines. *Cell Growth Differ* 9:545–555.

Iwato, M., Tachibana, O., Tohma, Y., Arakawa, Y., Nitta, H., Hasegawa, M., Yamashita, J., and Hayashi, Y. (2000). Alterations of the INK4a/ARF locus in human intracranial germ cell tumors. *Cancer Res* 60:2113–2115.

Jarmalaite, S., Kannio, A., Anttila, S., Lazutka, J. R., and Husgafvel-Pursiainen, K. (2003). Aberrant p16 promoter methylation in smokers and former smokers with nonsmall cell lung cancer. *Int. J. Cancer* 106(6):913–918.

Jia, L., Srinivasula, S. M., Liu, F. T., Newland, A. C., Fernandes-Alnemri, T., Alnemri, E. S., and Kelsey, S. M. (2001). Apaf-1 protein deficiency confers resistance to cytochrome c-dependent apoptosis in human leukemic cells. *Blood* 98414–421.

Johnstone, R. W., Ruefli, A. A., and Lowe, S. W. (2002). Apoptosis: a link between cancer genetics and chemotherapy. *Cell* 108:153–164.

Jolly, K. W., Malkin, D., Douglass, E. C., Brown, T. F., Sinclair, A. E., and Look, A. T. (1994). Splice-site mutation of the p53 gene in a family with hereditary breast-ovarian cancer. *Oncogene* 9:97–102.

Jones, S. N., Roe, A. E., Donehower, L. A., and Bradley, A. (1995). Rescue of embryonic lethality in Mdm2-deficient mice by absence of p53. *Nature* 378:206–208.

Kamb, A. (1995). Cell-cycle regulators and cancer. *Trends Genet* 11:136–140.

Kamijo, T., Zindy, F., Roussel, M. F., Quelle, D. E., Downing, J. R., Ashmun, R. A., Grosveld, G., and Sherr, C. J. (1997). Tumor suppression at the mouse INK4a locus mediated by the alternative reading frame product p19ARF. *Cell* 91:649–659.

Karuman, P., Gozani, O., Odze, R. D., Zhou, X. C., Zhu, H., Shaw, R., Brien, T. P., Bozzuto, C. D., Ooi, D., Cantley, L. C., and Yuan, J. (2001). The Peutz-Jegher gene product LKB1 is a mediator of p53-dependent cell death. *Mol Cell* 7:1307–1319.

Kemp, C. J., Wheldon, T., and Balmain, A. (1994). p53-deficient mice are extremely susceptible to radiation-induced tumorigenesis. *Nat Genet* 8:66–69.

Khatib, Z. A., Matsushime, H., Valentine, M., Shapiro, D. N., Sherr, C. J., and Look, A. T. (1993). Coamplification of the CDK4 gene with MDM2 and GLI in human sarcomas. *Cancer Res* 53:5535–5541.

Khosravi, R., Maya, R., Gottlieb, T., Oren, M., Shiloh, Y., and Shkedy, D. (1999). Rapid ATM-dependent phosphorylation of MDM2 precedes p53 accumulation in response to DNA damage. *Proc Natl Acad Sci USA* 96:14973–14977.

Kovach, J. S., Hartmann, A., Blaszyk, H., Cunningham, J., Schaid, D., and Sommer, S. S. (1996). Mutation detection by highly sensitive methods indicates that p53 gene mutations in breast cancer can have important prognostic value. *Proc Natl Acad Sci USA* 93:1093–1096.

Kovacs, A., Weber, M. L., Burns, L. J., Jacob, H. S., and Vercellotti, G. M. (1996). Cytoplasmic sequestration of p53 in cytomegalovirus-infected human endothelial cells. *Am J Pathol* 149:1531–1539.

Kubbutat, M. H., Jones, S. N., and Vousden, K. H. (1997). Regulation of p53 stability by Mdm2. *Nature* 387:299–303.

Kumar, R., Smeds, J., Lundh Rozell, B., and Hemminki, K. (1999). Loss of heterozygosity at chromosome 9p21 (INK4-p14 ARF locus): homozygous deletions and mutations in the p16 and p14 ARF genes in sporadic primary melanomas. *Melanoma Res* 2:138–147.

Kuperwasser, C., Pinkas, J., Hurlbut, G. D., Naber, S. P., and Jerry, D. J. (2000). Cytoplasmic sequestration and functional repression of p53 in the mammary epithelium is reversed by hormonal treatment. *Cancer Res* 60:2723–2729.

Kussie, P. H., Gorina, S., Marechal, V., Elenbaas, B., Moreau, J., Levine, A. J., and Pavletich, N. P. (1996). Structure of the MDM2 oncoprotein bound to the p53 tumor suppressor transactivation domain. [letter; comment.]. *Science* 274:948–953.

Labuhn, M., Jones, G., Speel, E. J., Maier, D., Zweifel, C., Gratzl, O., Van Meir, E. G., Hegi, M. E., and Merlo, A. (2001). Quantitative real-time PCR does not show selective targeting of p14(ARF) but concomitant inactivation of both p16(INK4A) and p14(ARF) in 105 human primary gliomas. *Oncogene* 20:1103–1109.

Ladanyi, M., Cha, C., Lewis, R., Jhanwar, S. C., Huvos, A. G., and Healey, J. H. (1993). MDM2 gene amplification in metastatic osteosarcoma. *Cancer Res* 53:16–18.

Lalle, P., Moyret-Lalle, C., Wang, Q., Vialle, J. M., Navarro, C., Bressac-de Paillerets, B., Magaud, J. P., and Ozturk, M. (1995). Genomic stability and wild-type p53 function of lymphoblastoid cells with germ-line p53 mutation. *Oncogene* 10:2447–2454.

Larsen, C. J. (1996). p16INK4a: a gene with a dual capacity to encode unrelated proteins that inhibit cell cycle progression. *Oncogene* 12:2041–2044.

Lavigueur, A., Maltby, V., Mock, D., Rossant, J., Pawson, T., and Bernstein, A. (1989). High incidence of lung, bone, and lymphoid tumors in transgenic mice overexpressing mutant alleles of the p53 oncogene. *Mol Cell Biol* 9:3982–3991.

Law, J. C., Strong, L. C., Chidambaram, A., and Ferrell, R. E. (1991). A germ line mutation in exon 5 of the p53 gene in an extended cancer family. *Cancer Res* 51:6385–6387.

Leach, F. S., Tokino, T., Meltzer, P., Burrell, M., Oliner, J. D., Smith, S., Hill, D. E., Sidransky, D., Kinzler, K. W., and Vogelstein, B. (1993). p53 Mutation and MDM2 amplification in human soft tissue sarcomas. *Cancer Res* 532231–2234.

Li, J., Yang, Y., Peng, Y., Austin, R. J., van Eyndhoven, W. G., Nguyen, K. C., Gabriele, T., McCurrach, M. E., Marks, J. R., Hoey, T., *et al.* (2002). Oncogenic properties of PPM1D located within a breast cancer amplification epicenter at 17q23. *Nat Genet* 31:133–134.

Livingstone, L. R., White, A., Sprouse, J., Livanos, E., Jacks, T., and Tlsty, T. D. (1992). Altered cell cycle arrest and gene amplification potential accompany loss of wild-type p53. *Cell* 70:923–935.

Lonardo, F., Ueda, T., Huvos, A. G., Healey, J., and Ladanyi, M. (1997). p53 and MDM2 alterations in osteosarcomas: correlation with clinicopathologic features and proliferative rate. *Cancer* 79:1541–1547.

Lou, M. A., Tseng, S. L., Chang, S. F., Yue, C. T., Chang, B. L., Chou, C. H., Yang, S. L., Teh, B. H., Wu, C. W., and Shen, C. Y. (1997). Novel patterns of p53 abnormality in breast cancer from Taiwan: experience from a low-incidence area. *Br J Cancer* 75:746–751.

Maher, V. M., Yang, J. L., and McCormick, J. J. (1990). Ability of adducts formed in a shuttle vector by reactive metabolites of 1-nitropyrene and benzo(a)pyrene to induce mutations when the plasmid replicates in human cells. *Acta Biol Hung* 41:173–186.

Malkin, D., Jolly, K. W., Barbier, N., Look, A. T., Friend, S. H., Gebhardt, M. C., Andersen, T. I., Borresen, A. L., Li, F. P., and Garber, J. (1992). Germline mutations of the p53 tumor-suppressor gene in children and young adults with second malignant neoplasms. [see comments.]. *New Engl J Med* 326:1309–1315.

Malkin, D., Li, F. P., Strong, L. C., Fraumeni, J. F., Jr., Nelson, C. E., Kim, D. H., Kassel, J., Gryka, M. A., Bischoff, F. Z., and Tainsky, M. A. (1990). Germ line p53 mutations in a familial syndrome of breast cancer, sarcomas, and other neoplasms. [see comments.]. *Science* 250:1233–1238.

Marchenko, N. D., Zaika, A., and Moll, U. M. (2000). Death signal-induced localization of p53 protein to mitochondria. A potential role in apoptotic signaling. *J Biol Chem* 275:16202–16212.

Metzger, A. K., Sheffield, V. C., Duyk, G., Daneshvar, L., Edwards, M. S., and Cogen, P. H. (1991). Identification of a germ-line mutation in the p53 gene in a patient with an intracranial ependymoma. *Proc Natl Acad Sci USA* 88, 7825–7829.

Mietz, J. A., Unger, T., Huibregtse, J. M., and Howley, P. M. (1992). The transcriptional transactivation function of wild-type p53 is inhibited by SV40 large T-antigen and by HPV-16 E6 oncoprotein. *EMBO J* 11:5013–5020.

Miller, C. W., Ikezoe, T., Krug, U., Hofmann, W. K., Tavor, S., Vegesna, V., Tsukasaki, K., Takeuchi, S., and Koeffler, H. P. (2002). Mutations of the CHK2 gene are found in some osteosarcomas, but are rare in breast, lung, and ovarian tumors. *Genes Chromosomes Cancer* 33:17–21.

Milner, J., and Medcalf, E. A. (1991). Cotranslation of activated mutant p53 with wild type drives the wild-type p53 protein into the mutant conformation. *Cell* 65:765–774.

Miyashita, T., and Reed, J. C. (1995). Tumor suppressor p53 is a direct transcriptional activator of the human bax gene. *Cell* 80:293–299.

Moll, U. M., Erster, S., and Zaika, A. (2001). p53, p63 and p73–solos, alliances and feuds among family members. *Biochim Biophys Acta* 1552:47–59.

Moll, U. M., LaQuaglia, M., Benard, J., and Riou, G. (1995). Wild-type p53 protein undergoes cytoplasmic sequestration in undifferentiated neuroblastomas but not in differentiated tumors. *Proc Natl Acad Sci USA* 92:4407–4411.

Moll, U. M., Ostermeyer, A. G., Haladay, R., Winkfield, B., Frazier, M., and Zambetti, G. (1996). Cytoplasmic sequestration of wild-type p53 protein impairs the G1 checkpoint after DNA damage. *Mol Cell Biol* 16:1126–1137.

Moll, U. M., Riou, G., and Levine, A. J. (1992). Two distinct mechanisms alter p53 in breast cancer: mutation and nuclear exclusion. *Proc Natl Acad Sci USA* 89:7262–7266.

Moll, U. M., and Zaika, A. (2000). Disrupting the p53-mdm2 interaction as a potential therapeutic modality. *Drug Resist Updat* 3:217–221.

Montes de Oca Luna, R., Wagner, D. S., and Lozano, G. (1995). Rescue of early embryonic lethality in mdm2-deficient mice by deletion of p53. *Nature* 378:203–206.

Montesano, R., Hainaut, P., and Wild, C. P. (1997). Hepatocellular carcinoma: from gene to public health. *J Natl Cancer Inst* 89:1844–1851.

Moroni, M. C., Hickman, E. S., Denchi, E. L., Caprara, G., Colli, E., Cecconi, F., Muller, H., and Helin, K. (2001). Apaf-1 is a transcriptional target for E2F and p53. *Nat Cell Biol* 3:552–558.

Nakagawa, T., Takahashi, M., Ozaki, T., Watanabe Ki, K., Todo, S., Mizuguchi, H., Hayakawa, T., and Nakagawara, A. (2002). Autoinhibitory regulation of p73 by Delta Np73 to modulate cell survival and death through a p73-specific target element within the Delta Np73 promoter. *Mol Cell Biol* 22:2575–2585.

Nakano, K., and Vousden, K. H. (2001). PUMA, a novel proapoptotic gene, is induced by p53. *Mol Cell* 7:683–694.

Nilbert, M., Rydholm, A., Willen, H., Mitelman, F., and Mandahl, N. (1994). MDM2 gene amplification correlates with ring chromosome in soft tissue tumors. *Genes Chromosomes Cancer* 9:261–265.

Oda, E., Ohki, R., Murasawa, H., Nemoto, J., Shibue, T., Yamashita, T., Tokino, T., Taniguchi, T., and Tanaka, N. (2000a). Noxa, a BH3-only member of the Bcl-2 family and candidate mediator of p53-induced apoptosis. *Science* 288:1053–1058.

Oda, K., Arakawa, H., Tanaka, T., Matsuda, K., Tanikawa, C., Mori, T., Nishimori, H., Tamai, K., Tokino, T., Nakamura, Y., and Taya, Y. (2000b). p53AIP1, a potential mediator of p53-dependent apoptosis, and its regulation by Ser-46-phosphorylated p53. *Cell* 102:849–862.

Okamura, S., Arakawa, H., Tanaka, T., Nakanishi, H., Ng, C. C., Taya, Y., Monden, M., and Nakamura, Y. (2001). p53DINP1, a p53-inducible gene, regulates p53-dependent apoptosis. *Mol Cell* 8:85–94.

Oliner, J. D., Kinzler, K. W., Meltzer, P. S., George, D. L., and Vogelstein, B. (1992). Amplification of a gene encoding a p53-associated protein in human sarcomas. [see comments.]. *Nature* 358:80–83.

Omura-Minamisawa, M., Diccianni, M. B., Chang, R. C., Batora, A., Bridgeman, L. J., Schiff, J., Cohn, S. L., London, W. B., and Yu, A. L. (2001). p16/p14 (ARF) cell cycle regulatory pathways in primary neuroblastoma: p16 expression is associated with advanced stage disease. *Clin Cancer Res* (11)3481–3490.

Ostermeyer, A. G., Runko, E., Winkfield, B., Ahn, B., and Moll, U. M. (1996). Cytoplasmically sequestered wild-type p53 protein in neuroblastoma is relocated to the nucleus by a C-terminal peptide. *Proc Natl Acad Sci USA* 93:15190–15194.

Ozturk, M. (1991). p53 mutation in hepatocellular carcinoma after aflatoxin exposure. *Lancet* 338:1356–1359.

Parshad, R., Price, F. M., Pirollo, K. F., Chang, E. H., and Sanford, K. K. (1993). Cytogenetic response to G2-phase X irradiation in relation to DNA repair and radiosensitivity in a cancer-prone family with Li-Fraumeni syndrome. *Radiat Res* 136:236–240.

Phillips, A. C., and Vousden, K. H. (2001). E2F-1 induced apoptosis. *Apoptosis* 6:173–182.

Poi, M. J., Yen, T., Li, J., Song, H., Lang, J. C., Schuller, D. E., Pearl, D. K., Casto, B. C., Tsai, M. D., and Weghorst, C. M. (2001). Somatic INK4a-ARF locus mutations: a significant mechanism of gene inactivation in squamous cell carcinomas of the head and neck *Mol. Carcinog* (1):26–36.

Polsky, D., Bastian, B. C., Hazan, C., Melzer, K., Pack, J., Houghton, A., Busam, K., Cordon-Cardo, C., and Osman, I. (2001). HDM2 protein overexpression, but not gene amplification, is related to tumorigenesis of cutaneous melanoma. *Cancer Res* 61:7642–7646.

Pomerantz, J., Schreiber-Agus, N., Liegeois, N. J., Silverman, A., Alland, L., Chin, L., Potes, J., Chen, K., Orlow, I., Lee, H. W., *et al.* (1998). The Ink4a tumor suppressor gene product, p19Arf, interacts with MDM2 and neutralizes MDM2's inhibition of p53. *Cell* 92:713–723.

Pozniak, C. D., Radinovic, S., Yang, A., McKeon, F., Kaplan, D. R., and Miller, F. D. (2000). An anti-apoptotic role for the p53 family member, p73, during developmental neuron death. [see comments.]. *Science* 289:304–306.

Prosser, J., Porter, D., Coles, C., Condie, A., Thompson, A. M., Chetty, U., Steel, C. M., and Evans, H. J. (1992). Constitutional p53 mutation in a non-Li-Fraumeni cancer family. *Br J Cancer* 65:527–528.

Prost, S., Ford, J. M., Taylor, C., Doig, J., and Harrison, D. J. (1998). Hepatitis B x protein inhibits p53-dependent DNA repair in primary mouse hepatocytes. *J Biol Chem* 273:33327–33332.

Rady, P., Scinicariello, F., Wagner, R. F., Jr., and Tyring, S. K. (1992). p53 mutations in basal cell carcinomas. *Cancer Res* 52:3804–3806.

Reddy, A., Yuille, M., Sullivan, A., Repellin, C., Bell, A., Tidy, J. A., Evans, D. J., Farrell, P. J., Gusterson, B., Gasco, M., and Crook, T. (2002). Analysis of CHK2 in vulval neoplasia. *Br J Cancer* 86:756– 760.

Rizos, H., Darmanian, A. P., Holland, E. A., Mann, G. J., and Kefford, R. F. (2001). Mutations in the INK4a/ARF melanoma suspectability locus functionally impair p14ARF. *J Biol Chem* (44) 41424–41434.

Rizos, H., Puig, S., Badenas, C., Malreky, J., Darmanian, A. P., Jimenez, L., Mila, M., and Kefford, R. F. (2001). A melanoma-associated germline mutation in exon 1 beta inactivates p14ARF. *Oncogene* (39)5543–5547.

Robles, A. I., Bemmels, N. A., Foraker, A. B., and Harris, C. C. (2001). APAF-1 is a transcriptional target of p53 in DNA damage-induced apoptosis. *Cancer Res* 61:6660–6664.

Rodriguez, M. S., Desterro, J. M., Lain, S., Midgley, C. A., Lane, D. P., and Hay, R. T. (1999). SUMO-1 modification activates the transcriptional response of p53. *EMBO J* 18:6455–6461.

Ruggeri, B., DiRado, M., Zhang, S. Y., Bauer, B., Goodrow, T., and Klein-Szanto, A. J. (1993). Benzo[a]pyrene-induced murine skin tumors exhibit frequent and characteristic G to T mutations in the p53 gene. *Proc Natl Acad Sci USA* 90:1013–1017.

Saito, S., Goodarzi, A. A., Higashimoto, Y., Noda, Y., Lees-Miller, S. P., Appella, E., and Anderson, C. W. (2002). ATM mediates phosphorylation at multiple p53 sites, including Ser(46), in response to ionizing radiation. *J Biol Chem* 277:12491–12494.

Saitoh, S., Cunningham, J., De Vries, E. M., McGovern, R. M., Schroeder, J. J., Hartmann, A., Blaszyk, H., Wold, L. E., Schaid, D., and Sommer, S. S. (1994). p53 gene mutations in breast cancers in midwestern US women: null as well as missense-type mutations are associated with poor prognosis. *Oncogene* 9:2869–2875.

Salomoni, P., and Pandolfi, P. P. (2002). The role of PML in tumor suppression. *Cell* 108:165–170.

Sanchez-Cespedes, M., Reed, A. L., Buta, M., Wu, L., Westra, W. H., Herman, J. G., Yang, S. C., Jen, J., and Sidransky, D. (1999). Inactivation of the INK4a/ARF locus frequently coexists with TP53 mutations in non-small cell lung cancer. *Oncogene* 18(43):5843–5849.

Sansome, C., Zaika, A., Marchenko, N. D., and Moll, U. M. (2001). Hypoxia death stimulus induces translocation of p53 protein to mitochondria. Detection by immunofluorescence on whole cells. *FEBS Lett* 488:110–115.

Sato, M., Nishigori, C., Zghal, M., Yagi, T., and Takebe, H. (1993). Ultraviolet-specific mutations in p53 gene in skin tumors in xeroderma pigmentosum patients. *Cancer Res.* 53:2944–2946.

Saunders, W. B. (1999). *Robbins Pathologic Basis of Disease*. Cotran, Kumar, Collins (ed). 6th Edition, pp. 741–753.

Sayan, A. E., Sayan, B. S., Findikli, N., and Ozturk, M. (2001). Acquired expression of transcriptionally active p73 in hepatocellular carcinoma cells. *Oncogene* 20:5111–5117.

Scheffner, M., Munger, K., Byrne, J. C., and Howley, P. M. (1991). The state of the p53 and retinoblastoma genes in human cervical carcinoma cell lines. *Proc Natl Acad Sci USA* 88:5523–5527.

Scheffner, M., Werness, B. A., Huibregtse, J. M., Levine, A. J., and Howley, P. M. (1990). The E6 oncoprotein encoded by human papillomavirus types 16 and 18 promotes the degradation of p53. *Cell* 63:1129–1136.

Schlamp, C. L., Poulsen, G. L., Nork, T. M., and Nickells, R. W. (1997). Nuclear exclusion of wild-type p53 in immortalized human retinoblastoma cells. [see comments.]. *J Natl Cancer Inst* 89:1530–1536.

Schmitt, C. A., McCurrach, M. E., de Stanchina, E., Wallace-Brodeur, R. R., and Lowe, S. W. (1999). INK4a/ARF mutations accelerate lymphomagenesis and promote chemoresistance by disabling p53. *Genes Dev* 13:2670–2677.

Schuler, M., and Green, D. R. (2001). Mechanisms of p53-dependent apoptosis. *Biochem Soc Trans* 29:684–688.

Sengupta, S., Vonesch, J. L., Waltzinger, C., Zheng, H., and Wasylyk, B. (2000). Negative cross-talk between p53 and the glucocorticoid receptor and its role in neuroblastoma cells. *EMBO J* 19:6051–6064.

Serrano, M., Lee, H., Chin, L., Cordon-Cardo, C., Beach, D., and DePinho, R. A. (1996). Role of the INK4a locus in tumor suppression and cell mortality. *Cell* 85:27–37.

Sharpless, N. E., Bardeesy, N., Lee, K. H., Carrasco, D., Castrillon, D. H., Aguirre, A. J., Wu, E. A., Horner, J. W., and DePinho, R. A. (2001). Loss of p16Ink4a with retention of p19Arf predisposes mice to tumorigenesis. *Nature* 413:86–91.

Shieh, S. Y., Ahn, J., Tamai, K., Taya, Y., and Prives, C. (2000). The human homologs of checkpoint kinases Chk1 and Cds1 (Chk2) phosphorylate p53 at multiple DNA damage-inducible sites. *Genes Dev* 14:289–300 [Erratum in: *Genes Dev* (2000) March 15; 14(6):750.].

Shimizu, Y., Zhu, J. J., Han, F., Ishikawa, T., and Oda, H. (1999). Different frequencies of p53 codon-249 hot-spot mutations in hepatocellular carcinomas in Jiang-su province of China. *Int J Cancer* 82:187–190.

Shirangi, T. R., Zaika, A., and Moll, U. M. (2002). Nuclear degradation of p53 occurs during down-regulation of the p53 response after DNA damage. *FASEB J* 16:420–422.

Sidransky, D., Tokino, T., Helzlsouer, K., Zehnbauer, B., Rausch, G., Shelton, B., Prestigiacomo, L., Vogelstein, B., and Davidson, N. (1992). Inherited p53 gene mutations in breast cancer. *Cancer Res* 52:2984–2986.

Siliciano, J. D., Canman, C. E., Taya, Y., Sakaguchi, K., Appella, E., and Kastan, M. B. (1997). DNA damage induces phosphorylation of the amino terminus of p53. *Genes Dev* 11:3471–3481.

Silva, J., Dominguez, G., Silva, J. M., Garcia, J. M., Gallego, I., Corbacho, C., Provencio, M., Espana, P., and Bonilla, F. (2001). Analysis of genetic and epigenetic processes that influence p14ARF expression in breast cancer. *Oncogene* 33:4586–4590.

Soengas, M. S., Capodieci, P., Polsky, D., Mora, J., Esteller, M., Opitz-Araya, X., McCombie, R., Herman, J. G., Gerald, W. L., Lazebnik, Y. A., *et al.* (2001). Inactivation of the apoptosis effector Apaf-1 in malignant melanoma. [see comments]. *Nature* 409:207–211.

Sommer, S. S., Cunningham, J., McGovern, R. M., Saitoh, S., Schroeder, J. J., Wold, L. E., and Kovach, J. S. (1992). Pattern of p53 gene mutations in breast cancers of women of the midwestern United States. *J Natl Cancer Inst* 84:246–252.

Speir, E., Modali, R., Huang, E. S., Leon, M. B., Shawl, F., Finkel, T., and Epstein, S. E. (1994). Potential role of human cytomegalovirus and p53 interaction in coronary restenosis. [see comments]. *Science* 265:391–394.

Sproston, A. R., Boyle, J. M., Heighway, J., Birch, J. M., and Scott, D. (1996). Fibroblasts from Li-Fraumeni patients are resistant to low dose-rate irradiation. *Int J Radiat Biol* 70:145–150.

Srivastava, S., Zou, Z. Q., Pirollo, K., Blattner, W., and Chang, E. H. (1990). Germ-line transmission of a mutated p53 gene in a cancer-prone family with Li-Fraumeni syndrome. [see comments]. *Nature* 348:747–749.

Stenmark-Askmalm M, Stal O, Sullivan S, Ferraud L, Sun XF, Carstensen J, and B., N. (1994). Cellular accumulation of p53 protein: an independent prognostic factor in stage II breast cancer. *Eur J Cancer* 30A(2):175–180.

Stiewe, T., Zimmermann, S., Frilling, A., Esche, H., and Putzer, B. M. (2002). Transactivation-deficient DeltaTA-p73 acts as an oncogene. *Cancer Res* 62:3598–3602.

Sun, X. F., Carstensen, J. M., Zhang, H., Arbman, G., and Nordenskjold, B. (1996). Prognostic significance of p53 nuclear and cytoplasmic overexpression in right and left colorectal adenocarcinomas. *Eur J Cancer* 32A:1963–1967.

Sun, X. F., Carstensen, J. M., Zhang, H., Stal, O., Wingren, S., Hatschek, T., and Nordenskjold, B. (1992). Prognostic significance of cytoplasmic p53 oncoprotein in colorectal adenocarcinoma. *Lancet* 340:1369–1373.

Symonds, H., Krall, L., Remington, L., Saenz-Robles, M., Lowe, S., Jacks, T., and Van Dyke, T. (1994). p53-dependent apoptosis suppresses tumor growth and progression in vivo. *Cell* 78:703–711.

Takada, N., Isogai, E., Kawamoto, T., Nakanishi, H., Todo, S., and Nakagawara, A. (2001). Retinoic acid-induced apoptosis of the CHP134 neuroblastoma cell line is associated with nuclear accumulation of p53 and is rescued by the GDNF/Ret signal. *Med Pediatr Oncol* 36:122–126.

Takekawa, M., Adachi, M., Nakahata, A., Nakayama, I., Itoh, F., Tsukuda, H., Taya, Y., and Imai, K. (2000). p53-inducible wip1 phosphatase mediates a negative feedback regulation of p38 MAPK-p53 signaling in response to UV radiation. *EMBO Journal* 19:6517–6526.

Tanaka, K., Zou, J. P., Takeda, K., Ferrans, V. J., Sandford, G. R., Johnson, T. M., Finkel, T., and Epstein, S. E. (1999). Effects of human cytomegalovirus immediate-early proteins on p53-mediated apoptosis in coronary artery smooth muscle cells. *Circulation* 99:1656–1659.

Tannapfel, A., Busse, C., Weinans, L., Benicke, M., Katalinic, A., Geissler, F., Hauss, J., and Wittekind, C. (2001). INK4a-ARF alterations and p53 mutations in hepatocellular carcinomas. *Oncogene* 20(48):7104–7109.

Tavor, S., Takeuchi, S., Tsukasaki, K., Miller, C. W., Hofmann, W. K., Ikezoe, T., Said, J. W., and Koeffler, H. P. (2001). Analysis of the CHK2 gene in lymphoid malignancies. *Leuk Lymphoma* 42:517–520.

Toguchida, J., Yamaguchi, T., Dayton, S. H., Beauchamp, R. L., Herrera, G. E., Ishizaki, K., Yamamuro, T., Meyers, P. A., Little, J. B., and Sasaki, M. S. (1992). Prevalence and spectrum of germline mutations of the p53 gene among patients with sarcoma. [see comments.]. *New Engl J Med* 326:1301–1308.

Tommasino, M., Accardi, R., Caldeira, S., Dong, W., Malanchi, I., Smet, A., Zehbe, I. (2003). The role of TP53 in Cervical carcinogenesis. *Hum Mutat* 3:307–312.

Ueda, H., Ullrich, S. J., Gangemi, J. D., Kappel, C. A., Ngo, L., Feitelson, M. A., and Jay, G. (1995). Functional inactivation but not structural mutation of p53 causes liver cancer. *Nat Genet* 9:41–47.

Vahteristo, P., Bartkova, J., Erola, H., Syrjakoski, K., Ojala, S., Kilpivaara, O., Tamminen, A., Kononen, J., Aittomaki K., Heikkila, P., Holli, K., Blomqvist, C., Bartek, J., Kallioniemi, O.P., Nevanlinna, H. (2002). A CHEK2 genetic variant contributing to a substantial fraction of familial breast cancer. *Am J Hum Genet* 71(2):432–438.

Vahteristo, P., Tamminen, A., Karvinen, P., Eerola, H., Eklund, C., Aaltonen, L. A., Blomqvist, C., Aittomaki, K., and Nevanlinna, H. (2001). p53, CHK2, and CHK1 genes in Finnish families with Li-Fraumeni syndrome: further evidence of CHK2 in inherited cancer predisposition. *Cancer Res* 61:5718–5722.

Wagner, A. J., Kokontis, J. M., and Hay, N. (1994). Myc-mediated apoptosis requires wild-type p53 in a manner independent of cell cycle arrest and the ability of p53 to induce p21waf1/cip1. *Genes Dev* 8, 2817–2830.

Wang, J., Belcher, J. D., Marker, P. H., Wilcken, D. E., Vercellotti, G. M., and Wang, X. L. (2001). Cytomegalovirus inhibits p53 nuclear localization signal function. *J Mol Med* 78, 642–647.

Wang, X. W., Forrester, K., Yeh, H., Feitelson, M. A., Gu, J. R., and Harris, C. C. (1994). Hepatitis B virus X protein inhibits p53 sequence-specific DNA binding, transcriptional activity, and association with transcription factor ERCC3. *Proc Nat Acad Sci USA* 91:2230–2234.

Weber, J. D., Taylor, L. J., Roussel, M. F., Sherr, C. J., and Bar-Sagi, D. (1999). Nucleolar Arf sequesters Mdm2 and activates p53. *Nat Cell Biol* 1:20–26.

Weinberg, R. A. (1995). The retinoblastoma protein and cell cycle control. *Cell* 81:323–330.

Williams, K. J., Boyle, J. M., Birch, J. M., Norton, J. D., and Scott, D. (1997). Cell cycle arrest defect in Li-Fraumeni syndrome: a mechanism of cancer predisposition? *Oncogene* 14:277–282.

Williams, K. J., Heighway, J., Birch, J. M., Norton, J. D., and Scott, D. (1996). No defect in G1/S cell cycle arrest in irradiated Li-Fraumeni lymphoblastoid cell lines. *Br J Cancer* 74:698–703.

Wolf, D., and Rotter, V. (1985). Major deletions in the gene encoding the p53 tumor antigen cause lack of p53 expression in HL-60 cells. *Proc Natl Acad Sci* 82:790–794.

Wu, L., and Levine, A. J. (1997). Differential regulation of the p21/WAF-1 and mdm2 genes after high-dose UV irradiation: p53-dependent and p53-independent regulation of the mdm2 gene. *Mol Med* 3:441–451.

Wu, X., Bayle, J. H., Olson, D., and Levine, A. J. (1993). The p53-mdm-2 autoregulatory feedback loop. *Genes Dev* 7:1126–1132.

Wu, Y., Mehew, J. W., Heckman, C. A., Arcinas, M., and Boxer, L. M. (2001). Negative regulation of bcl-2 expression by p53 in hematopoietic cells. *Oncogene* 20:240–251.

Yew, P. R., and Berk, A. J. (1992). Inhibition of p53 transactivation required for transformation by adenovirus early 1B protein. *Nature* 357:82–85.

Yin, Y., Tainsky, M. A., Bischoff, F. Z., Strong, L. C., and Wahl, G. M. (1992). Wild-type p53 restores cell cycle control and inhibits gene amplification in cells with mutant p53 alleles. *Cell* 70:937–948.

Yoo, L. I., Chung, D. C., and Yuan, J. (2002). LKB1–a master tumour suppressor of the small intestine and beyond. *Nat Rev Cancer* 2:529–535.

Yu, J., Zhang, L., Hwang, P. M., Kinzler, K. W., and Vogelstein, B. (2001). PUMA induces the rapid apoptosis of colorectal cancer cells. *Mol Cell* 7:673–682.

Zaika, A., Marchenko, N., and Moll, U. M. (1999). Cytoplasmically "sequestered" wild type p53 protein is resistant to Mdm2-mediated degradation.. *J Biol Chem* 274:27474–27480 [Erratum in: *J Biol Chem* (2000) April 14; 275(15):11538.].

Zaika, A., Slade, N., Erster, S.H., Sansome, C., Joseph, T.W., Pearl, M., Chalas, E., Moll, U.M. (2002) DNp73, a Dominant negative inhibitor of wild type p53 and TAp73, is upregulated in human tumours. *J Exp Med* 6:765–80.

Zhao, R., Gish, K., Murphy, M., Yin, Y., Notterman, D., Hoffman, W. H., Tom, E., Mack, D. H., and Levine, A. J. (2000). Analysis of p53-regulated gene expression patterns using oligonucleotide arrays. *Genes Dev* 14:981–993.

Ziegler, A., Leffell, D. J., Kunala, S., Sharma, H. W., Gailani, M., Simon, J. A., Halperin, A. J., Baden, H. P., Shapiro, P. E., and Bale, A. E. (1993). Mutation hotspots due to sunlight in the p53 gene of nonmelanoma skin cancers. *Proc Natl Acad Sci USA* 90:4216–4220.

7

MDM2 and MDMX Regulators of p53 Activity

Jamil Momand, Paul Joseph Aspuria, and Saori Furuta

SUMMARY

MDM2 possesses three activities that, together, effectively inhibit the p53 tumor suppressor. First, it binds to p53 and sterically blocks p53 interaction with TATA box protein accessory factors thereby shutting down its transcriptional transactivation function. Second, MDM2 shuttles p53 from its site of action within the nucleus into the cytoplasm. Third, MDM2 is an E3 ligase that transfers ubiquitin onto lysine residues of p53. Ubiquitinated p53 is rapidly degraded by the 26S proteosome. Because the MDM2 oncoprotein mediates three progressive stages of inhibition, it is the principal regulator of p53 activity. The *MDM2* gene is located on chromosome 12q14.3-q15 and is amplified in several types of neoplasms, most of which are of mesenchymal tissue origin. MDM2 binding to p53 can be inhibited by phosphorylation of either MDM2 or p53. The kinases responsible for this phosphorylation are activated by cell stressors in general (hypoxia, nitric oxide, hydrogen peroxide) and DNA damaging agents in particular (ionizing radiation, UV-light). MDM2 can be inhibited by RAS or MYC oncoproteins. RAS and MYC activate the tumor suppressor protein p19[Arf] which sequesters MDM2 into the nucleolus and, in doing so, allows p53 levels to rise. The *MDM2* gene is activated by p53, which means that, in effect, p53 inhibits itself through MDM2. The cell requires a fine balance of MDM2 and p53 to maintain cell growth

J. MOMAND, P. J. ASPURIA, AND S. FURUTA • Department of Chemistry and Biochemistry, California State University at Los Angeles, Los Angeles, CA 90032, USA

The p53 Tumor Suppressor Pathway and Cancer, edited by Zambetti.
Springer Science+Business Media, New York, 2005.

and a rapid response to stressors. MDMX is a paralog of MDM2 that has retained the ability to inhibit p53 binding to TATA box protein accessory factors. MDMX does not possess other p53 inhibitory activities. There has been recent progress in the development of small molecules that block MDM2 from binding to p53. Although these molecules are in the early stages of development, it is hoped that they will contribute to the war on cancer. This chapter summarizes the key studies that have increased our understanding of the interplay between p53, MDM2, and MDMX.

7.1. INTRODUCTION

Every sophisticated engineering system allows for regulation via feedback. Feedback mechanisms ensure that systems do not spiral out of control. In the area of cell growth one can consider tumor suppression as a system in which p53 is the central operator and that MDM2 and MDMX are essential proteins in the feedback mechanism that controls p53. The *mdm2* gene was initially characterized as an amplified oncogene by Dr. Donna George at the University of Pennsylvania (Cahilly-Snyder et al., 1987; Fakharzadeh et al., 1991). Its protein product, MDM2 (sometimes named HDM2 when referring to the human protein), downregulates p53 in distinct stages. First, p53 forms a complex with MDM2 and, through this interaction, is prevented from increasing the transcription of its effector genes. Second, MDM2 shuttles p53 away from the genome out to the cytoplasm. Finally, MDM2 marks p53 for degradation. MDMX (also known as MDM4) was later discovered because its sequence is similar to MDM2 (Shvarts et al., 1996). MDMX does not promote p53 degradation nor transport it from the nucleus to the cytoplasm. However, like MDM2, MDMX inhibits p53-mediated transactivation by masking the transactivation domain. Both oncoproteins have been shown to be abnormally upregulated in human tumors. In this chapter, we will discuss the molecules that regulate the ability of MDM2 and MDMX to control p53. We will explore avenues that may lead to anticancer therapies based on MDM2/p53 interactions and highlight areas that will likely be investigated in the near future.

7.2. MDM2 AND MDMX STRUCTURE/FUNCTION RELATIONSHIPS

From its mRNA sequence, MDM2 is predicted to be composed of 491 amino acid residues and its gene is located on human chromosome 12q13-14 (Oliner et al., 1992). Note that we will use the term MDM2 to refer to the human ortholog unless otherwise noted. MDM2 is highly phosphorylated (Momand et al., 1992) and is quickly turned over with a half-life of approximately 20 minutes (Hinds et al., 1990; Olson et al., 1993). A common method used to identify important functional regions of proteins is sequence alignment and previous studies that compared MDM2 and MDMX amino acid sequences pinpointed three regions of high similarity (Momand et al., 2000; Piette et al., 1997). Figure 7.1A shows an alignment of 8 MDM2 and 2 MDMX sequences using the Multialign algorithm (Barton and Sternberg, 1987) where these three similar regions are denoted with the abbreviations CR1, CR2, and CR3 ("CR" is an abbreviation for conserved region). Figure 7.1A also shows the sites where MDM2 is posttranslationally modified and the regions that control its subcellular localization. Analysis of CR1, CR2, and CR3 predicts that MDM2 binds p53, partakes in regulating the transport of molecules in and out of the nucleus, and transfers ubiquitin onto protein substrates.

The functions of the three conserved regions accurately describe the biochemical activities of MDM2. CR1 (amino acid residues 27–94) is the portion of MDM2 that binds to p53. Most of the binding takes place through van der Waals forces (Kussie et al., 1996). CR2 (residues 301–329) contains a sequence that is similar to Ran-1, a protein that controls nuclear export via a zinc-binding sequence (Yaseen and Blobel, 1999). MDM2 shuttles between the nucleus and the cytoplasm and controls the export of p53 from nucleus, but CR2 has not been identified as a necessary domain for this process. CR3 (residues 437–483) binds zinc (Lai et al., 1998) and contains a critical cysteine residue at position 464 that is necessary for transferring ubiquitin onto p53 (Honda et al., 1997). Ubiquitination of several lysine residues near the p53 carboxyl terminus is required for its degradation by the 26S proteosome (Kubbutat et al., 1999; Lai et al., 2001). In some cases, Cys 464 is also necessary to transport p53 from the nucleus to the cytoplasm (Boyd et al., 2000; Geyer et al., 2000). Interestingly, MDMX, which also contains a conserved cysteine residue within CR3, fails to transfer ubiquitin onto p53 making it likely that regions outside of CR3 are necessary for MDM2 ubiquitin ligase activity. Figure 7.1B shows a schematic diagram of the linear sequence of MDM2 and some of the key regions that control its activity. Structural analysis of MDM2 and MDMX should help to further elucidate their functions.

At the moment, we know the structure of the interface between CR1 of MDM2 and p53. X-ray diffraction studies performed by Dr. Nikola Pavletich's group at Columbia University followed by nuclear magnetic resonance spectroscopy studies showed that CR1 creates a hydrophobic cleft for p53 (Kussie et al., 1996; Stoll et al., 2001). Analysis of the interface region has provided insight into the design of molecules that bind MDM2 and release p53 for the purpose of preventing the spread of cancer. Figure 7.2A shows a portion of MDM2 bound to p53 depicted as filled van der Waals surfaces. Electron-rich regions are red, electron-poor regions are blue, and neutral regions are white. The p53 peptide is presented as a red α-helix with five side chains in the shapes of balls and sticks. The peptide bears the sequence Glu[17]-Thr-Phe-Ser-Asp-Leu-Trp-Lys-Leu-Leu-Pro-Glu-Asn[29], but only side chains of Phe19, Leu22, Trp23, Leu25, and Leu26 are shown for clarity. Phe19, Trp23, and Leu26 side chains lie on one face of the alpha-helix and direct their hydrophobic atoms into the pocket of MDM2. On the opposite face lies two other hydrophobic side chains, Leu22 and Leu25. It appears that at least one of these, Leu25, may interact with His73 of MDM2. Thus, MDM2 interacts with residues on opposite faces of the p53 alpha-helix that constitutes the transactivation domain.

A

Block positions 1–50–100:

```
                    1         50                                                              100
Human    MDM2  MCNTNNMSVPT DGAVTTSQIP ASEQETLVRP KPLLLKLLKS VGAQKDTVTM KEVLFYLGQY IMTKRLYDEK QQHIVYCSND LLGDLFGVPS FSVKEHRKIY
Horse    MDM2  MCNTNNMSVST DGAVSTSQIP ASEQETLVRP KPLLLKLLKS VGAQKDTVTM KEVIFYLGQY IMTKRLYDEK QQHIVYCSND LLGDLFGVPS FSVKEHRKIY
Dog      MDM2  MCNTNNMSVST GGAVSTSQIP ASEQETLVRP KPLLLKLLKS VGAQKDTYTM KEVIFYLGQY IMTKRLYDEK QQHIVYCSND LLGDLFGVPS FSVKEHRKIY
Mouse    MDM2  MCNTNNMSVST EGAASTSQIP ASEQETLVRP KPFLFKLLKS VGAQNDTYTM KEVIFYLGQY IMTKRLYDEK QQHIVYCSND LLGDLFGVPS FSVKEHRKIH
Hamster  MDM2  .......ST   DGAEGTSQIP ASEQETLVRP KPLFLKLLKS KEIILS.WQY IMTKRLYDEK QQHIVHCSND PLGELFGVQE FSVKEPRRLY
Xenopus  MDM2  ...MNLJ.ST  TNCLENNHIS TSDQEKLVQP TPLLLSLLKS AGAQKETFFM KEVIVHLGQY IMAKQLYDEK QQHIVHCSND LLGDLFGVQE FSVKEPRRLF
Zebrafish MDM2 ...MAT      ESCLSSSQIS KVDNEKLVRP            AGADKDTFFM KEVMFYLGKY IMSKELYDKQ EQHIVHCAND LLGDLFGVTS FSVKEHRRIY
Chicken  MDM2  MCNTEMTSLT  DG....SPVS ASEQEALVKP KPLLLKLLKL AGAQRDTFFM KEVIFYLGQY IMSKELYDKQ EQHIVHCAND LLGDLFGVTS FSVKEHRRIY
Human    MDMX  MTSFSTSAQC  STSDSACRIS .PQQINQVRP KLPLLKILHA AGAQBEMFTV KEVMHYLGQY IMVKQLYDQQ EQHMYYCGGD LLGEILGRQS FSVKNPSPLY
Mouse    MDMX  MTSHSTSAQC  SASDSACRIS .SEQISQVRP KLQLLKILHA AGAQGEVFTM KEVMHYLGQY IMVKQLYDQQ EQHMVYCGSD LLGDLLGCQS FSVKDPSPLY
Consensus      M---------  ------V-P- ----L----- -GA---T-   KE-----G-Y IM-K-LYD-- -QH-V-C--D -LG--G----  -FSVK-
```

CR I

Block positions 101–148–197:

```
                   101        148                                                            197
Human    MDM2  TMIYRNLVVV .NQQESSDS  GTSVSENRCH LEGGSDQKDL VQELQEEKPS SSHLVSRPST EE.NSDELSG ERQRKRHKSD SISLSFDESL
Horse    MDM2  TMIYRNLVVV .SQQEPSDS  GTSVSENRCH LEGGSNQKDL VQELQEEKPS SSDMVSRPST EE.NSDELPG ERQRKRHKSD NISLSFDESL
Dog      MDM2  TMIYRNLVVV .NQHEPSDS  GTSVSENSCH REGGSDQKDL VQELQBEKPS SSDLISRPST EE.HADDLPG           SISLSFDESL
Mouse    MDM2  AMIYRNLVAV .SQQ...DS  GTSLSESRRQ PEGGSDLKDP LQAPPEEKPS SSDLVSRPST EE.NTDELPG ERHRKRRRS. ..LSFDPSL
Hamster  MDM2  IMIYRNLVVV .SQQETLQS  GTSVSESSRQ            VQEPQOEK.S SDSVSRPST  EE.NADELPG DRQRKRHRS. ..LSFDESL
Xenopus  MDM2  AMISRNLVSA .NVKESSED  ..IFGNVCCF PDKQSQKEK  LQELPD.KLI APASDSKPCN LSQRKSSNET EEISSVDHPA SFSLIFDESL
Zebrafish MDM2 ALINRNLVTV .KNPE.SQS  TFSEPRSQSE PDRGPGDTD. .SDSRSSTSQ QQRRRRRSD  PESSSAEDES RERRKRHKSD SFSLTFDSL
Chicken  MDM2  SMISRNLIAI .NQQDSTLA  VPPEMMPNFG LKKKMPKRKS MQELEEKQ.  TSSNATSQPT TSRRRTHSES EENSSDDLHS .DRRRKRHKSD SISLTFDESL
Human    MDMX  DMLRKNLVTL ATATTDAAQT LALAQDHSMD IPSQDOLKQS AEBSSTSRKR TTEDDIPTLP TSEHKCIHSR EDEDLIENLA QDETSRLDLG .FEEWDVAGL
Mouse    MDMX  DMLRKNLVTS ASNNTDAAQT LALAQDHTMD FPSQDRLKHG ATEYSNPRKR TEEEDTHTLP TSRHKCRDSR ADEDLIEHLS QDETSRLDLD .FEEWDVAGL
Consensus      ------NL--                                                                       -----R----          --------L
```

NLS

Block positions 198–240–287:

```
                   198        240                                                            287
Human    MDM2  ALCVIREICC ....ERSSS  SESTGTPSNP DLDAGVSEHS GD..WLDQDS VSDQFSVEFE VESLDSEDYS LSEBGQELSD ED...DEVYQ VTVYQAGESD
Horse    MDM2  ALCVIREICC ....ERSSS  SESTGTPSNP DLDAGVSEHS GD..WLDQDS VSDQFSVEFE VESLDSEDYS LSEBGQELSD ED...DEVYR VTVYQAGESD
Dog      MDM2  ALCVIREICC ....ERSSS  SESTGTPSNP DLDDGVSEHS GD..WLDQDS VSDQFSVEFE VESLDSEDYS LSEBGQELSD ED...DEVYR VTVYQAGESD
Mouse    MDM2  GLCELRFMCS GGSSSSSSSS SESTETPSNQ DLDDGVSEHS GD..CLDQDS VSDQFSVEFE VESLDSEDYS LSDEGHELSD ED...DEVYR VTVYQTGESD
Hamster  MDM2  ALCVLREICC ....ERSSS  SESTDTPSNQ DLDDGVSEHS GD..WLDQDS VSDQFSVEFE VESLDSEDYS LSEGGQELSD ED...DEVYR VTVYQSGESD
Xenopus  MDM2  SWWVISGLRC ....DRNS   SESTDSSSNS DPE...RHST ND..NSEHDS .DQFSVEFE  VESVCSDDYS PSGDEHGVSE EEEINDEVYQ VTIYETEESE
Zebrafish MDM2 SWCVIGGLH. ....RERGN  SESSDANSNS DVGISRSEGS EE..SEDSDS DSDNFSVEFE VESINSDAYS ENDVDSVFGE ...NEIYE   VTIFAEDE.
Chicken  MDM2  SWCVVSGLCR ....DRSNS  SDSTDSVSIP DLDASSLSEN SD..WFDHGS VSDQFSVEFE VESIYSEDYS HNEBGQELTD ED...DEVYQ LITIYQDEDSD
Human    MDMX  PWWFLGNLRS ....NTYPRS NGSTDLQTNQ DVGTAIVSDT TDDLWFLANE VSEQLGVGIK VEAADTEQTS EEVGKVSDKV V...IEVGK  NDDLEDSKSL
Mouse    MDMX  PWWFLGNLRN ....NCIPKS NGSTDLQTNQ DIGTAIVSDT TDDLWFLNET VSEQLGVGIK VEAANSEQTS E.VGKTSNKK T...VEVGK  DDDLEDSRSL
Consensus      ------NL--                                  ----D----  ----------- -----V---- --------E- ---------- ------
```

NES

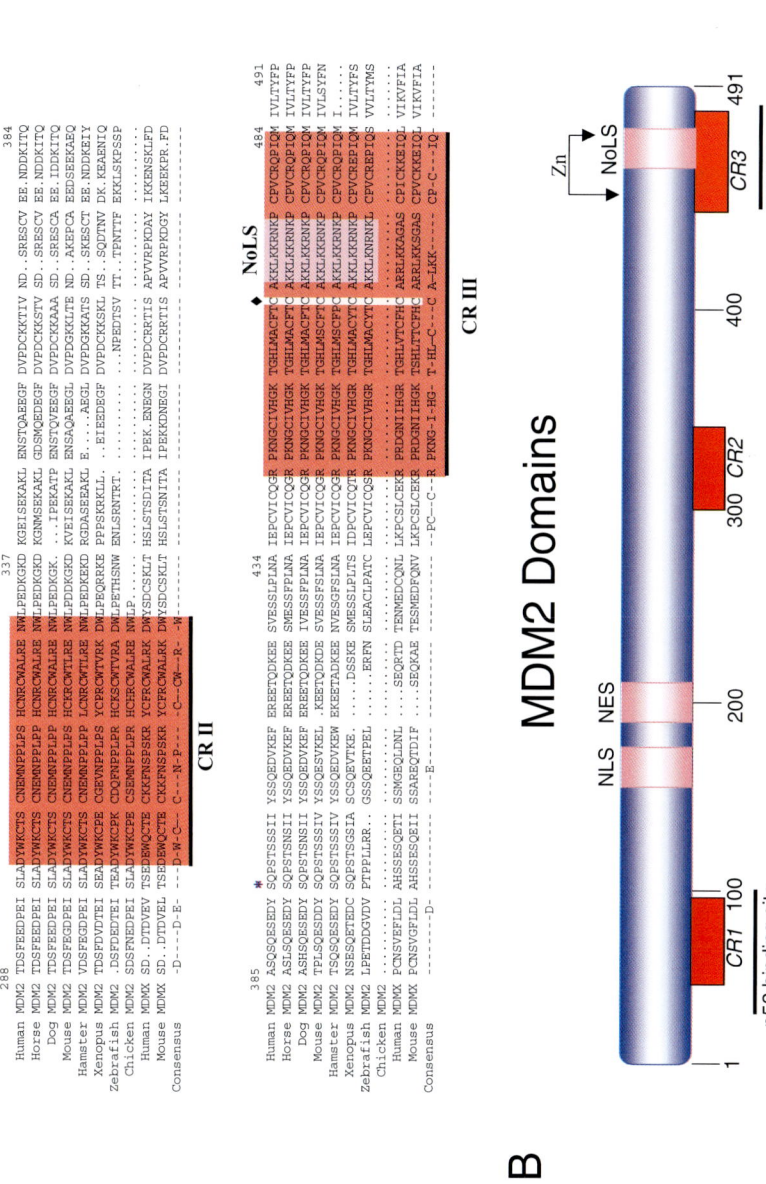

Figure 7.1. MDM2 and MDMX protein domains. **A:** Sequence alignment of MDM2 and MDMX protein sequences. CR1, conserved region 1; CR2, conserved region 2; CR3, conserved region 3; NLS, nuclear localization sequence; NES, nuclear export sequence; NoLS, nucleolus localization sequence; black diamond, cysteine required for ubiquitin transfer to p53; blue asterisk, phosphorylation site. Sequences were aligned with the MultiAlign program. **B:** Schematic diagram of MDM2 protein domains (see text for details). Known Zn binding sites and the RING finger domain are indicated.

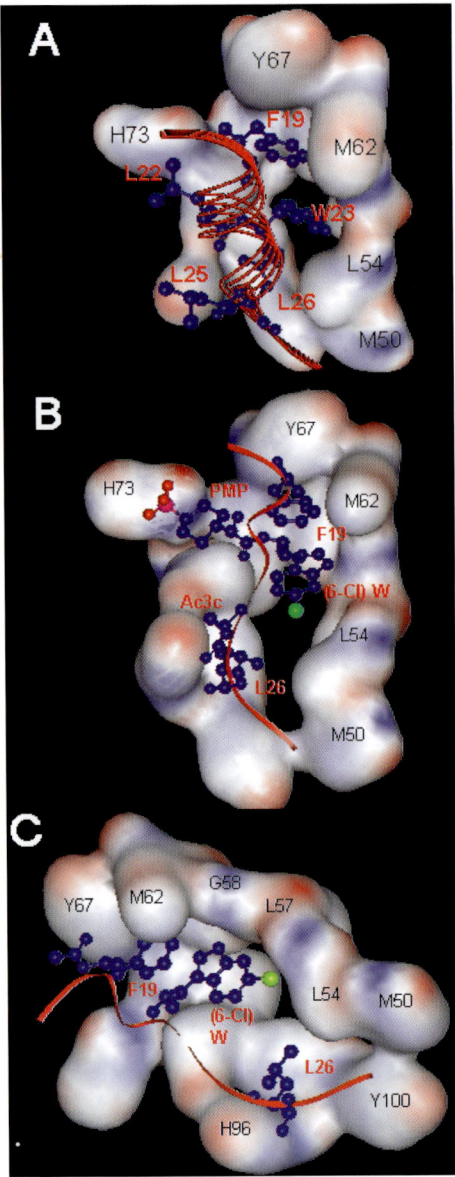

Figure 7.2. Peptide inhibitors bound to MDM2 interactions. **A:** p53 peptide bound to MDM2. The red ribbon represents the backbone peptide alpha-helix of p53. The three p53 residues (Phe 19, Trp 23, Leu 26) on one face of the alpha-helix form van der Waals interactions with the MDM2. Two other p53 residues (Leu 22 and Leu 25) on the opposite face of the alpha-helix also appear to interact with MDM2. MDM2 residues are depicted as filled van der Waals surfaces, with red indicating high electron density and blue indicating low electron density. The MDM2 residues involved in binding p53 are labeled with black font. **B:** Model of CGP 84700 bound to MDM2. The CGP 84700 peptide model was created by substituting the side chains of CGP 84700 for the p53 peptide side chains. **C:** Different orientation of CGP 84700 bound to MDM2. All models were generated using WebLab ViewerPro©. Coordinates for MDM2 and p53 residues were obtained from data deposited in the Protein Data Bank (Kussie et al., 1996). Peptide atom color code: blue, carbon; pink, phosphorous; red, oxygen; green, chlorine.

Aside from knowing the shape of p53 when it is bound to MDM2 it is also important to understand how its conformation changes upon binding. A synthetic p53 peptide containing residues 14–28 has the propensity to form a two-β-turn structure stabilized by Phe19, Leu22, Trp23, and Leu25 (Botuyan et al., 1997). The conformation of this two β-turn structure is similar to the structure of the p53 peptide when it is bound to MDM2. This scenario opens the possibility that this portion of p53 is "primed" for MDM2 binding. The structure of the MDM2-p53 complex has paved the way for the design of potent inhibitors of binding.

7.2.1. Artificial Modulation of MDM2-p53 Complex Formation

Dr. Bert Vogelstein and his colleagues at Johns Hopkins University presented the first evidence that MDM2 inactivates p53 in human tumors (Oliner et al., 1992). They demonstrated that MDM2 is overexpressed in a large percentage of sarcomas. This result gave the impetus for scientists to design inhibitors of MDM2 as a possible therapeutic to reestablish p53 tumor suppressor activity in these tumors. In one general approach, small molecules have been synthesized (or isolated) that can release p53 from MDM2. In a second approach, oligonucleotides have been designed to bind and destroy MDM2 mRNA. Table 7.1 shows patents and patent-pending applications related to MDM2. The majority of these patents cover approaches to inhibit MDM2 activity, and are discussed below.

To create small molecules that prevent MDM2 from binding p53, it was necessary to accurately map the regions of interaction. To map these sites, truncated MDM2 transcripts were translated in the presence of p53 and complexes were captured by coimmunoprecipitation (Chen et al., 1993). Using this method, it was discovered that the two proteins interact through regions near their amino termini. Confirmation that these two domains interact came from yeast two-hybrid assays (Oliner et al., 1993). Next, peptide libraries were used to narrow the binding region within p53 to residues 18–23 ([18]TFSDLW[23]) (Picksley et al., 1994). Amino acid replacements at Leu22 and Trp23 effectively prevented its ability to bind MDM2 (Lin et al., 1994). This "double mutant" p53 was also incapable of activating genes driven by a p53-responsive promoter, indicating that the MDM2-binding domain and the domain responsible for transactivation overlap.

7.2.2. High Affinity Molecules that Dissociate MDM2 and p53

With the sites required for interaction accurately mapped, reagents were developed to inhibit MDM2-p53 complex formation. Five of these inhibitors and their respective IC_{50} values are presented in Table 7.2. IC_{50} values represent the amount of inhibitor necessary to reduce the binding of p53 to MDM2 by 50%. A compound developed by Novartis, named CGP 84700, has the exceptionally low IC_{50} of 5 nM, and is sufficiently hydrophobic to penetrate cultured tumor cells and induce accumulation of p53 (Chene et al., 2000). Treatment of cancer cells with CGP 84700 leads to p53-mediated transactivation and promotes cell suicide (apoptosis), indicating its potential as a therapeutic. Two views of a model of CGP 84700 bound to MDM2

Table 7.1. List of patents and patent-pending applications related to MDM2.

Assignee or inventor	Year granted or applied	Patent or application number	Title
University of Johns Hopkins	1995	US5,411,860; WO9320238A3	Amplification of human MDM2 gene in human tumors
University of Johns Hopkins	1997	US5,618,921	Antibodies for detection of human MDM2 protein
University of Dundee	1998	US5,770,377	Interruption of binding of MDM2 and p53 protein and therapeutic application thereof
Ludwig Institute for Cancer Research	1998 2001	WO9813064A1; US6,204,253	Factors which interact with oncoproteins
Cancer research campaign technology limited	1998	WO9801467A2;	Inhibitors of the interaction between p53 and MDM2
University of Dundee	1998	WO9847525A1	Materials and methods relating to inhibiting the interaction of p53 and MDM2
Ruiwen Zhang	1999	WO9910486A3	MDM2-specific antisense oligonucleotides
Zeneca Limited	2000	WO0015657	Piperizine-4-phenyl derivatives as inhibitors of the interaction between MDM2 and p53
David Lane et al.	2001	App. No. 20010018511 (US)	Inhibitors of the interaction between p53 and MDM2
Yijia Bao et al.	2001	App. No. 20010018183	Simultaneous measurement of gene expression and genomic abnormalities using nucleic acid microarrays
Loren J. Miraglia et al.	2001	App. No. 20010016575	Antisense modulation of human MDM2 expression

Table 7.2. Small molecule inhibitors of p53-MDM2 interaction.

Common Name	Molecular structure	IC$_{50}$ μM	Reference
p53 peptide	Ac-Thr[18]-Phe-Ser-Asp-Leu-Trp[26]-NH$_2$	286–1000	Picksley et al. (1994)
CGP 84700	Ac-Phe[19]-Met-Aib-Pmp-(6-Cl)Trp-Glu-Ac$_3$c-Leu[26]-NH$_2$[a]	0.005	Chene et al. (2000)
Chalcone	1,3-diphenyl-2-propen-1-one, compound B-1	117	Stoll et al. (2001)
Chlorofusin (B-1)	C$_{66}$H$_{99}$O$_{19}$N$_{12}$Cl	4.6	Duncan et al. (2001)
Nutlin-3	C$_{34}$H$_{29}$O$_4$N$_4$Cl$_2$	0.09	Vassilev et al. (2004)

[a]Where Aib is α-aminoisobutyric acid, Pmp is phosphonophenylalanine, (6-Cl)Trp is 6-chorotryptophan, Ac$_3$c is 1-amino-cyclopropane carboxylic acid.

are shown in Figures 7.2B and C. The model was created by using the structure of the p53 peptide in Figure 7.2A as a scaffold to build the compound in an alpha helical form with minimal alteration of the MDM2 cleft. Chalcones, a class of molecules derived from 1,3-diphenyl-2-propen-1-one, also show promise as inhibitors of MDM2-p53 complex formation (Stoll et al., 2001). These molecules were originally isolated from plants, and have a wide variety of anticancer effects. Of the chalcones tested, the most effective at inhibiting MDM2-p53 complex formation is derivative B-1, which binds to the region of MDM2 that normally binds Trp 23 of p53. Another molecule that releases MDM2 from p53 is chlorofusin, a circular peptide isolated from the fungus *Fusarum* (Duncan et al., 2001). This nine amino acid peptide was isolated by screening over 53,000 compounds that could potentially inhibit MDM2-p53 interaction. It will be exciting to see if any of these inhibitors can be further developed as an anticancer therapeutic with minimal toxic side effects. Scientists at Hoffman-La Roche and Pharma developed a series of *cis*-imidazoline analogs named Nutlin-1, -2, and -3 (Vassilev et al., 2004). These small organic compounds dissociated recombinant p53 from MDM2 with the median IC$_{50}$ in the 100 to 300 nM range. The compounds inhibit cell cycle progression and promote apoptosis in a p53-dependent manner. Initial studies of Nutlin-3 on a human osteosarcoma cell-line xenograft in nude mice demonstrated that it was as effective as doxorubicin in reducing tumor volume although the effective dose of Nutlin-3 was 20-fold higher.

7.2.3. Antisense Therapy

A second potential therapeutic route to inhibit MDM2 function is to use antisense oligodeoxyribonucleotides (ODNs). In this approach, the ODN binds the transcript and forms a localized duplex, which then becomes a target for degradation by the endogenous nuclease RNAse H. Two research groups have reported that ODNs reduce the levels of MDM2 transcript and MDM2 in cultured cells (Chen et al., 1998; Teoh et al., 1997). Using multiple myeloma cells, Teoh et al (1997) showed that the ODN 5'-dGACATGTTGGTATTGCACAT-3' (complements nucleotides 1–20 of the coding sequence within the transcript) reduces MDM2 expression, increases p53 protein levels, and inhibits DNA synthesis. A second group created the ODN 5'-dGATCACTCCCACCTTCAAGG-3', which is complementary to nucleotides 714–733 of MDM2 transcript (Chen et al., 1998). This ODN contains a phosphoramidite backbone that is designed to decrease its degradation by nucleases, and appears to be very effective in destroying MDM2 transcripts.

Strategies to treat cancers with MDM2 targeting agents are in the initial stages of exploration. To proceed with their development, one should be cognizant of their potential uses and limitations. Generally, such agents are proposed to release p53 and thus cause cancers cells to undergo apoptosis. From the perspective of the p53 autoregulatory loop, there are four molecular scenarios that could benefit from anti-MDM2 therapy: (1) The *mdm2* gene is amplified and MDM2 is overexpressed; (2) MDM2 mRNA is overexpressed in the absence of gene amplification; (3) The *p53* gene is wild type but MDM2 is not overexpressed; and (4) p53 is not properly phosphorylated for activation. The third and fourth scenarios may not be obvious candidates for anti-MDM2 therapy. In the case of the third scenario, one must consider a cancer cell where the signaling pathway that *inactivates* MDM2 is defective. An example of this scenario is the ARF signaling pathway (see ARF-MDM2 complex formation below). The ARF protein normally activates p53 by inhibiting MDM2. ARF is inactivated in a large number of cancers allowing MDM2 to constitutively inhibit p53. In such cancers anti-MDM2 therapy may be able to activate p53. The fourth scenario that may benefit from anti-MDM2 therapy is one where p53 is not properly modified by kinases and acetylases—modifications known to increase p53 activity (Giaccia and Kastan, 1998). Overexpression of unmodified p53 by anti-MDM2 therapy may suppress tumors because the high level will be sufficient to activate target genes. While most of the benefit of anti-MDM2 therapy to cancers will likely be derived from p53 activation, such therapy may also benefit cancers with low Retinoblastoma tumor suppressor activity. MDM2 has been shown to inhibit RB's ability to prevent tumor cell growth (Xiao et al., 1995). A major challenge currently facing scientists is to develop an efficient anti-MDM2 therapeutic delivery vehicle that spares potential toxic side effects.

7.2.4. SUMO-1 Modification

In the past few years there has been some progress in our understanding of how MDM2 is regulated by posttranslational modifications. A list of these modifications and their functional consequences is presented in Table 7.3. One of the major modifications is the covalent attachment of a polypeptide to a lysine residue of MDM2. Based on the number of amino acids encoded by the *mdm2* gene, MDM2 should have a molecular size of 54 kDal. However, MDM2 isolated from mammalian cells has a relative molecular size of 90 kDal as determined by denaturing polyacrylamide gel electrophoresis (Barak and Oren, 1992; Momand et al., 1992). This paradox of varying molecular sizes was partially solved when it was discovered that the majority of MDM2 is covalently bound to the small-ubiquitin-like modifier protein (SUMO-1),

Table 7.3. Modifiers of MDM2 protein and putative outcomes.

MDM2 modification (molecule/site)	Modifier of MDM2	General characteristics of modifier	Stressor/effect on modification	Effect on MDM2	Effect on p53	References
Phosphoryl group/ser17	DNA-dependent protein kinase	Phosphorylates proteins when bound to double stranded DNA molecules-often as a result of DNA damage	Data not available	Decrease binding to p53	Increase	Mayo et al. (1997)
Phosphoryl group/Ser166	Data not available	Data not available	Mitogen activation of PI3-kinase and Akt-PKB serine-threonine kinase/increase phosphorylation	Promotes translocation to nucleus	Lowers levels of p53	Mayo and Donner, (2001)
Phosphoryl group/Ser186	Data not available	Data not available	Mitogen activation of PI3-kinase and Akt/PKB serine-threonine kinase/increase phosphorylation	Promotes translocation to nucleus	Lowers levels of p53	Mayo and Donner (2001)
Phosphoryl group/ser269	Creatine Kinase 2	Associates with general transcription factors	Data not available	Data not available	Data not available	Gotz et al. (1999)

Table 7.3. (Continued)

MDM2 modification (molecule/site)	Modifier of MDM2	General characteristics of modifier	Stressor/effect on modification	Effect on MDM2	Effect on p53	References
Phosphoryl group/ser395	Ataxia Telangiectasia a Mutated protein	DNA-damage responsive kinase; mutated in AT patients leading to radiosensitivity	IR/increase phosphorylation of MDM2	Data not available	Correlates with increase in p53 protein levels	Khosravi et al. (1999); Maya et al. (2001)
SUMO-1/unknown	E1 activating complex (Aos1-Uba2); E2 (Ubc9); E3 (unknown)	Modifier of lysine residues on proteins	UV/decrease level IR/decrease level	Increase protein level	Lower protein level	Buschmann et al. (2000); Melchior and Hengst (2000); Buschmann et al. (2001)
Ubiquitin/lys446	E1 activating complex (?); E2 (?); E3 (MDM2)	Modifier of lysine residues on proteins	Data not available	Decrease protein level	Increase protein level	Buschmann et al. (2000)
Ubiquitin/cys464	E2 (Ubc9)	Transfer ubiquitin from E2 to E3	Data not available	Part of transfer reaction to ubiquitinate p53	Transfer of Ub to Lys residues of p53	Honda et al. (1997)

which consists of 101 amino acid residues (Buschmann et al., 2000). SUMO-1 attachment increases the size of MDM2 and enhances the ubiquitin ligase activity, which likely hastens p53 proteolysis.

A hallmark of p53 activation after DNA damage is stabilization through decreased degradation. Because MDM2 is responsible for p53 degradation, one would predict that DNA damage would reduce its ability to ligate ubiquitin onto p53. In line with this prediction, MDM2 modification by SUMO-1 is inhibited in response to DNA damage and correlates with higher p53 levels (Buschmann et al., 2000). The mechanism controlling SUMO-1 modification in response to DNA damage will likely be another major research front in the future.

7.2.5. Nuclear-Cytoplasmic Translocation

Localization of MDM2 within the cell is tightly regulated, as is suggested by the high number of subcellular localization signals it contains. Within its primary amino acid sequence are regions responsible for nuclear import and export (Roth et al., 1998), and nucleolar import (Lohrum et al., 2000) (see Fig. 7.1). These sequences help MDM2 control the transport of itself and the transport of p53. First, one must place MDM2's control of p53 transport within the context of what we know about p53 subcellular trafficking. After p53 synthesis in the cytoplasm it is transported to the nucleus. Under nonstressed conditions p53 is quickly exported from the nucleus whereupon it meets its destruction by the 26S proteosome in the cytoplasm. This cycle of nuclear import, nuclear export, and destruction is broken when the cell is stressed. Upon stressor treatment, p53 export from the nucleus is blocked and it accumulates. Once the concentration reaches a threshold level it binds target genes that execute its tumor suppressor function. Work in Dr. Arnold Levine's laboratory at Princeton University showed that MDM2 is the transporter that carries p53 from the nucleus to the cytoplasm (Roth et al., 1998; Tao and Levine, 1999). Interestingly, a few clinical cancer cases have shown that high levels of p53 reside in the cytoplasm but not in the nucleus. This phenomenon, termed nuclear exclusion, suggests that there is defect in the p53 accumulation process in the nucleus of these cancer cells. Cancers showing this phenotype include inflammatory breast carcinomas, retinoblastomas, neuroblastomas, and colorectal carcinomas (Moll et al., 1992, 1995; Domagala et al., 1993; Schlamp et al., 1997). Recent studies suggest that MDM2 may play a role in p53 nuclear exclusion. For example, experimental overexpression of MDM2 in cells expressing wild-type p53 can, in some cases, lead to p53 nuclear exclusion (Rodriguez-Lopez et al., 2001). Furthermore, some tumor cells displaying nuclear exclusion require MDM2 to maintain p53 in the cytoplasm (Lu et al., 2000). It is possible that these cells express a highly active MDM2 that exports p53 and a defect in p53 degradation mechanism. Therefore, tumor cells displaying p53 nuclear exclusion may be good candidates for anti-MDM2 therapy.

Exactly how MDM2 shuttles p53 out of the nucleus has been the subject of several recent studies. MDM2-mediated export of p53 requires CRM1, a mammalian export receptor that recognizes a leucine-rich nuclear export signal (Freedman and

Levine, 1998). Research has shown that Cys 464 within the RING finger domain of MDM2 is required for p53 nuclear export (Boyd et al., 2000; Geyer et al., 2000). This cysteine residue is also essential for ubiquitin ligase activity but it is not clear if p53 is ubiquitinated prior to export. There appear to be subtle nuances in the control of p53 nuclear export. Complicating the export issue is the fact that both p53 and MDM2 contain nuclear export sequences. Two studies have shown that p53 nuclear export requires a nuclear export sequence located within residues 340–351 but does not require the MDM2 nuclear export sequence (Boyd et al., 2000; Geyer et al., 2000). Another study has shown that export of p53 requires nuclear export sequences from both p53 and MDM2 (Tao and Levine, 1999). A third study has shown that p53 is capable of nuclear export in the absence of MDM2 (Stommel et al., 1999). How can we account for these apparent differences? Perhaps the mechanism of p53 nuclear export depends on other parameters that include species type, cell type, MDM2/p53 expression levels, and cell growth state.

An interesting twist on MDM2 control of p53 nuclear export was recently discovered (Zhang and Xiong, 2001). The first p53 nuclear export sequence discovered was located near its carboxyl terminus (Roth et al., 1998) far away from the MDM2 binding domain. However, Dr. Yue Xiong's laboratory at the University of North Carolina uncovered a second p53 nuclear export sequence located near the N-terminus, where MDM2 binds. Under nonstressed circumstances the N-terminal p53 export sequence is functional; p53 forms a complex with MDM2 and is shuttled out of the nucleus. Phosphorylation of p53 within this sequence, however, prevents MDM2 binding and, in addition, blocks its export. In this instance, kinases activated by genome damage can release p53 from MDM2 and at the same time retain p53 within the nucleus. In sum, p53 nuclear export is strongly influenced by MDM2.

7.3. STRESSOR INDUCED REGULATION OF MDM2–p53 INTERACTION

7.3.1. Phosphorylation of p53

Activation of p53 is characterized by an increase in protein level, nuclear accumulation, and the attainment of posttranslational modifications that enhance its ability to transactivate effector genes. Two well-characterized stressors that lead to p53 activation are ionizing radiation and inappropriate oncogene activation. Each stressor inhibits MDM2 activity but in unique ways. In 1991, Dr. Michael Kastan and his colleagues at Johns Hopkins University showed that ionizing radiation promotes upregulation of p53 activity and halts cellular proliferation (Kastan et al., 1991). Since this seminal discovery, several investigators have showed that p53 upregulation in response to DNA damage correlates with certain specific posttranslational modifications (see Giaccia and Kastan, 1998) for review). Phosphorylation of p53 occurs at Ser 15, Thr 18, and Ser 20 and prevents its ability to bind to MDM2 either by direct interference (Thr 18, Ser 20) or by alteration of its local conformation

(Ser 15) (Shieh et al., 1997; Unger et al., 1999a, b; Bean and Stark, 2001, 2002; Hirao et al., 2000; Sakaguchi et al., 2000). Recently, DNA damage has also been shown to result in MDM2 phosphorylation (Khosravi et al., 1999; Maya et al., 2001).

7.3.2. Phosphorylation of MDM2

One kinase that phosphorylates MDM2 is the *Ataxia Telangiectasia Mutated* gene product, ATM (Khosravi et al., 1999; Maya et al., 2001), the same one that modifies p53 in response to some forms of DNA damage. ATM appears to be responsible for phosphorylating MDM2 at Ser 395. Although Ser 395 does not reside in a known subcellular localization motif, its phosphorylation does appear to inhibit MDM2 export and assist in upregulating p53 activity. Another kinase that phosphorylates MDM2 is DNA-dependent protein kinase (Mayo et al., 1997). DNA-dependent kinase is activated by genome damage but the site of phosphorylation on MDM2 is not yet clear. It is likely that, with the development of phosphopeptide-specific antibodies, MDM2 phosphorylation regulation will be explored extensively in the next few years.

7.3.3. ARF–MDM2 Complex Formation

Classical yeast genetic studies showed that DNA damage promotes cell cycle arrest (reviewed in Hartwell and Weinert, 1989). Later, mammalian cell culture studies demonstrated that DNA damage could also lead to apoptosis (Clarke et al., 1993; Lowe et al., 1993). More recently, oncogene activation has been shown to lead to these cellular outcomes as well (reviewed in Sherr, 2001). Activation of oncogenes can be deleterious to the organism by signaling cells to divide at inappropriate times. If the cell receives a signal from an oncogene to divide when DNA is damaged, it risks the chance of sustaining a mutation. When this cell stress pathway was uncovered it did not take long to show that oncogene activation triggered a p53 response. Early studies demonstrated that *myc* oncogene activation led to apoptosis and that this process required p53 (Ramqvist et al., 1993; Wang et al., 1993). A key discovery was that the *Ink4a/ARF* tumor suppressor gene was required for oncogene signaling to p53 (de Stanchina et al., 1998; Zindy et al., 1998). The *Ink4a/ARF* gene produces two transcripts, each possessing tumor suppressor activity. In humans the first transcript produces p16[Ink4a], which inhibits cyclin D1/Cdk4 complex, a negative inhibitor of Rb. The second transcript of this gene produces ARF (in humans this is sometimes known as p14[Arf] and in mice as p19[Arf]), which binds and inhibits MDM2 activity. ARF also binds and prevents MDMX from interacting with p53 (Jackson et al., 2001). The current view of the mechanism of MDM2 and MDMX inactivation by ARF is presented in Figure 7.3. In the absence of oncogene activation p53 is removed from the nucleus by MDM2 and degraded by the 26S proteasome. A separate p53 subpopulation is directly bound to MDMX in the nucleus and is prevented from activating its effector genes. Upon oncogene activation ARF levels increase and bind MDM2 and MDMX, releasing p53 to transactivate its appropriate target genes. ARF then sequesters both MDM2 and MDMX into the nucleolus. Oncoproteins that activate

NORMAL

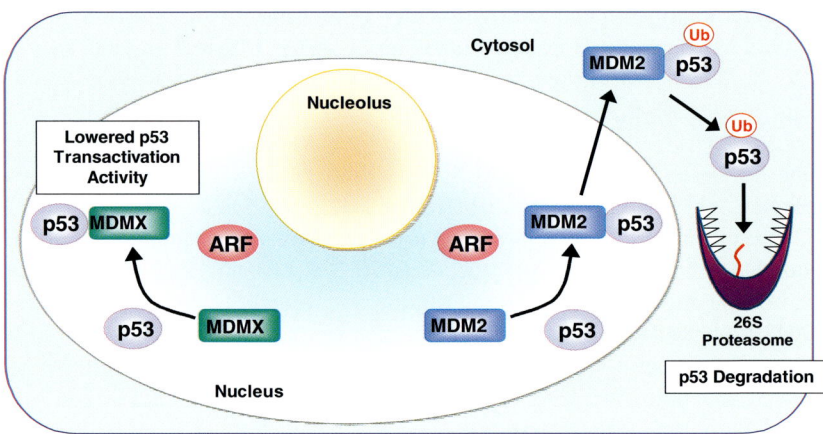

ONCOGENE ACTIVATION

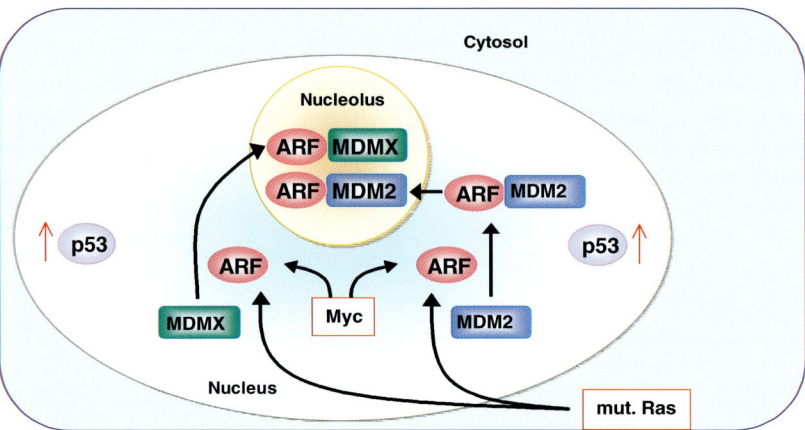

Figure 7.3. Oncogene activation of MDM2 and MDMX. Under normal conditions p53 is bound to MDMX in an inactive complex or shuttled to the cytoplasm by MDM2 where it is degraded by the 26S proteasome. When the Myc or mutant Ras oncoprotein is abnormally activated, ARF levels increase. Upon binding ARF, MDMX is released from p53 and shuttled into the nucleolus. When MDM2 binds ARF it fails to ubiquitinate p53 and is shuttled to the nucleolus. The released p53 transactivates genes that arrests cell proliferation or induces apoptosis.

p53 through the ARF pathway include myc (Pomerantz et al., 1998; Zindy et al., 1998), polyoma virus middle T-antigen (Lomax and Fried, 2001), E2F1 (Bates et al., 1998), viral oncoprotein E1A (de Stanchina et al., 1998), and mutant Ras (Ries et al., 2000). The list of mitogenic molecules that activate p53 through this pathway is likely to expand in the near future.

Recent studies suggest that ARF actually downregulates MDM2 activity by two mechanisms. First, MDM2 bound to ARF is inhibited in its ability to ubiquitinate p53 (Honda and Yasuda, 1999; Xirodimas et al., 2001). Second, ARF sequesters MDM2 into the nucleolus. These would seem to be redundant inhibitory mechanisms but redundancy appears to be the rule when dealing with MDM2 biochemical activities, not the exception. It is likely that the combination of the two mechanisms of MDM2 inactivation results in a more rapid and a more efficient response than a single one. Perhaps in the absence of nucleolar sequestration, upon initial binding to ARF, MDM2 is inactivated but remains in the vicinity of p53. Because binding is reversible, some MDM2 can dissociate from ARF and bind to p53 again. To be a more effective inhibitor ARF removes MDM2 to the nucleolus and allows newly synthesized ARF to bind and inhibit other MDM2 molecules. In sum, the two major upstream signals that activate p53 do so by distinct mechanisms. DNA damage modulates phosphorylation of MDM2 and p53. Oncogene activation controls ARF binding to MDM2. A fertile area of future research will be the delineation of more mechanisms that control MDM2. Other stressors known to control p53 activity include hypoxia, hyperoxia, and chemical carcinogens. These stressors may modulate MDM2 and MDMX through either of the routes listed above or through other posttranslational mechanisms such as oxidation, reduction, oligomerization, acetylation, SUMO-1 modification, and dephosphorylation.

7.3.4. Oligomerization and Acetylation

Aside from phosphorylation and ARF binding, other posttranslational events control MDM2–p53 complex formation. For example, p53 must form a dimer for efficient MDM2 binding (Maki, 1999) and p53 is incapable of being acetylated when it is bound to MDM2 (Ito et al., 2001; Kobet et al., 2000). Acetylation helps recruit transcriptional coactivators to p53 to help it increase its ability to transactivate genes (Barlev et al., 2001). Lysine residues near the C-terminus of p53 are substrates for acetylation. It is likely that the RING finger domain of MDM2 is involved in blocking acetylation because this domain is necessary for ubiquitinating lysine residues near the C-terminus of p53. If p53 deacetylation is inhibited it becomes more stable (Ito et al., 2001) suggesting that acetylation protects lysines from ubiquitination.

7.4. GENETICS OF MDM2 AND MDMX

Mouse genetics can be a powerful tool to test hypotheses derived from experiments performed with cultured mammalian cells. Prior to the mouse genetic studies, experiments using cultured cells led to the prediction that a mouse lacking *mdm2* would overexpress p53. Overexpression of p53 should lead to cell death or cell cycle arrest. This hypothesis was dramatically confirmed when it was shown that *mdm2 –/–* mouse embryos fail to survive after day 6.5–7.5 postgestation (Jones et al., 1995; Montes de Oca Luna et al., 1995). Tissues recovered from aborted embryos showed signs of having undergone p53-mediated programmed cell death (Montes de Oca Luna et al., 1995). Interestingly, the *mdm2 –/–* mouse was rescued when placed in a *p53* null genetic background. The double knockout mouse exhibits a phenotype almost identical to a mouse that expresses no p53. Both mice are born with a normal phenotype but tumors arise within the same timeframe and the type of tumors that develop are similar. Thus, *mdm2* appears to mainly function to maintain p53 protein in check.

To determine the physiological significance of MDMX an *mdmx* knockout mouse was created (Parant et al., 2001). Like the *mdm2 –/–* mouse, the *mdmx –/–* mouse was not viable and aborted at approximately 6.5 days postgestation. However, death was not due to excessive programmed cell death. Instead, cells in the aborted embryo simply failed to proliferate. It is possible that p53 was unchecked and that the cell cycle arrest genes it normally regulates were overexpressed. To partially test this hypothesis, a mouse with a knockout in both *mdmx* and *p53* genes was created. Surprisingly, the double knockout mouse survived and appeared normal. The oldest mouse at the time of the publication was four months and failed to show any signs of abnormality (It is predicted that the mouse will develop tumors similar to the p53 knockout mouse.) The mouse genetic studies indicate three important points: (1) MDM2 is a negative regulator of p53-mediated apoptosis activity; (2) MDMX is a negative regulator of p53-mediated cell cycle arrest activity; (3) MDM2 cannot substitute for MDMX and vice versa during mouse development.

It is intriguing that MDM2 and MDMX cannot substitute for one another given that both molecules bind to the same region within p53. It is possible that there may be cell-specific or time-specific expression of each p53 inhibitor, both being independently critical to embryo survival. It is also possible that control of p53's ability to transactivate its many effector genes is divided between MDM2 and MDMX. In the developing embryo, MDM2 may control proapoptotic genes while MDMX may control growth arrest genes.

7.5. THE AUTOREGULATORY LOOP

Soon after the discovery of MDM2 it was observed that its level was increased when p53 was experimentally overexpressed (Barak et al., 1993). This increase was due to a higher level of MDM2 transcript and was mediated by the binding of p53 to two specific sequences within intron 1 (Juven et al., 1993; Wu et al., 1993). DNA damage also leads to p53-mediated upregulation of the *mdm2* gene (Chen et al., 1994). These studies gave rise to the notion that p53 could regulate its own inhibition through an autoregulatory loop (reviewed in Zambetti and Levine, 1993). In this loop, it is thought that p53 activates its own destruction by transcribing *mdm2*, thus maintaining itself at a low level. Stressors activate signaling pathways that prevent p53 from binding MDM2, thus allowing its level to increase and transactivate effector genes that promote apoptosis and cell cycle arrest. When cell cycle arrest is the outcome, cells may eventually resume proliferation once the genome is repaired. Once p53 activates *mdm2*, its levels are lowered to a point where cells can proliferate again. Interestingly, p53 does not regulate *mdmx*. Presented in Figure 7.4 are the major discoveries of the molecules involved in the p53 and MDM2 pathways and the autoregulatory loop. Each arrow that connects two molecules in the pathway corresponds to a discovery made by scientists. The width of each arrow is proportional to the number of citations each discovery received per unit time (see Appendix 7.1 for details on how the width of each arrow was calculated). The black arrows correspond to the molecular events that are active in the autoregulatory loop that controls p53 activity. The yellow arrows show the molecular events that occur when the cell genome is damaged. The pink arrows show the molecular events that are activated by oncogene stimulation. The

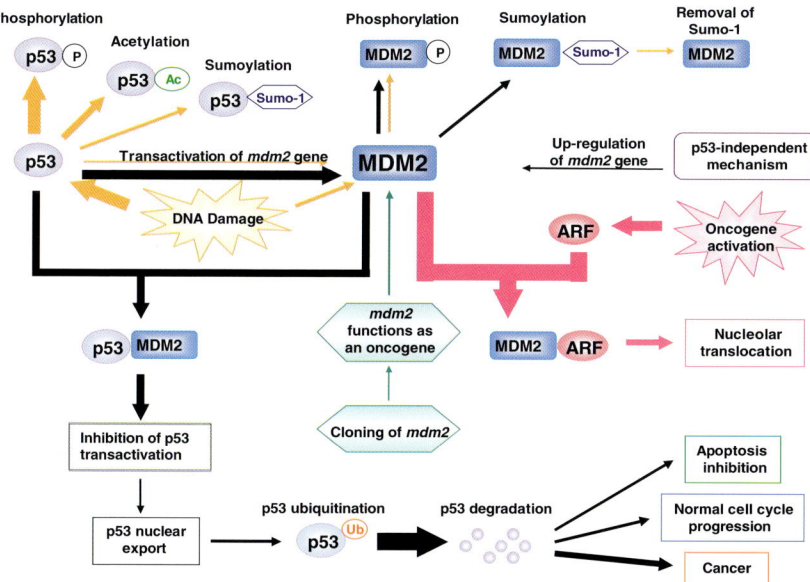

Figure 7.4. Major milestones of p53 autoregulation. The p53 protein activates *mdm2* resulting in an increase in MDM2 protein levels. In the absence of DNA damage (depicted with black arrows) p53 transactivates the *mdm2* gene. MDM2 binds p53 and ultimately assists in its degradation. In doing so, p53 is prevented from promoting apoptosis or inhibiting the cell cycle. Excess production of MDM2 lowers the level of p53 and can ultimately lead to cancer. DNA damage activates p53 (depicted in yellow arrows) by promoting p53 phosphorylation, p53 acetylation, SUMO-1 conjugation to p53. Subsequently, the *mdm2* gene is activated. DNA damage also promotes MDM2 phosphorylation and removal of SUMO-1 from MDM2. A second route to MDM2 inhibition is via oncogene activation (depicted in pink arrows). Oncogene activation increases the level of ARF, which in turn sequesters MDM2 into the nucleolus. Green arrows depict the discovery of the *mdm2* oncogene. The thickness of the arrows correlates with the number of citations received for each part of the pathway. (See Appendix 7.1 for details of calculations of arrow thicknesses.)

green arrows depict the major steps made in the discovery of the *mdm2* gene. This mode of p53 activation allows it to act quickly after application of the stressor.

7.6. *mdm2* GENE STRUCTURE AND TRANSCRIPTION

Murine *mdm2* consists of 12 exons and spans approximately 25 kb (Montes de Oca Luna et al., 1996; Jones et al., 1995). We now know that regulation of *mdm2* takes place at the transcriptional level, posttranscriptional level, and the posttranslational level. In this section we describe how differential promoter usage and alternative splicing regulate the MDM2 transcript. The first two exons of the *mdm2* gene do not code for protein, and two physically distinct promoters within *mdm2*, named P1 and P2, have been identified in both humans and mice (Barak et al., 1994; Landers et al., 1997; Zauberman et al., 1995). The P1 promoter is located 5′ to exon 1 and the P2 promoter is located just 5′ to exon 2. The P2 promoter contains two p53-responsive elements but elements that control the P1 promoter have not been identified. Studies on *mdm2* transcriptional regulation have yielded a complex pattern of mRNA production.

In adult murine tissues MDM2 transcripts initiated from P1 are approximately five times more abundant than transcripts initiated from P2 (Mendrysa and Perry, 2000). As expected, when a mouse is exposed to ionizing radiation, P2 initiated transcripts increase in a p53-dependent fashion. In the *p*53 knockout mouse, P2-initiated transcripts, named S-MDM2 transcripts, are present at low levels, but fail to increase after ionizing radiation treatment. The fact that S-MDM2 transcripts are detected in the *p*53 knockout mouse indicates that transcription factors other than p53 can express the S-MDM2 transcript. Interestingly, P2 becomes active upon removing cells from normal mice embryos and placing them in culture. This suggests that P2 becomes p53 dependent only in a state of constant stress. The process of culturing promotes that stress and implies that perhaps all cultured cells contain an active p53 autoregulatory loop. The corollary to this hypothesis is p53 is not downregulated by MDM2 under nonstressed conditions. This has implications for anti-MDM2 therapy. It was previously thought that a major impediment to anti-MDM2 therapy could be the inappropriate activation of p53 in noncancer tissue. This would elicit a cell death response in normal cells. If, however, the autoregulatory loop is active only after a stress event then anti-MDM2 therapy becomes attractive because there may be no p53-elicited cell death in nonstressed normal tissues. The transcription factors that maintain basal levels of MDM2 mRNA in nonstressed tissues are not known.

Transcription factors other than p53 can increase MDM2 mRNA. Dr. Moshe Oren and his colleagues at the Weizmann Institute in Rehovot have shown that DNA sequences corresponding to AP-1, EtsA, and EtsB transcription factor binding sites are observed within intron 1 (Ries et al., 2000). Ras oncoprotein activates a signaling pathway that leads to upregulation of AP-1, EtsA, and EtsB and, ultimately, cell proliferation. The existence of this pathway indicates that Ras can mediate its oncogenic function by MDM2 upregulation. The factors linking Ras and the DNA binding proteins are Raf, MEK, and MAPK. Combined with other studies previously discussed in this chapter, this observation indicates that Ras can activate two opposing pathways culminating with either upregulation or downregulation of MDM2. In one pathway, Ras activates ARF, which inhibits MDM2 through complex formation. In the other pathway, Ras activation increases transcription of *mdm2*. How can we explain these two seemingly contradictory outcomes? It is known that p53 levels are extremely low in proliferating cells and it may be that in such cells the Ras pathway that upregulates *mdm2* may be active, leading to low p53 levels. In this cell growth phase, Ras fails to activate ARF. However, if Ras is activated at an inappropriate time, it triggers ARF activation and upregulates p53. The switch that modulates these pathways is unknown. In cells that have lost the ability to properly express the *ARF* gene the repercussions of the *mdm2* activation by Ras becomes magnified. In such cells, Ras is predicted to decrease the level of p53 protein and drive them toward cancer formation.

An alternatively spliced form of MDM2 transcript affects its ability to inhibit p53. Truncated versions of MDM2 have been demonstrated to exist for almost 10 years (Olson et al., 1993; Haines et al., 1994). But, until recently, little was known as to how such truncated forms could control p53 activity. A murine *mdm2* transcript lacking exon 3 has been observed in cell lines (Saucedo et al., 1999). This transcript codes for a truncated version of MDM2, named p76^{MDM2}, which is translated beginning at codon 50. Interestingly, p76^{MDM2} increases the stability of full-length MDM2 and enhances p53 transactivation capacity (Perry et al., 2000). One hypothesis as to the mechanism of p76^{MDM2}-mediated upregulation of p53 is that it may compete for factors that normally bind or modify full-length MDM2 (perhaps E2 of the ubiquitin activating pathway). Binding these factors prevents MDM2 from properly inhibiting p53. The ratio of full-length MDM2 to p76^{MDM2} is not uniform amongst tissues and, when the ratio of p76^{MDM2} to full-length MDM2 is high, it is predicted that the p53 protein will be able to quickly stabilize after DNA damage.

Like its murine counterpart, the human *mdm2* gene produces two transcripts named L-MDM2 and S-MDM2 (Landers et al., 1997; Zauberman et al., 1995). Figure 7.5 delineates how these transcripts are generated. The L-MDM2 transcript is the product of the constitutive promoter, P1. Exon 2 encoded sequences are removed from the mature L-MDM2 transcript resulting in an RNA sequence that consists of exon 1 encoded sequence juxtaposed to exon 3 encoded sequence. The S-MDM2 transcript is the product of the p53 inducible promoter, P2. The S-MDM2 transcript is translated at eightfold higher efficiency than the L-MDM2 transcript, resulting in high expression levels of its protein product (Landers et al., 1997). Both S-MDM2 and L-MDM2 transcripts encode the full-length MDM2 protein.

7.7. MDM2 AND MDMX INVOLVEMENT IN CANCERS

Gene amplification is one of several mechanisms by which cancer cells can overexpress oncogenes. In fact, the *mdm2* gene was originally observed to be amplified in a mouse cell line that had spontaneously become tumorigenic (Cahilly-Snyder et al., 1987; Fakharzadeh et al., 1991). *mdm2* gene amplification has been detected in 14 types of tissues displaying abnormal growth patterns (Momand et al., 1998). The overall frequency of *mdm2* amplification in these tissues is 7%. Based on the

Human *mdm2* Transcripts and Protein Products

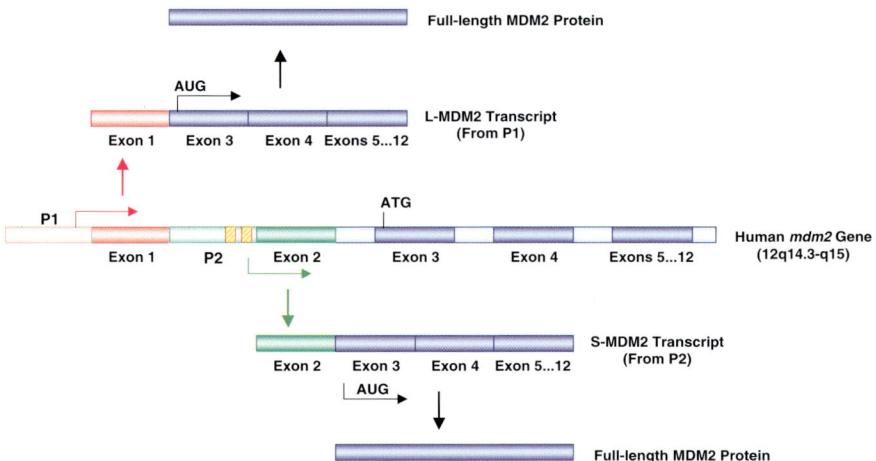

Figure 7.5. Alternative splice forms of human MDM2 transcripts. Human *mdm2* consists of 12 exons. The first two exons do not code for protein. Promoter P1 is located upstream of exon 1 and P2 is located within intron 1. Two yellow hatched regions represent the two p53-response elements within P2. L-MDM2 is transcribed beginning at exon 1 but splices out exon 2 to form the mature transcript. S-MDM2 is transcribed starting at exon 2. Full-length MDM2 is produced from both transcripts.

phenotype of the *mdmx* knockout mouse one would predict that the human *mdmx* gene might be upregulated in some cancers. This prediction was recently confirmed when in a study of 208 gliomas it was discovered that *mdmx* was amplified in five tumors (Riemenschneider et al., 1999). Because MDM2 and MDMX control p53 one would expect that mutations in the *p53* gene would be an infrequent event in cancers with *mdm2* and *mdmx* gene amplification. In all cases of cancers with *mdm2 or mdmx* amplification there was virtually no evidence for *p53* gene mutations, indicating that genetic alterations in *mdm2/mdmx* and *p53* rarely occur in the same tissue (Momand et al., 1998; Riemenschneider et al., 1999). Because ARF lies directly upstream of MDM2 within the oncoprotein activation pathway one would predict that the *ARF* gene would be mutated in high percentage of tumors; studies now attest to this (Gardie et al., 1998; Gazzeri et al., 1998). Taken together, we now know that the chance of observing an inactivating mutation in *ARF*, an activating mutation in *mdm2/mdmx*, or an inactivating mutation in *p53* is very high in human cancers.

7.8. CONCLUDING REMARKS

A striking take-home lesson presented by the p53 autoregulatory loop is that the ability of MDM2 to inhibit the tumor suppressor activity of p53 is biochemically redundant. MDM2 appears to target p53 for destruction in three progressive stages—targeting, removal, and destruction. In the targeting stage, the specific domain of p53 required for transcription is bound by MDM2 allowing for immediate cessation of p53 activity. Fulfillment of this stage allows p53 to bind MDM2 while it is bound to DNA, immediately shutting down transcription. In the removal stage, MDM2 continuously escorts p53 away from the nucleus; thus, maintaining a low concentration of p53 in the vicinity of its target genes. In the destruction stage, the initial step of p53 degradation is MDM2-mediated ubiquitination. In the absence of any of these stages MDM2 becomes a less efficient inhibitor and, therefore, may allow p53 to be active at inopportune times. The last two stages of MDM2 inhibition harkens back to the thermodyanamic principles that control product formation in chemical reactions. The change in free energy of the reaction depends, in large part, on efficient removal of products which drives the reaction forward. This biochemical redundancy is necessary for efficient responses to cellular needs. It is likely that redundant forms of inhibition are waiting to be found in other biochemical systems.

ACKNOWLEDGEMENTS

We would like to thank the support of the Minorities in Biomedical Research Sciences Program (NIGMS08101) and the Bridges-to-the-Future-Program (NIGMS49001).

APPENDIX 7.1.   CALCULATIONS AND REFERENCES FOR CITATIONS USED TO SET ARROW THICKNESS IN FIGURE 7.4.

Original communications were identified and placed under separate subject headings. The number of citations for each communication until November 2001 was obtained from the ISI Web of Science ® internet site. The number of citations for each communication was divided by the number of months elapsed since publication to obtain the citation rate. The citation rate was added to other citation rates within the same category and listed as "Total". This Total number divided by 2 was set as the point width of each arrow used in Figure 7.4.

p53 phosphorylation:
Phosphorylation at S15 upon DNA damage (Siliciano et al., 1997) **Dec (47 Months); # Cited: 227; Correction = 227/47 = 4.83**
Phosphorylation at S20 upon DNA damage
 (Chehab et al., 1999) **Nov (24 Months); # Cited: 67; Correction = 67/24 = 2.79**
 (Unger et al., 1999) **Apr (31 Months); # Cited: 88; Correction = 88/31 = 2.84**
Phosphorylation at S37 upon DNA damage
 (Shieh et al., 1997) **Oct (49 Months); # Cited: 388; Correction = 388/49 = 7.92**
Phosphorylation at T18 upon DNA damage
 (Sakaguchi et al., 2000) **Mar (20 Months); # Cited: 26; Correction = 26/20 = 1.3**

| **Total = 19.68** |

p53 acetylation:

(Gu and Roeder, 1997) **Aug (51 Months); # Cited: 532; Correction = 532/51 = <u>10.43</u>**

| Total = 10.43 |

p53 Sumoylation:

(Rodriguez et al., 1999) **Nov (24 Months); # Cited: 63; Correction = <u>2.63</u>**

(Gostissa et al., 1999) **Nov (24 Months); # Cited: 55; Correction = 55/24 = <u>2.29</u>**

| Total = 4.92 |

MDM2 phosphorylation:

S17 by DNA-PK (Mayo et al., 1997) **Nov (48 Months); # Cited: 91; Correction = 91/48 = <u>1.90</u>**

S395 by ATM (Khosravi et al., 1999) **Dec (23 Months); # Cited: 61; Correction = 61/23 = <u>2.65</u>**

(Maya et al., 2001) **May (6 Months); # Cited: 4; Correction = 4/6 = <u>0.67</u>**

S269 by CK2 (Gotz et al., 1999) **Dec (23 Months); # Cited: 5; Correction = 5/23 = <u>0.22</u>**

S166 by PI3K/Akt(PKB) (Mayo and Donner, 2001) **Sep (2 Months); # Cited: 1; Correction = $^1/_2$ = <u>0.50</u>**

S186 by PI3K/Akt(PKB) (Mayo and Donner, 2001) **(2 Months); # Cited: 1; Correction = $^1/_2$ = <u>0.50</u>**

| Total = 6.44 |

MDM2 Sumoylation:

(Buschmann et al., 2000) **Jun (17 Months); # Cited: 37; Correction = 37/17 = <u>2.18</u>**

(Melchior and Hengst, 2000) **Sep (14 Months); # Cited: 1; Correction = 1/14 = <u>0.07</u>**

| Total = 2.25 |

Cloning of MDM2:

(Cahilly-Snyder et al., 1987) **May (174 Months); # Cited = 153; Correction = 153/174 = <u>0.88</u>**

| Total = 0.88 |

Identification of MDM2 as an oncogene:

(Fakharzadeh et al., 1991) **Jun (125 Months); # Cited = 336; Correction = 336/125 = <u>2.69</u>**

| Total = 2.69 |

p53 up-regulation by DNA damage:

(Kastan et al., 1991) **Dec (119 Months); # Cited = 2266; Correction = 2266/119 = <u>19.09</u>**

| Total = 19.09 |

Transactivation of MDM2 gene by p53:

(Wu et al., 1993) **Jul (100 Months): # Cited: 728; Correction = 738/100 = <u>7.38</u>**

(Barak et al., 1993) **Feb (105 Months): # Cited: 558; Correction = 558/105 = <u>5.31</u>**

(Juven et al., 1993) **Dec (95 Months): # Cited: 216; Correction = 216/95 = <u>2.27</u>**

| Total = 14.96 |

Transcriptional up-regulation of MDM2 independent of p53:
    (Perry et al., 2000) **May (18 Months); # Cited: 8; Correction $= 8/18 = \underline{0.44}$**
    (Barak et al., 1994) **1994 Aug (51 Months); # Cited 129; Correction $= 129/51$**
$= \underline{2.53}$ $\boxed{\textbf{Total} = \textbf{2.97}}$

MDM2/p53 complex formation:
    (Momand et al., 1992) **Jun (113 Months); # Cited: 1358; Correction $= 1358/113$**
$= \underline{12.02}$ $\boxed{\textbf{Total} = \textbf{12.02}}$

p53 transactivation inhibition by MDM2:
    (Momand et al., 1992) **Jun (113 Months); # Cited: 1358; Correction $= 1358/113$**
$= \underline{12.02}$ $\boxed{\textbf{Total} = \textbf{12.02}}$

p53 export by MDM2:
    (Tao and Levine, 1999) **Jun (29 Months); # Cited: 81; Correction $= 81/29 =$**
$\underline{2.79}$ $\boxed{\textbf{Total} = \textbf{2.79}}$

Ubiquitin ligase activity of MDM2:
    (Honda et al., 1997) **Dec (47 Months); # Cited: 216; Correction $= 216/47 =$**
$\underline{4.60}$ $\boxed{\textbf{Total} = \textbf{4.60}}$

p53 degradation by MDM2:
    (Haupt et al., 1997) **May (54 Months); # Cited: 705; Correction $= 705/54 =$**
$\underline{13.06}$
    (Kubbutat et al., 1997) **May (54 Months); # Cited: 629; Correction $= 629/54 =$**
$\underline{11.65}$
    (Midgley and Lane, 1997) **Sep (50 Months); # Cited: 106; Correction $= 106/$**
$50 = \underline{2.12}$ $\boxed{\textbf{Total} = \textbf{26.83}}$

Cancer progression upon p53 inhibition by MDM2:
    (Oliner et al., 1992) **Jul (112 Months); # Cited: 1053; Correction $= 1053/112$**
$= \underline{9.40}$ $\boxed{\textbf{Total} = \textbf{9.40}}$

Normal cell cycle progression upon p53 degradation by MDM2:
    (Chen et al., 1994) **March (92 Months); # Cited: 226; Correction $= 226/92 =$**
$\underline{2.46}$ $\boxed{\textbf{Total} = \textbf{2.46}}$

DNA damage induced upregulation of MDM2 gene:
    (Chen et al., 1994) **March (92 Months); # Cited: 226; Correction $= 226/92 =$**
$\underline{2.46}$ $\boxed{\textbf{Total} = \textbf{2.46}}$

Apoptosis inhibition upon p53 degradation by MDM2:
    (Haupt et al., 1996) **Feb (69 Months); # Cited: 150; Correction $= 150/69 = \underline{2.17}$**
    (Chen et al., 1996) **May (66 Months); # Cited: 145; Correction $= 145/66 = \underline{2.20}$**
                            $\boxed{\textbf{Total} = \textbf{4.37}}$

Oncogene Activation of ARF:

(Kamijo et al., 1998) **Jul (40 Months); # Cited : 198; Correction = 198/40 = 4.95**
(de Stanchina et al., 1998) **Aug (39 Months); # Cited: 188; Correction = 188/39 = 4.82**
(Zindy et al., 1998) **Aug (39 Months); # Cited: 238; Correction = 238/39 = 6.10**

$$\boxed{\text{Total} = 15.87}$$

MDM2/ARF complex formation:

(Pomerantz et al., 1998) **Mar (44 Months); # Cited: 755; Correction = 755/ 44 = 17.16**
(Zhang et al., 1998) **Mar (44 Months); # Cited: 395; Correction = 395/44 = 8.98**

$$\boxed{\text{Total} = 26.14}$$

MDM2/ARF complex nucleolus translocation:

(Weber et al., 1999) **May (29 Months); # Cited: 140; Correction = 140/29 = 4.83**
(Tao and Levine, 1999) **June (27 Months); # Cited: 145; Correction = 145/29 = 5.00**

$$\boxed{\text{Total} = 9.83}$$

DNA damage induced MDM2 Phosphorylation changes:

S395 by ATM (Khosravi et al., 1999) **Dec (23 Months); # Cited: 61; Correction = 61/23 = 2.65**
(Maya et al., 2001) **May (6 Months); # Cited: 4; Correction = 4/6 = 0.67**

$$\boxed{\text{Total} = 3.32}$$

DNA damage induced MDM2 SUMO-1 removal:

(Buschmann et al., 2000) **Jun (17 Months); # Cited: 37; Correction = 37/17 = 2.18**

$$\boxed{\text{Total} = 2.18}$$

REFERENCES

Barak, Y., Gottlieb, E., Juven-Gershon, T., and Oren, M. (1994). Regulation of mdm2 expression by p53: alternative promoters produce transcripts with nonidentical translation potential. *Genes Dev* 8:1739–1749.

Barak, Y., Juven, T., Haffner, R., and Oren, M. (1993). mdm2 expression is induced by wild type p53 activity. *EMBO J* 12:461–468.

Barak, Y., and Oren, M. (1992). Enhanced binding of a 95 kDa protein to p53 in cells undergoing p53-mediated growth arrest. *EMBO J* 11:2115–2121.

Barlev, N. A., Liu, L., Chehab, N. H., Mansfield, K., Harris, K. G., Halazonetis, T. D., and Berger, S. L. (2001). Acetylation of p53 Activates Transcription through Recruitment of Coactivators/Histone Acetyltransferases. *Mol Cell* 8:1243–1254.

Barton, G. J., and Sternberg, M. J. (1987). A strategy for the rapid multiple alignment of protein sequences. Confidence levels from tertiary structure comparisons. *J Mol Biol* 198:327–337.

Bates, S., Phillips, A. C., Clark, P. A., Stott, F., Peters, G., Ludwig, R. L., and Vousden, K. H. (1998). p14ARF links the tumour suppressors RB and p53. *Nature* 395:124–125.

Bean, L. J., and Stark, G. R. (2001). Phosphorylation of serines 15 and 37 is necessary for efficient accumulation of p53 following irradiation with UV. *Oncogene* 20:1076–1084.

Bean, L. J., and Stark, G. R. (2002). Regulation of the accumulation and function of p53 by phosphorylation of two residues within the domain that binds to Mdm2. *J Biol Chem* 277:1864–1871.

Botuyan, M. V., Momand, J., and Chen, Y. (1997). Solution conformation of an essential region of the p53 transactivation domain. *Fold Des* 2:331–342.

Boyd, S. D., Tsai, K. Y., and Jacks, T. (2000). An intact HDM2 RING-finger domain is required for nuclear exclusion of p53. *Nat Cell Biol* 2:563–568.

Buschmann, T., Fuchs, S. Y., Lee, C. G., Pan, Z. Q., and Ronai, Z. (2001). Erratum: SUMO-1 modification of Mdm2 prevents its self-ubiquitination and increases Mdm2 ability to ubiquitinate p53. *Cell* 107:549.

Buschmann, T., Fuchs, S. Y., Lee, C. G., Pan, Z. Q., and Ronai, Z. (2000). SUMO-1 modification of Mdm2 prevents its self-ubiquitination and increases Mdm2 ability to ubiquitinate p53. *Cell* 101:753–762.

Cahilly-Snyder, L., Yang-Feng, T., Francke, U., and George, D. L. (1987). Molecular analysis and chromosomal mapping of amplified genes isolated from a transformed mouse 3T3 cell line. *Somat Cell Mol Genet* 13:235–244.

Chehab, N. H., Malikzay, A., Stavridi, E. S., and Halazonetis, T. D. (1999). Phosphorylation of Ser-20 mediates stabilization of human p53 in response to DNA damage. *Proc Natl Acad Sci USA* 96:13777–13782.

Chen, C. Y., Oliner, J. D., Zhan, Q., Fornace, A. J., Jr., Vogelstein, B., and Kastan, M. B. (1994). Interactions between p53 and MDM2 in a mammalian cell cycle checkpoint pathway. *Proc Natl Acad Sci USA* 91:2684–2688.

Chen, J., Marechal, V., and Levine, A. J. (1993). Mapping of the p53 and mdm-2 interaction domains. *Mol Cell Biol* 13:4107–4114.

Chen, J., Wu, X., Lin, J., and Levine, A. J. (1996). mdm-2 inhibits the G1 arrest and apoptosis functions of the p53 tumor suppressor protein. *Mol Cell Biol* 16:2445–2452.

Chen, L., Agrawal, S., Zhou, W., Zhang, R., and Chen, J. (1998). Synergistic activation of p53 by inhibition of MDM2 expression and DNA damage. *Proc Natl Acad Sci USA* 95:195–200.

Chene, P., Fuchs, J., Bohn, J., Garcia-Echeverria, C., Furet, P., and Fabbro, D. (2000). A small synthetic peptide, which inhibits the p53-hdm2 interaction, stimulates the p53 pathway in tumour cell lines. *J Mol Biol* 299:245–253.

Clarke, A. R., Purdie, C. A., Harrison, D. J., Morris, R. G., Bird, C. C., Hooper, M. L., and Wyllie, A. H. (1993). Thymocyte apoptosis induced by p53-dependent and independent pathways. *Nature* 362:849–852.

de Stanchina, E., McCurrach, M. E., Zindy, F., Shieh, S. Y., Ferbeyre, G., Samuelson, A. V., Prives, C., Roussel, M. F., Sherr, C. J., and Lowe, S. W. (1998). E1A signaling to p53 involves the p19(ARF) tumor suppressor. *Genes Dev* 12:2434–42.

Domagala, W., Harezga, B., Szadowska, A., Markiewski, M., Weber, K., and Osborn, M. (1993). Nuclear p53 protein accumulates preferentially in medullary and high- grade ductal but rarely in lobular breast carcinomas. *Am J Pathol* 142:669–674.

Duncan, S. J., Gruschow, S., Williams, D. H., McNicholas, C., Purewal, R., Hajek, M., Gerlitz, M., Martin, S., Wrigley, S. K., and Moore, M. (2001). Isolation and structure elucidation of Chlorofusin, a novel p53-MDM2 antagonist from a Fusarium sp. *J Am Chem Soc* 123:554–560.

Fakharzadeh, S. S., Trusko, S. P., and George, D. L. (1991). Tumorigenic potential associated with enhanced expression of a gene that is amplified in a mouse tumor cell line. *EMBO J* 10:1565–1569.

Freedman, D. A., and Levine, A. J. (1998). Nuclear export is required for degradation of endogenous p53 by MDM2 and human papillomavirus E6. *Mol Cell Biol* 18:7288–7293.

Gardie, B., Cayuela, J. M., Martini, S., and Sigaux, F. (1998). Genomic alterations of the p19ARF encoding exons in T-cell acute lymphoblastic leukemia. *Blood* 91:1016–1020.

Gazzeri, S., Della Valle, V., Chaussade, L., Brambilla, C., Larsen, C. J., and Brambilla, E. (1998). The human p19ARF protein encoded by the beta transcript of the p16INK4a gene is frequently lost in small cell lung cancer. *Cancer Res* 58:3926–3931.

Geyer, R. K., Yu, Z. K., and Maki, C. G. (2000). The MDM2 RING-finger domain is required to promote p53 nuclear export. *Nat Cell Biol* 2:569–573.

Giaccia, A. J., and Kastan, M. B. (1998). The complexity of p53 modulation: emerging patterns from divergent signals. *Genes Dev* 12:2973–2983.

Gostissa, M., Hengstermann, A., Fogal, V., Sandy, P., Schwarz, S. E., Scheffner, M., and Del Sal, G. (1999). Activation of p53 by conjugation to the ubiquitin-like protein SUMO-1. *EMBO J* 18:6462–6471.

Gotz, C., Kartarius, S., Scholtes, P., Nastainczyk, W., and Montenarh, M. (1999). Identification of a CK2 phosphorylation site in mdm2. *Eur J Biochem* 266:493–501.

Gu, W., and Roeder, R. G. (1997). Activation of p53 sequence-specific DNA binding by acetylation of the p53 C-terminal domain. *Cell* 90:595–606.

Haines, D. S., Landers, J. E., Engle, L. J., and George, D. L. (1994). Physical and functional interaction between wild-type p53 and mdm2 proteins. *Mol Cell Biol* 14:117i–1178.

Hartwell, L. H., and Weinert, T. A. (1989). Checkpoints: controls that ensure the order of cell cycle events. *Science* 246:629–634.

Haupt, Y., Barak, Y., and Oren, M. (1996). Cell type-specific inhibition of p53-mediated apoptosis by mdm2. *EMBO J* 15:1596–1606.

Haupt, Y., Maya, R., Kazaz, A., and Oren, M. (1997). Mdm2 promotes the rapid degradation of p53. *Nature* 387:296–299.

Hinds, P. W., Finlay, C. A., Quartin, R. S., Baker, S. J., Fearon, E. R., Vogelstein, B., and Levine, A. J. (1990). Mutant p53 DNA clones from human colon carcinomas cooperate with ras in transforming primary rat cells: a comparison of the "hot spot" mutant phenotypes. *Cell Growth Differ* 1: 571–580.

Hirao, A., Kong, Y. Y., Matsuoka, S., Wakeham, A., Ruland, J., Yoshida, H., Liu, D., Elledge, S. J., and Mak, T. W. (2000). DNA damage-induced activation of p53 by the checkpoint kinase Chk2. *Science* 287:1824–1827.

Honda, R., Tanaka, H., and Yasuda, H. (1997). Oncoprotein MDM2 is a ubiquitin ligase E3 for tumor suppressor p53. *FEBS Lett* 420:25–27.

Honda, R., and Yasuda, H. (1999). Association of p19(ARF) with Mdm2 inhibits ubiquitin ligase activity of Mdm2 for tumor suppressor p53. *EMBO J* 18:22–27.

Ito, A., Lai, C. H., Zhao, X., Saito, S., Hamilton, M. H., Appella, E., and Yao, T. P. (2001). p300/CBP-mediated p53 acetylation is commonly induced by p53-activating agents and inhibited by MDM2. *EMBO J* 20:1331–1340.

Jackson, M. W., Lindstrom, M. S., and Berberich, S. J. (2001). MdmX binding to ARF affects Mdm2 protein stability and p53 transactivation. *J Biol Chem* 276:25336–25341.

Jones, S. N., Roe, A. E., Donehower, L. A., and Bradley, A. (1995). Rescue of embryonic lethality in Mdm2-deficient mice by absence of p53. *Nature* 378:206–208.

Juven, T., Barak, Y., Zauberman, A., George, D. L., and Oren, M. (1993). Wild type p53 can mediate sequence-specific transactivation of an internal promoter within the mdm2 gene. *Oncogene* 8:3411–3416.

Kamijo, T., Weber, J. D., Zambetti, G., Zindy, F., Roussel, M. F., and Sherr, C. J. (1998). Functional and physical interactions of the ARF tumor suppressor with p53 and Mdm2. *Proc Natl Acad Sci USA* 95:8292–8297.

Kastan, M. B., Onyekwere, O., Sidransky, D., Vogelstein, B., and Craig, R. W. (1991). Participation of p53 protein in the cellular response to DNA damage. *Cancer Res* 51:6304–6311.

Khosravi, R., Maya, R., Gottlieb, T., Oren, M., Shiloh, Y., and Shkedy, D. (1999). Rapid ATM-dependent phosphorylation of MDM2 precedes p53 accumulation in response to DNA damage. *Proc Natl Acad Sci USA* 96:14973–14977.

Kobet, E., Zeng, X., Zhu, Y., Keller, D., and Lu, H. (2000). MDM2 inhibits p300-mediated p53 acetylation and activation by forming a ternary complex with the two proteins. *Proc Natl Acad Sci USA* 97:12547–12552.

Kubbutat, M. H., Jones, S. N., and Vousden, K. H. (1997). Regulation of p53 stability by Mdm2. *Nature* 387:299–303.

Kubbutat, M. H., Ludwig, R. L., Levine, A. J., and Vousden, K. H. (1999). Analysis of the degradation function of Mdm2. *Cell Growth Differ* 10:87–92.

Kussie, P. H., Gorina, S., Marechal, V., Elenbaas, B., Moreau, J., Levine, A. J., and Pavletich, N. P. (1996). Structure of the MDM2 oncoprotein bound to the p53 tumor suppressor transactivation domain. *Science* 274:948–953.

Lai, Z., Ferry, K. V., Diamond, M. A., Wee, K. E., Kim, Y. B., Ma, J., Yang, T., Benfield, P. A., Copeland, R. A., and Auger, K. R. (2001). Human mdm2 mediates multiple mono-ubiquitination of p53 by a mechanism requiring enzyme isomerization. *J Biol Chem* 276:31357–31367.

Lai, Z., Freedman, D. A., Levine, A. J., and McLendon, G. L. (1998). Metal and RNA binding properties of the hdm2 RING finger domain. *Biochemistry* 37:7005–7015.

Landers, J. E., Cassel, S. L., and George, D. L. (1997). Translational enhancement of mdm2 oncogene expression in human tumor cells containing a stabilized wild-type p53 protein. *Cancer Res* 57:3562–3568.

Lin, J., Chen, J., Elenbaas, B., and Levine, A. J. (1994). Several hydrophobic amino acids in the p53 amino-terminal domain are required for transcriptional activation, binding to mdm-2 and the adenovirus 5 E1B 55-kD protein. *Genes Dev* 8:1235–1246.

Lohrum, M. A., Ashcroft, M., Kubbutat, M. H., and Vousden, K. H. (2000). Identification of a cryptic nucleolar-localization signal in MDM2. *Nat Cell Biol* 2:179–181.

Lomax, M., and Fried, M. (2001). Polyoma virus disrupts ARF signaling to p53. *Oncogene* 20:4951–4960.

Lowe, S. W., Schmitt, E. M., Smith, S. W., Osborne, B. A., and Jacks, T. (1993). p53 is required for radiation-induced apoptosis in mouse thymocytes. *Nature* 362:847–849.

Lu, W., Pochampally, R., Chen, L., Traidej, M., Wang, Y., and Chen, J. (2000). Nuclear exclusion of p53 in a subset of tumors requires MDM2 function. *Oncogene* 19:232–240.

Maki, C. G. (1999). Oligomerization is required for p53 to be efficiently ubiquitinated by MDM2. *J Biol Chem* 274:16531–16535.

Maya, R., Balass, M., Kim, S. T., Shkedy, D., Leal, J. F., Shifman, O., Moas, M., Buschmann, T., Ronai, Z., Shiloh, Y., Kastan, M. B., Katzir, E., and Oren, M. (2001). ATM-dependent phosphorylation of Mdm2 on serine 395: role in p53 activation by DNA damage. *Genes Dev* 15:1067–1077.

Mayo, L. D., and Donner, D. B. (2001). A phosphatidylinositol 3-kinase/Akt pathway promotes translocation of Mdm2 from the cytoplasm to the nucleus. *Proc Natl Acad Sci USA* 98:11598–11603.

Mayo, L. D., Turchi, J. J., and Berberich, S. J. (1997). Mdm-2 phosphorylation by DNA-dependent protein kinase prevents interaction with p53. *Cancer Res* 57:5013–5016.

Melchior, F., and Hengst, L. (2000). Mdm2-SUMO1: is bigger better? *Nat Cell Biol* 2:E161–3.

Mendrysa, S. M., and Perry, M. E. (2000). The p53 tumor suppressor protein does not regulate expression of its own inhibitor, MDM2, except under conditions of stress. *Mol Cell Biol* 20:2023–20230.

Midgley, C. A., and Lane, D. P. (1997). p53 protein stability in tumour cells is not determined by mutation but is dependent on Mdm2 binding. *Oncogene* 15:1179–1189.

Moll, U. M., LaQuaglia, M., Benard, J., and Riou, G. (1995). Wild-type p53 protein undergoes cytoplasmic sequestration in undifferentiated neuroblastomas but not in differentiated tumors. *Proc Natl Acad Sci USA* 92:4407–4411.

Moll, U. M., Riou, G., and Levine, A. J. (1992). Two distinct mechanisms alter p53 in breast cancer: mutation and nuclear exclusion. *Proc Natl Acad Sci USA* 89:7262–7266.

Momand, J., Jung, D., Wilczynski, S., and Niland, J. (1998). The MDM2 gene amplification database. *Nucleic Acids Res* 26:3453–3459.

Momand, J., Wu, H. H., and Dasgupta, G. (2000). MDM2–master regulator of the p53 tumor suppressor protein. *Gene* 242:15–29.

Momand, J., Zambetti, G. P., Olson, D. C., George, D., and Levine, A. J. (1992). The mdm-2 oncogene product forms a complex with the p53 protein and inhibits p53-mediated transactivation. *Cell* 69:1237–1245.

Montes de Oca Luna, R., Wagner, D. S., and Lozano, G. (1995). Rescue of early embryonic lethality in mdm2-deficient mice by deletion of p53. *Nature* 378:203–206.

Montes de Oca Luna, R., Tabor, A. D., Eberspaecher, H., Hulboy, D. L., Worth, L. L., Colman, M. S., Finlay, C. A., and Lozano, G. (1996). The organization and expression of the mdm2 gene. *Gene* 33:352–357.

Oliner, J. D., Kinzler, K. W., Meltzer, P. S., George, D. L., and Vogelstein, B. (1992). Amplification of a gene encoding a p53-associated protein in human sarcomas. *Nature* 358:80–83.

Oliner, J. D., Pietenpol, J. A., Thiagalingam, S., Gyuris, J., Kinzler, K. W., and Vogelstein, B. (1993). Oncoprotein MDM2 conceals the activation domain of tumour suppressor p53. *Nature* 362:857–860.

Olson, D. C., Marechal, V., Momand, J., Chen, J., Romocki, C., and Levine, A. J. (1993). Identification and characterization of multiple mdm-2 proteins and mdm-2-p53 protein complexes. *Oncogene* 8:2353–2360.

Parant, J., Chavez-Reyes, A., Little, N. A., Yan, W., Reinke, V., Jochemsen, A. G., and Lozano, G. (2001). Rescue of embryonic lethality in Mdm4-null mice by loss of Trp53 suggests a nonoverlapping pathway with MDM2 to regulate p53. *Nat Genet* 29:92–95.

Perry, M. E., Mendrysa, S. M., Saucedo, L. J., Tannous, P., and Holubar, M. (2000). p76(MDM2) inhibits the ability of p90(MDM2) to destabilize p53. *J Biol Chem* 275:5733–5738.

Picksley, S. M., Vojtesek, B., Sparks, A., and Lane, D. P. (1994). Immunochemical analysis of the interaction of p53 with MDM2;–fine mapping of the MDM2 binding site on p53 using synthetic peptides. *Oncogene* 9:2523–2529.

Piette, J., Neel, H., and Marechal, V. (1997). Mdm2: keeping p53 under control. *Oncogene* 15:1001–1010.

Pomerantz, J., Schreiber-Agus, N., Liegeois, N. J., Silverman, A., Alland, L., Chin, L., Potes, J., Chen, K., Orlow, I., Lee, H. W., Cordon-Cardo, C., and DePinho, R. A. (1998). The Ink4a tumor suppressor gene product, p19Arf, interacts with MDM2 and neutralizes MDM2's inhibition of p53. *Cell* 92:713–723.

Ramqvist, T., Magnusson, K. P., Wang, Y., Szekely, L., Klein, G., and Wiman, K. G. (1993). Wild-type p53 induces apoptosis in a Burkitt lymphoma (BL) line that carries mutant p53. *Oncogene* 8:1495–1500.

Riemenschneider, M. J., Buschges, R., Wolter, M., Reifenberger, J., Bostrom, J., Kraus, J. A., Schlegel, U., and Reifenberger, G. (1999). Amplification and overexpression of the MDM4 (MDMX) gene from 1q32 in a subset of malignant gliomas without TP53 mutation or MDM2 amplification. *Cancer Res* 59:6091–6096.

Ries, S., Biederer, C., Woods, D., Shifman, O., Shirasawa, S., Sasazuki, T., McMahon, M., Oren, M., and McCormick, F. (2000). Opposing effects of Ras on p53: transcriptional activation of mdm2 and induction of p19ARF. *Cell* 103:321–330.

Rodriguez, M. S., Desterro, J. M., Lain, S., Midgley, C. A., Lane, D. P., and Hay, R. T. (1999). SUMO-1 modification activates the transcriptional response of p53. *EMBO J* 18:6455–6461.

Rodriguez-Lopez, A. M., Xenaki, D., Eden, T. O., Hickman, J. A., and Chresta, C. M. (2001). MDM2 mediated nuclear exclusion of p53 attenuates etoposide-induced apoptosis in neuroblastoma cells. *Mol Pharmacol* 59:135–143.

Roth, J., Dobbelstein, M., Freedman, D. A., Shenk, T., and Levine, A. J. (1998). Nucleo-cytoplasmic shuttling of the hdm2 oncoprotein regulates the levels of the p53 protein via a pathway used by the human immunodeficiency virus rev protein. *EMBO J* 17:554–564.

Sakaguchi, K., Saito, S., Higashimoto, Y., Roy, S., Anderson, C. W., and Appella, E. (2000). Damage-mediated phosphorylation of human p53 threonine 18 through a cascade mediated by a casein 1-like kinase. Effect on Mdm2 binding. *J Biol Chem* 275:9278–9283.

Saucedo, L. J., Myers, C. D., and Perry, M. E. (1999). Multiple murine double minute gene 2 (MDM2) proteins are induced by ultraviolet light. *J Biol Chem* 274:8161–8168.

Schlamp, C. L., Poulsen, G. L., Nork, T. M., and Nickells, R. W. (1997). Nuclear exclusion of wild-type p53 in immortalized human retinoblastoma cells. *J Natl Cancer Inst* 89:1530–1536.

Sherr, C. J. (2001). The ink4a/arf network in tumour suppression. *Nat Rev Mol Cell Biol* 2:731–737.

Shieh, S. Y., Ikeda, M., Taya, Y., and Prives, C. (1997). DNA damage-induced phosphorylation of p53 alleviates inhibition by MDM2. *Cell* 91:325–334.

Shvarts, A., Steegenga, W. T., Riteco, N., van Laar, T., Dekker, P., Bazuine, M., van Ham, R. C., van der Houven van Oordt, W., Hateboer, G., van der Eb, A. J., and Jochemsen, A. G. (1996). MDMX: a novel p53-binding protein with some functional properties of MDM2. *EMBO J* 15:5349–5357.

Siliciano, J. D., Canman, C. E., Taya, Y., Sakaguchi, K., Appella, E., and Kastan, M. B. (1997). DNA damage induces phosphorylation of the amino terminus of p53. *Genes Dev* 11:3471–3481.

Stoll, R., Renner, C., Hansen, S., Palme, S., Klein, C., Belling, A., Zeslawski, W., Kamionka, M., Rehm, T., Muhlhahn, P., Schumacher, R., Hesse, F., Kaluza, B., Voelter, W., Engh, R. A., and Holak, T. A. (2001). Chalcone derivatives antagonize interactions between the human oncoprotein MDM2 and p53. *Biochemistry* 40:336–344.

Stommel, J. M., Marchenko, N. D., Jimenez, G. S., Moll, U. M., Hope, T. J., and Wahl, G. M. (1999). A leucine-rich nuclear export signal in the p53 tetramerization domain: regulation of subcellular localization and p53 activity by NES masking. *EMBO J* 18:1660–1672.

Tao, W., and Levine, A. J. (1999). Nucleocytoplasmic shuttling of oncoprotein Hdm2 is required for Hdm2-mediated degradation of p53. *Proc Natl Acad Sci USA* 96:3077–3080.

Tao, W., and Levine, A. J. (1999). P19(ARF) stabilizes p53 by blocking nucleo-cytoplasmic shuttling of Mdm2. *Proc Natl Acad Sci USA* 96:6937–6941.

Teoh, G., Urashima, M., Ogata, A., Chauhan, D., DeCaprio, J. A., Treon, S. P., Schlossman, R. L., and Anderson, K. C. (1997). MDM2 protein overexpression promotes proliferation and survival of multiple myeloma cells. *Blood* 90:1982–1992.

Unger, T., Juven-Gershon, T., Moallem, E., Berger, M., Vogt Sionov, R., Lozano, G., Oren, M., and Haupt, Y. (1999)a. Critical role for Ser20 of human p53 in the negative regulation of p53 by Mdm2. *EMBO J* 18:1805–1814.

Unger, T., Sionov, R. V., Moallem, E., Yee, C. L., Howley, P. M., Oren, M., and Haupt, Y. (1999)b. Mutations in serines 15 and 20 of human p53 impair its apoptotic activity. *Oncogene* 18:3205–3212.

Vassilev, L. T., Vu, B. T., Graves, B., Carvajal, D., Podlaski, F., Filipovic, Z., Kong, N., Kamlott, U., Lukacs, C., Klein, C., Fotouhi, N. and Liu, E. A. (2004). In vivo activation of the p53 pathway by small-molecule antagonists of MDM2. *Science* 303:844–848.

Wang, Y., Ramqvist, T., Szekely, L., Axelson, H., Klein, G., and Wiman, K. G. (1993). Reconstitution of wild-type p53 expression triggers apoptosis in a p53- negative v-myc retrovirus-induced T-cell lymphoma line. *Cell Growth Differ* 4:467–473.

Weber, J. D., Taylor, L. J., Roussel, M. F., Sherr, C. J., and Bar-Sagi, D. (1999). Nucleolar Arf sequesters Mdm2 and activates p53. *Nat Cell Biol* 1:20–26.

Wu, X., Bayle, J. H., Olson, D., and Levine, A. J. (1993). The p53-mdm-2 autoregulatory feedback loop. *Genes Dev* 7:1126–1132.

Xiao, Z. X., Chen, J., Levine, A. J., Modjtahedi, N., Xing, J., Sellers, W. R., and Livingston, D. M. (1995). Interaction between the retinoblastoma protein and the oncoprotein MDM2. *Nature* 375:694–698.

Xirodimas, D., Saville, M. K., Edling, C., Lane, D. P., and Lain, S. (2001). Different effects of p14ARF on the levels of ubiquitinated p53 and Mdm2 in vivo. *Oncogene* 20:4972–4983.

Yaseen, N. R., and Blobel, G. (1999). Two distinct classes of Ran-binding sites on the nucleoporin Nup-358. *Proc Natl Acad Sci USA* 96:5516–5521.

Zambetti, G. P., and Levine, A. J. (1993). A comparison of the biological activities of wild-type and mutant p53. *FASEB J* 7:855–865.

Zauberman, A., Flusberg, D., Haupt, Y., Barak, Y., and Oren, M. (1995). A functional p53-responsive intronic promoter is contained within the human mdm2 gene. *Nucleic Acids Res* 23:2584–2592.

Zhang, Y., and Xiong, Y. (2001). A p53 amino-terminal nuclear export signal inhibited by DNA damage-induced phosphorylation. *Science* 292:1910–1915.

Zhang, Y., Xiong, Y., and Yarbrough, W. G. (1998). ARF promotes MDM2 degradation and stabilizes p53: ARF-INK4a locus deletion impairs both the Rb and p53 tumor suppression pathways. *Cell* 92:725–734.

Zindy, F., Eischen, C. M., Randle, D. H., Kamijo, T., Cleveland, J. L., Sherr, C. J., and Roussel, M. F. (1998). Myc signaling via the ARF tumor suppressor regulates p53-dependent apoptosis and immortalization. *Genes Dev* 12:2424–2433.

8

p53 Family Members: p63 and p73

Elsa R. Flores and Tyler Jacks

SUMMARY

Almost two decades subsequent to the discovery of p53, two homologues p63 and p73 were revealed. Much excitement erupted in the p53 field due to the fact that these genes bear significant homology to p53 primarily in the DNA binding domain. Although the structure of these genes is quite complex, p63 and p73 have been shown to have similar functions to p53 transcriptionally activating a number of known p53 target genes. In mouse models deficient for p63 and p73, developmental roles for these genes have been unveiled. To date, no clear evidence has shown that these genes have tumor suppressive functions similar to those seen for p53, but p63 and p73 have been shown to play a role in apoptosis, an important antitumorigenic pathway. A previously unrecognized connection between p53 and its family members has recently been revealed. p53 depends on p63 and p73 for the induction of apoptosis in response to DNA damage. These new p53 family members appear to be necessary for p53 to bind to DNA and transactivate target genes involved in the apoptotic response. While p63 and p73 were only recently discovered, many new functions of these genes have been found that have important implications for the p53 pathway and cancer therapy.

E. R. FLORES • The University of Texas M.D. Anderson Cancer Center and the University of Texas Graduate School of Biomedical Sciences, Department of Molecular and Cellular Oncology, Houston, TX 77030, USA   T. JACKS • Massachusetts Institute of Technology, Department of Biology and Center for Cancer Research, Cambridge, MA 02139, USA, and Howard Hughes Medical Institute, 4000 Jones Bridge Road, Chevy Chase, MD 20185, USA

The p53 Tumor Suppressor Pathway and Cancer, edited by Zambetti.
Springer Science+Business Media, New York, 2005.

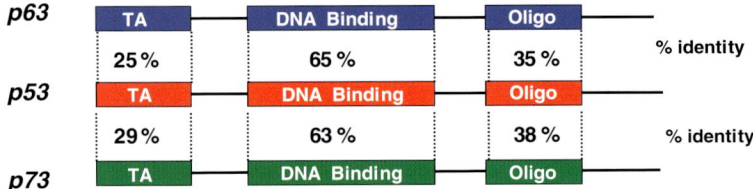

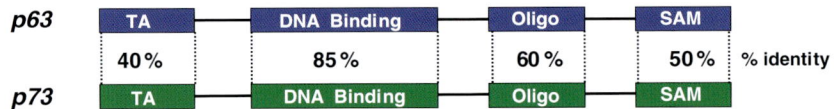

Figure 8.1. Comparison of the structure of the p53 family members, p63 and p73. **A**: p63 and p73 members share ~65% identity to p53 in the DNA binding domain. They share ~25% identity with p53 in the transactivation (TA) and ~35% in the oligomerization (Oligo) domains. **B**: p63 and p73 are more similar to each other than to p53. They share ~85% identity in the DNA binding domain, ~40% in the transactivation domain, ~60% in the oligomerization domain, and 50% in the steric alpha motif (SAM) domain.

8.1. INTRODUCTION TO THE p53 FAMILY

p53 has long been known to be a cellular sensor for DNA damage and has been dubbed the guardian of the genome because of its ability to protect the cell by responding to cellular insults by inducing cell cycle arrest or apoptosis (Vogelstein et al., 2000). In 1997, almost 20 years after the discovery of p53, two genes named p63 and p73 were discovered (Augustin et al., 1998; Kaghad et al., 1997; Osada et al., 1998; Yang et al., 1998). This finding created much excitement, as there are various regions within p63 and p73 that bear significant homology to the well-studied p53. Like p53, p63 and p73 have a transactivation domain, a DNA binding domain, and an oligomerization domain. Both share significant homology with p53 particularly in the DNA binding domain (Fig. 8.1A), though p63 and p73 are more homologous to one another than to p53 (Fig. 8.1B). Because these genes are similar to p53, much speculation grew about their ability to behave like their well-known sibling p53 in their ability to act as tumor suppressor genes. Moreover, these genes were found to reside on chromosomes that are frequently lost or mutated in human cancers. These genes are quite complex and unraveling their functions may help us understand unresolved mysteries about p53.

8.2. p63 AND p73: ORIGIN AND STRUCTURE

p63 (KET, p51, p40, p73L) and p73 were cloned from cDNA libraries (Augustin et al., 1998; Kaghad et al., 1997; Osada et al., 1998; Yang et al., 1998). p63 was found to reside on chromosome 3q27 while p73 is located on chromosome 1p36. The location of these genes gave rise to much conjecture as to whether these genes are involved in tumorigenesis. Some evidence supports this idea. Chromosome 3q27 is amplified in advanced cervical carcinoma and other squamous cell carcinomas, and chromosome 1p36 is frequently lost in neuroblastoma, breast, and colorectal cancer (Kaghad et al., 1997). Thus, it is possible that these genes play a role in the development of cancer.

Unlike p53, many isoforms of p63 and p73 exist (Melino et al., 2002; Yang et al., 2002). Two distinct promoters within p63 and p73 give rise to the full length transactivation forms (TA) of each and the truncated forms lacking the transactivation domain (ΔN) (Fig. 8.2) (Augustin et al., 1998; Osada et al., 1998; Yang et al., 1998, 2000). All isoforms share a common core domain containing the DNA binding domain. Alternative splicing at the carboxy terminus gives rise to multiple isoforms. The isoforms identified for p63 include TAp63α, β, γ and ΔNp63α, β, .γ (Fig. 8.2A) (Yang et al., 1998) Many more isoforms have been identified for p73: the C-terminal isoforms include TAp73α, β, γ, δ, ε, ζ, η the N-terminal isoforms include: the splice variants, Δ2, Δ3, Δ2/3, and variants arising from an internal promoter, ΔNp73α, β, γ, δ, ε, ζ, η (Fig. 8.2B) (De Laurenzi et al., 1998, 1999; Kaghad et al., 1997; Melino et al., 2002; Ueda et al., 1999). The alpha isoforms of both p63 and p73 contain a SAM (steric alpha motif) domain (Arrowsmith, 1999; Chi et al., 1999; Thanos and Bowie, 1999). SAM domains may have important biological function as they mediate protein homodimerization, thus these domains may be important for protein interactions among the family members or with other proteins. The gamma isoforms of p63 and p73, which lack a SAM domain most closely resemble p53. These isoforms are expressed at various levels in different tissues. For example, the ΔN isoforms are generally found at higher levels in murine tissues than the TA isoforms (Yang et al., 1999, 2000). This family of genes also shares homology in the oligomerization domain, a region of the protein important for tetramerization of the p53 molecule necessary for DNA binding and subsequent transactivation of target genes (Vogelstein et al., 2000).

8.3. THE p53 FAMILY TREE

Although p63 and p73 were discovered nearly 20 years after p53, it is thought that p63 and p73 are ancestors of p53 (Yang et al., 2002). As a geologist examines layers of prehistoric rock and the context in which it is found to learn more about the origins of our solar system, the sequencing of genomes from primitive organisms can provide clues about the origins of the p53 family. Many questions have arisen as to whether p63 and p73 evolved from p53 or whether p53 evolved from p63 and p73. There is evidence to support both hypotheses although it is now becoming clear that p53 evolved from p73, which evolved from the ancient ancestor, p63. The p53-like molecule found in *Drosophila melangaster* has a genoprotective role (Brodsky et al., 2000). Based on such findings, p53 is thought to be the ancestor of p63 and p73. An additional piece of evidence that supports this model is that *Drosophila* p53

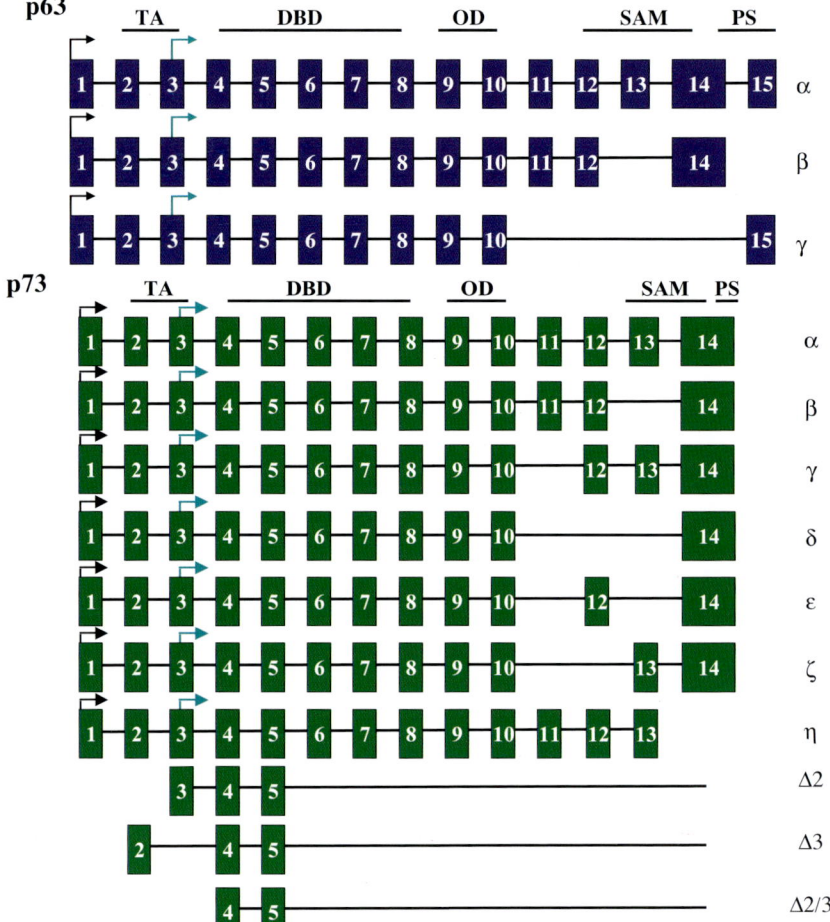

Figure 8.2. Isoforms of p63 and p73. Many isforms of both p63 and p73 have been identified. **A**: The isoforms dubbed TAp63α, β, and γ contain a transactivation domain encompassing exons 2 and 3. TAp63α contains 15 exons while TAp63β and γ are carboxy-terminus splice form derivatives. Only TAp63α+ ΔN$_p$63α contain a SAM and post-SAM (PS) domain. Isoforms lacking a transactivation domain ΔN of each of these exist. The ΔN isoforms are transcribed from an intronic promoter shown by an arrow. **B**: Many isoforms for p73 have been found. TAp73α contains 14 exons, a SAM, and post-SAM (PS) domain. TAp73β, γ, δ, ε, ζ, η are C-terminal splice variants of TAp73α. As with p63, the ΔN isoforms of p73 are transcribed from an internal promoter shown by an arrow.

lacks a SAM domain. Since mammals have p63 and p73 containing a SAM domain, this suggests that isoforms with a SAM domain evolved later (Schultz et al., 1997). Evidence from sequence analysis tells a different story. A sequence comparison of the three family members indicates that p63 is the ancient ancestor. The p53 molecule present in less evolved species such as *Drosophila* and *Caenorrhabditis elegans* more closely resembles p63 and p73 (Yang et al., 2002). The earliest p53 family member was TAp63 with a SAM domain (Fig. 8.2A). Orthologues of p53 became apparent at the vertebrate branch of the evolutionary tree (Yang et al., 2002). At this transition point, there was a gain of the intronic promoter giving rise to the ΔN isoforms of p63. Moreover, C-terminal splicing isoforms appeared lacking the SAM domain. p73 and subsequently p53 arose from gene duplication. Thus, it seems that p73 and p53 coevolved with the ΔN isoforms of p63 and p73.

8.4. PHENOTYPES OF THE p63 AND p73 KNOCKOUT MICE

Shortly after the discovery of p63 and p73, mice were created with deletions in the DNA binding domains of the respective genes resulting in the deletion of all isoforms of either p63 or p73 (Mills et al., 1999; Yang et al., 1999, 2000). Both the p63 and p73 knockout mice exhibit striking developmental defects. Mice lacking p63 die shortly after birth (Mills et al., 1999; Yang et al., 1999). They are born with craniofacial abnormalities and have severe limb truncations. The death of the mice was shown to be due to dessication as these mice are born with abnormal epithelium. Upon staining with various epithelial specific differentiation markers, the p63 deficient skin lacked a normal staining pattern. Embryos deficient for p63 did not stain positively for keratin 5, a marker for the basal compartment of the epithelium. Patches of differentiated skin were apparent on these p63 deficient embryos and were positive for loricrin, a marker of the stratum corneum of the epithelium. These data suggest that p63 is not required for epithelial differentiation yet it may be essential for the renewal of the basal cells of the epithelium. The defects found in p63 deficient mice mimic the defects exhibited in the human disease, ectodermal dysplastic syndrome (EEC) (Celli et al., 1999; Mills et al., 1999; van Bokhoven et al., 2001; Yang et al., 1999). These patients have point mutations in the DNA binding domain of p63 giving rise to electrodactaly and cleft palates.

p73 deficient (p73–/–) mice also have developmental abnormalities (Yang et al., 2000). Unlike p63–/– mice, p73–/– mice can survive to adulthood but are more prone to bacterial infections and exhibit many secretory defects. In some cases, these mice can live to be over a year old with no increased incidence of tumorigenesis. They are born with hydrocephalus due to the increased secretion of cerebral spinal fluid by the choroid plexus. They also exhibit many other developmental abnormalities including increased sympathetic neuronal cell death, hippocampal defects, hyper-inflammatory responses, defects in pheromone detection, and a runted apprearance (Pozniak et al., 2000; Yang et al., 2000). The phenotype of the p73 deficient mice is so complex that it is difficult to assign a single function to p73.

The study of the p63 and p73 deficient mice has provided some evidence that p63 and p73 may not be involved in cancer. Unlike the p53 knockout mouse which develop thymic lymphomas, fibrosarcomas, and several other tumor types, the p73 knockout mice can live into adulthood without the evidence of tumors (Donehower et al., 1992; Jacks et al., 1994). These data suggest that p63 and p73 evolved as essential genes for development while p53 evolved as a cellular sensor for genotoxic stress.

8.5. p63 AND p73 EXHIBIT p53 – LIKE PROPERTIES

The new p53 family members have many p53 like properties. They can transactivate p53 target genes and induce apoptosis (Jost et al., 1997; Yang et al., 1998). The transactivation competent isoforms of p63 and p73 share similar functions to p53. Some isoforms of TAp63 and TAp73 have been reported to transactivate the p53 target genes, p21, bax, GADD45, and mdm2, to varying degrees, in some cases, more effectively than p53 itself (Lee and La Thangue, 1999; Zhu et al., 1998). Also, some isoforms of these family members can induce apoptosis. For example, TAp63γ has been the most effective p63 isoform in inducing apoptosis (Yang et al., 1998). ΔNp63γ also induced a modest amount of apoptosis. These isoforms of p63 could also induce luciferase plasmids containing a p53 response element. While the gamma isoforms of p63 induce apoptosis, both TAp63α and ΔNp63α could not induce apoptosis. By sequence analysis, the gamma isoforms of the family members most closely resemble p53 because they lack the SAM domain. Consequently, it is not surprising that these are the isoforms that induce apoptosis. Both TAp73α and TAp73β are potent transactivators of p53 target genes. These isoforms of p73 also induce apoptosis while the ΔN versions of these isoforms do not (Jost et al., 1997). Much like E2F-1 can regulate p53, it has also been shown to regulate p73. Its action on p73 is quite different from that on p53. Promoter analysis of p73 has revealed three E2F-1 binding sites upstream of exon 1 (Seelan et al., 2002). E2F-1 was found to transactivate p73 and lead to a p53-independent mechanism of apoptosis (Irwin et al., 2000; Lissy et al., 2000).

Some of the isoforms of p63 and p73 have antagonistic effects on p53 and the transactivation competent isoforms of p63 and p73. The ΔN isoforms of both p63 and p73 have been found to have antagonistic effects on p53 (Pozniak et al., 2000; Yang et al., 1998, 2000). These isoforms can hetero-oligomerize with p53, TAp63, and TAp73 and presumably compete with the transactivation competent forms for DNA binding sites, thus impairing transactivation by p53, TAp63, and TAp73.

p63 and p73 also share distinct properties from p53. While mdm2 associates with p53 and targets it for ubiquitin-mediated degradation, the same is not true for p73 (Zeng et al., 1999). mdm2 binds to p73 and blocks its ability to transactivate target genes by suppressing the association of p73 with p300-CBP. Viral oncoproteins have been shown to have different activities on p53 and its family members. These viral oncoproteins disrupt the normal function of cellular proteins like p53 to make the cell a more hospitable environment for viral replication that can lead to cellular transformation. While SV40 large T antigen and E6 from human papillomavirus type 18 (HPV-18) interact and inactivate p53, they do not inactivate p63 or p73 (Marin et al., 1998; Roth and Dobbelstein, 1999; Roth et al., 1998). In contrast, both adenovirus E1A inactivates p73, and the Tax protein from human T-cell lymphotropic virus 1 (HTLV1) bind and inactivate p63 and p73 (Kaida et al., 2000). These data suggest that the p53 family members play different roles in some of these viruses' life cycles and cellular transformation.

8.6. p63 AND p73 IN THE DNA DAMAGE RESPONSE

p53 binds to DNA as a tetramer and transactivates a multitude of genes involved in cell cycle arrest or apoptosis in response to DNA damage (Vogelstein et al., 2000; Vousden and Lu, 2002). In this manner, p53 elicits its antitumorigenic activities. Given the homology between p53 and its newly discovered family members, p63 and p73, it was an intriguing possibility that these family members may play a role in the response to DNA damage. p73 has previously been shown to be induced in response to specific DNA damaging agents such as cisplatin (Gong et al., 1999). This induction requires the c-abl tyrosine kinase which phosphorylates p73 following DNA damage (Agami et al., 1999; Gong et al., 1999; Yuan et al., 1999). This result demonstrated that p73 is indeed involved in the DNA damage response. To test the possibility that all three p53 family members play a role in the DNA damage response, developing mouse embryos and mouse embryo fibroblasts (MEFs) deficient for the p53 family members individually and in combination were treated with DNA damaging agents: doxorubicin, cisplatin, or gamma irradiation. Cells deficient for the p53 family members were found to be defective for the apoptotic response (Flores et al., 2002). Embryos and MEFs deficient for p63 or p73 individually exhibited an intermediate resistance while cells deficient for both p63 and p73 were as resistant to apoptosis as cells lacking p53 itself. This result was surprising and unexpected given that p53 was present in these cells and indicated that p63 and p73 are required for p53-dependent apoptosis by compromising the ability of p53 to induce apoptosis.

In cells lacking p63 and p73, p53 was induced to the same levels detected in wild-type MEFs (Flores et al., 2002). Northern and western blot analysis revealed that p53 was defective in its ability to induce genes involved specifically in apoptosis while genes involved in cell cycle arrest were unaffected. Using the chromatin immunoprecipitation technique, in cells lacking both p63 and p73, p53 was found to occupy its binding sites on the promoters of cell cycle arrest genes only. The ability of p53 to bind to promoters of apoptotic target genes, such as bax, PERP, and NOXA, was greatly impaired in cells deficient for p63 and p73 indicating that they are needed for p53 to occupy these promoters.

These data have provided a piece of the puzzle of a question that has long plagued the p53 field: how does p53 induce cell cycle arrest in some conditions and apoptosis in others? Perhaps, the critical players involved in the cell death decision process are p63 and p73. In a cell cycle arrest mode, p53 acts alone in response to stress and contacts the promoters of genes such as p21 to signal the cell to stop proliferating. In an apoptotic mode, p53 depends on the presence of both p63 and p73 to contact the promoters of apoptotic genes and subsequently induce them (Fig. 8.3). The action of p63 and p73 on p53 in apoptosis is unclear at this time, but may have to do with the ability of p53 to occupy promoter elements.

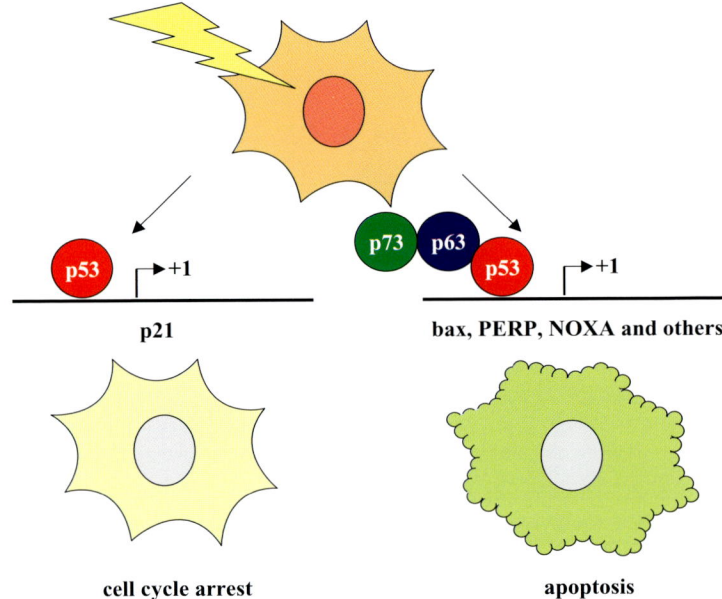

Figure 8.3. p63 and p73 in the DNA damage response pathway. p53 responds to cellular insults by inducing cell cycle arrest (left) or apoptosis (right). Cell cycle arrest is induced by the transcriptional activation of genes like p21 by p53. Apoptosis is induced by the transactivation of many genes, such as, bax, PERP, and NOXA. The presence of p63 and p73 are required for p53 to occupy the binding sites on these promoters.

How might p63 and p73 be acting to help p53 occupy its binding sites? Some p53 target genes have been found to have low affinity p53 binding sites. Other sites are known to bind p53 quite strongly like p21. The paper by Flores et al. (2002) showed that within the PERP promoter containing two p53 consensus binding sites, p53 binds preferentially to the site at –218 while p63 preferentially binds site at –2097 (Fig. 8.4). The preferential binding by different p53 family members may be explained by various scenarios. Some possibilities are that p63 and p73 could be recruiting p53 to binding sites. In the case of the PERP promoter, p63 could be serving the purpose of binding to and recruiting or anchoring p53 to the neighboring low affinity site. Alternatively, p63, which binds to its own site, could be needed for full transcriptional activation of the PERP gene. More experiments need to be performed on multiple p53 target genes involved in cell cycle arrest and apoptosis to get a clear answer of the interplay of the p53 family at various promoters. While we have data indicating p63's possible action on p53, the role of p73 has yet to be investigated. Clearly, there are caveats to these models. p63 and p73 have been shown to weakly interact with wild-type p53, but perhaps weak interactions are all that is needed to bring this family together at the promoters of apoptotic genes (Gaiddon et al., 2001). Moreover, under conditions of stress where levels of all three proteins have been shown to increase, these interactions

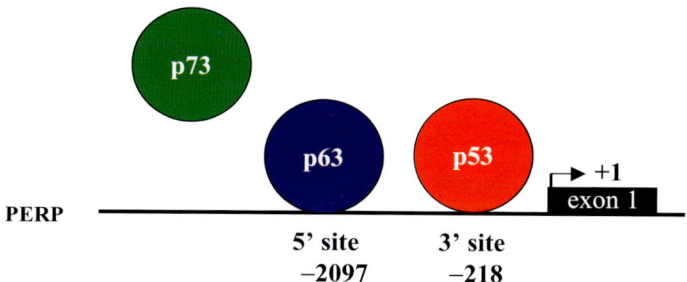

Figure 8.4. Promoter occupancy of p53 and p63. The two putative p53 binding sites on the PERP promoter at –2097 and –218 were analyzed by chromatin immunoprecipitation (ChIP). p53 preferentially binds to the site at –218 while p63 preferentially binds site at –2096. The role of p73 in promoter occupancy has yet to be determined.

could take place more robustly (Flores et al., 2002; Zaika et al., 2001; Zeng et al., 2002). All of these possibilities require more investigation.

These data place p63 and p73 in the critically important p53 antitumorigenic pathway. They play essential roles in p53-dependent apoptosis. This raises important questions about p63 and p73 in vivo. Are they involved in cancer and what does this say about the phenotype of the p63 and p73 knockout mice? Recall that the p63 deficient mice could not be aged to determine its role in tumorigenesis while the p73 deficient mice showed no increase in tumorigenesis. Based on the apoptosis data, it is possible that p63 and p73 act in concert in the development of tumors. Perhaps, both need to be deleted in order for tumors to form. These questions have yet to be answered, but will yield important data on how these genes may act in cancer. Ageing p63/p73 double heterozygous mice and other models where these genes are deleted together in specific tissues will yield important insight into these genes' possible contributions to tumorigenesis.

8.7. IMPLICATIONS FOR CANCER AND THE FUTURE

When p63 and p73 were first discovered, much excitement erupted because these genes were thought to be tumor suppressor genes like p53. After thorough analyses, p63 and p73 were found to be rarely mutated in human cancers (Irwin and Kaelin, 2001). Do they really have a role in cancer? This is an important question and much work needs to be done to investigate this.

Many studies have been conducted to find mutations in the p63 and p73 in human tumors. p73 is located on a region of chromosome 1 that is frequently lost in neuroblastoma, breast, and colorectal cancer (Kaghad et al., 1997). p63 is located on a region of chromosome 3 that is altered in cancers of the lung cervix and ovary (Yang et al., 1998). In addition, the p73 knockout mice do not show an increased susceptibility to tumorigenesis (Yang et al., 2000). While the p63 knockout mice die too early to assess its role in tumorigenesis, p63 heterozygous mice do not appear to be more susceptible to tumorigenesis (Yang et al., 1999). How then could these genes be involved in cancer? These genes have a clear role in the DNA damage response. Based on this finding alone, taking another look at the role of p63 and p73 in cancer is worth the effort. It is possible that p63 and p73 do not act as classical tumor suppressor genes and that through their loss they promote tumorigenesis through another mechanism. One possibility that has been put forth is that perhaps p63 and p73 may play a role in p53 tumorigenesis. The data supporting this idea is that p63 and p73 can interact with certain mutants of p53 while interactions with wild-type p53 are weak (Gaiddon et al., 2001; Strano et al., 2002). These interactions were shown to inhibit the activities of p63 and p73. This implies that mutant p53 could inactivate p63 and p73 through these interactions blunting p63 and p73 apoptotic response. This finding is significant, because cancer prone patients with the human disease Li Fraumeni Syndrome (LFS) have point mutations within the DNA binding domain of p53 (Vousden and Lu, 2002). It is precisely these types of p53 mutations that have been found in complex with p63 and p73 (Gaiddon et al., 2001; Strano et al., 2002). Through this mechanism, p63 and p73 may be assigned a role in the genesis of tumors.

Mutant p53, like that found in LFS is thought to induce chemoresistance through binding and neutralization of p73. Recent studies have indeed demonstrated that p73 plays a key role in chemosensitivity (Bergamaschi et al., 2003; Irwin et al., 2003). p73 was shown to accumulate in many cell lines including squamous cell carcinomas after treatment with various chemotherapeutic agents. In these studies, knockdown of mutant p53 using siRNA technology enhances the effectiveness of chemotherapeutic agents. This enhanced chemosensitivity is thought to be due to the loss of the neutralization of p73 by mutant p53. These studies could lead to a new anti-tumor therapy via inhibition of mutant p53.

One of the most well studied genes welcomed two new members into its family. In the first five years after the discovery of p63 and p73, much new knowledge has been accumulated not only about these genes but also about the p53 family as a whole. p63 and p73 are both much more complex than p53, and it is clear that there is interplay between these family members. Much more investigation is needed to understand this family as a whole in processes such as development and cancer.

REFERENCES

Agami, R., Blandino, G., Oren, M., and Shaul, Y. (1999). Interaction of c-Abl and p73alpha and their collaboration to induce apoptosis. *Nature* 399:809–813.

Arrowsmith, C. H. (1999). Structure and function in the p53 family. *Cell Death Differ* 6:1169–1173.

Augustin, M., Bamberger, C., Paul, D., and Schmale, H. (1998). Cloning and chromosomal mapping of the human p53-related KET gene to chromosome 3q27 and its murine homolog Ket to mouse chromosome 16. *Mamm Genome* 9:899–902.

Bergamaschi, D., Gasco, M., Hiller, L., Sullivan, A., Syed, N., Trigiante, G., Yulug, I., Merlano, M., Numico, G., Comino, A., et al. (2003). p53 polymorphism influences response in cancer chemotherapy via modulation of p73-dependent apoptosis. *Cancer Cell* 3:387–402.

Brodsky, M. H., Nordstrom, W., Tsang, G., Kwan, E., Rubin, G. M., and Abrams, J. M. (2000). Drosophila p53 binds a damage response element at the reaper locus. *Cell* 101:103–113.

Celli, J., Duijf, P., Hamel, B. C., Bamshad, M., Kramer, B., Smits, A. P., Newbury-Ecob, R., Hennekam, R. C., van Buggenhout, G., van Haeringen, A., et al. (1999). Heterozygous germline mutations in the p53 homolog p63 are the cause of EEC syndrome. *Cell* 99:143–153.

Chi, S. W., Ayed, A., and Arrowsmith, C. H. (1999). Solution structure of a conserved C-terminal domain of p73 with structural homology to the SAM domain. *EMBO J* 18:4438–4445.

De Laurenzi, V., Costanzo, A., Barcaroli, D., Terrinoni, A., Falco, M., Annicchiarico-Petruzzelli, M., Levrero, M., and Melino, G. (1998). Two new p73 splice variants, gamma and delta, with different transcriptional activity. *J Exp Med* 188:1763–1768.

De Laurenzi, V. D., Catani, M. V., Terrinoni, A., Corazzari, M., Melino, G., Costanzo, A., Levrero, M., and Knight, R. A. (1999). Additional complexity in p73: induction by mitogens in lymphoid cells and identification of two new splicing variants epsilon and zeta. *Cell Death Differ* 6:389–390.

Donehower, L. A., Harvey, M., Slagle, B. L., McArthur, M. J., Montgomery, C. A., Jr., Butel, J. S., and Bradley, A. (1992). Mice deficient for p53 are developmentally normal but susceptible to spontaneous tumours. *Nature* 356:215–221.

Flores, E. R., Tsai, K. Y., Crowley, D., Sengupta, S., Yang, A., McKeon, F., and Jacks, T. (2002). p63 and p73 are required for p53-dependent apoptosis in response to DNA damage. *Nature* 416: 560–564.

Gaiddon, C., Lokshin, M., Ahn, J., Zhang, T., and Prives, C. (2001). A subset of tumor-derived mutant forms of p53 down-regulate p63 and p73 through a direct interaction with the p53 core domain. *Mol Cell Biol* 21:1874–1887.

Gong, J. G., Costanzo, A., Yang, H. Q., Melino, G., Kaelin, W. G., Jr., Levrero, M., and Wang, J. Y. (1999). The tyrosine kinase c-Abl regulates p73 in apoptotic response to cisplatin-induced DNA damage. *Nature* 399:806–809.

Irwin, M., Marin, M. C., Phillips, A. C., Seelan, R. S., Smith, D. I., Liu, W., Flores, E. R., Tsai, K. Y., Jacks, T., Vousden, K. H., and Kaelin, W. G., Jr. (2000). Role for the p53 homologue p73 in E2F-1-induced apoptosis. *Nature* 407:645–648.

Irwin, M. S., and Kaelin, W. G., Jr. (2001). Role of the newer p53 family proteins in malignancy. *Apoptosis* 6:17–29.

Irwin, M. S., Kondo, K., Marin, M. C., Cheng, L. S., Hahn, W. C., and Kaelin, W. G. (2003). Chemosensitivity linked to p73 function. *Cancer Cell* 3:403–410.

Jacks, T., Remington, L., Williams, B. O., Schmitt, E. M., Halachmi, S., Bronson, R. T., and Weinberg, R. A. (1994). Tumor spectrum analysis in p53-mutant mice. *Curr Biol* 4:1–7.

Jost, C. A., Marin, M. C., and Kaelin, W. G., Jr. (1997). p73 is a simian [correction of human] p53-related protein that can induce apoptosis. *Nature* 389:191–194.

Kaghad, M., Bonnet, H., Yang, A., Creancier, L., Biscan, J. C., Valent, A., Minty, A., Chalon, P., Lelias, J. M., Dumont, X., et al. (1997). Monoallelically expressed gene related to p53 at 1p36, a region frequently deleted in neuroblastoma and other human cancers. *Cell* 90:809–819.

Kaida, A., Ariumi, Y., Ueda, Y., Lin, J. Y., Hijikata, M., Ikawa, S., and Shimotohno, K. (2000). Functional impairment of p73 and p51, the p53-related proteins, by the human T-cell leukemia virus type 1 Tax oncoprotein. *Oncogene* 19:827–830.

Lee, C. W., and La Thangue, N. B. (1999). Promoter specificity and stability control of the p53-related protein p73. *Oncogene* 18:4171–4181.

Lissy, N. A., Davis, P. K., Irwin, M., Kaelin, W. G., and Dowdy, S. F. (2000). A common E2F-1 and p73 pathway mediates cell death induced by TCR activation. *Nature* 407:642–645.

Marin, M. C., Jost, C. A., Irwin, M. S., DeCaprio, J. A., Caput, D., and Kaelin, W. G., Jr. (1998). Viral oncoproteins discriminate between p53 and the p53 homolog p73. *Mol Cell Biol* 18:6316–6324.

Melino, G., De Laurenzi, V., and Vousden, K. H. (2002). p73: Friend or foe in tumorigenesis. *Nat Rev Cancer* 2:605–615.

Mills, A. A., Zheng, B., Wang, X. J., Vogel, H., Roop, D. R., and Bradley, A. (1999). p63 is a p53 homologue required for limb and epidermal morphogenesis. *Nature* 398:708–713.

Osada, M., Ohba, M., Kawahara, C., Ishioka, C., Kanamaru, R., Katoh, I., Ikawa, Y., Nimura, Y., Naka-gawara, A., Obinata, M., and Ikawa, S. (1998). Cloning and functional analysis of human p51, which structurally and functionally resembles p53. *Nat Med* 4:839–843.

Pozniak, C. D., Radinovic, S., Yang, A., McKeon, F., Kaplan, D. R., and Miller, F. D. (2000). An anti-apoptotic role for the p53 family member, p73, during developmental neuron death. *Science* 289: 304–306.

Roth, J., and Dobbelstein, M. (1999). Failure of viral oncoproteins to target the p53-homologue p51A. *J Gen Virol* 80 (12):3251–3255.

Roth, J., Konig, C., Wienzek, S., Weigel, S., Ristea, S., and Dobbelstein, M. (1998). Inactivation of p53 but not p73 by adenovirus type 5 E1B 55-kilodalton and E4 34-kilodalton oncoproteins. *J Virol* 72:8510–8516.

Schultz, J., Ponting, C. P., Hofmann, K., and Bork, P. (1997). SAM as a protein interaction domain involved in developmental regulation. *Protein Sci* 6:249–253.

Seelan, R. S., Irwin, M., van der Stoop, P., Qian, C., Kaelin, W. G., Jr., and Liu, W. (2002). The human p73 promoter: characterization and identification of functional E2F binding sites. *Neoplasia* 4:195–203.

Strano, S., Fontemaggi, G., Costanzo, A., Rizzo, M. G., Monti, O., Baccarini, A., Del Sal, G., Levrero, M., Sacchi, A., Oren, M., and Blandino, G. (2002). Physical interaction with human tumor-derived p53 mutants inhibits p63 activities. *J Biol Chem* 277:18817–18826.

Thanos, C. D., and Bowie, J. U. (1999). p53 Family members p63 and p73 are SAM domain-containing proteins. *Protein Sci* 8:1708–1710.

Ueda, Y., Hijikata, M., Takagi, S., Chiba, T., and Shimotohno, K. (1999). New p73 variants with altered C-terminal structures have varied transcriptional activities. *Oncogene* 18:4993–4998.

van Bokhoven, H., Hamel, B. C., Bamshad, M., Sangiorgi, E., Gurrieri, F., Duijf, P. H., Vanmolkot, K. R., van Beusekom, E., van Beersum, S. E., Celli, J., et al. (2001). p63 Gene mutations in eec syndrome, limb-mammary syndrome, and isolated split hand-split foot malformation suggest a genotype-phenotype correlation. *Am J Hum Genet* 69:481–492.

Vogelstein, B., Lane, D., and Levine, A. J. (2000). Surfing the p53 network. *Nature* 408:307–310.

Vousden, K. H., and Lu, X. (2002). Live or let die: the cell's response to p53. *Nat Rev Cancer* 2:594–604.

Yang, A., Kaghad, M., Caput, D., and McKeon, F. (2002). On the shoulders of giants: p63, p73 and the rise of p53. *Trends Genet* 18:90–95.

Yang, A., Kaghad, M., Wang, Y., Gillett, E., Fleming, M. D., Dotsch, V., Andrews, N. C., Caput, D., and McKeon, F. (1998). p63, a p53 homolog at 3q27-29, encodes multiple products with transactivating, death-inducing, and dominant-negative activities. *Mol Cell* 2:305–316.

Yang, A., Schweitzer, R., Sun, D., Kaghad, M., Walker, N., Bronson, R. T., Tabin, C., Sharpe, A., Caput, D., Crum, C., and McKeon, F. (1999). p63 is essential for regenerative proliferation in limb, craniofacial and epithelial development. *Nature* 398:714–718.

Yang, A., Walker, N., Bronson, R., Kaghad, M., Oosterwegel, M., Bonnin, J., Vagner, C., Bonnet, H., Dikkes, P., Sharpe, A., et al. (2000). p73-deficient mice have neurological, pheromonal and inflam-matory defects but lack spontaneous tumours. *Nature* 404:99–103.

Yuan, Z. M., Shioya, H., Ishiko, T., Sun, X., Gu, J., Huang, Y. Y., Lu, H., Kharbanda, S., Weichselbaum, R., and Kufe, D. (1999). p73 is regulated by tyrosine kinase c-Abl in the apoptotic response to DNA damage. *Nature* 399:814–817.

Zaika, A., Irwin, M., Sansome, C., and Moll, U. M. (2001). Oncogenes induce and activate endogenous p73 protein. *J Biol Chem* 276:11310–11316.

Zeng, S. X., Dai, M. S., Keller, D. M., and Lu, H. (2002). SSRP1 functions as a co-activator of the transcriptional activator p63. *EMBO J* 21:5487–5497.

Zeng, X., Chen, L., Jost, C. A., Maya, R., Keller, D., Wang, X., Kaelin, W. G., Jr., Oren, M., Chen, J., and Lu, H. (1999). MDM2 suppresses p73 function without promoting p73 degradation. *Mol Cell Biol* 19:3257–3266.

Zhu, J., Jiang, J., Zhou, W., and Chen, X. (1998). The potential tumor suppressor p73 differentially regulates cellular p53 target genes. *Cancer Res* 58:5061–5065.

9

The Oncogenic Activity of p53 Mutants

Alex Sigal and Varda Rotter

SUMMARY

Single mutations in the DNA binding domain of p53 cause a radical shift in function from tumor suppressor to oncogene. The mutated proteins lose the negative feedback regulation mediated by MDM2. Their oncogenic activity consists of a dominant negative inhibition of the remaining wild-type p53 protein, and a gain of function (GOF) activity independent of wild-type p53 inhibition. An understanding of the properties of these very common oncogenes is yielding promising therapeutic approaches, and is predicted to offer more clinical applications as the field develops.

9.1. INTRODUCTION

The p53 tumor suppressor gene is mutated in the majority of cancer cases, and this has spurred intense interest in the multifunctional protein. The mutations, about 95% of which fall within the DNA binding domain (DBD), cause both a loss of wild-type p53 function and a gain of oncogenic function.

The majority of mutations in p53 are missense, which result in full-length, albeit mutant proteins. The majority of mutations in other common tumor suppressors such as ATM are frameshift, nonsense, and other types, that lead to a truncated protein or no protein at all (Hussain and Harris, 1998b) (Fig. 9.1). This implies that the presence

A. SIGAL AND V. ROTTER • Department of Molecular Cell Biology, Weizmann Institute of Science, Rehovot, 76100, Israel

The p53 Tumor Suppressor Pathway and Cancer, edited by Zambetti.
Springer Science+Business Media, New York, 2005.

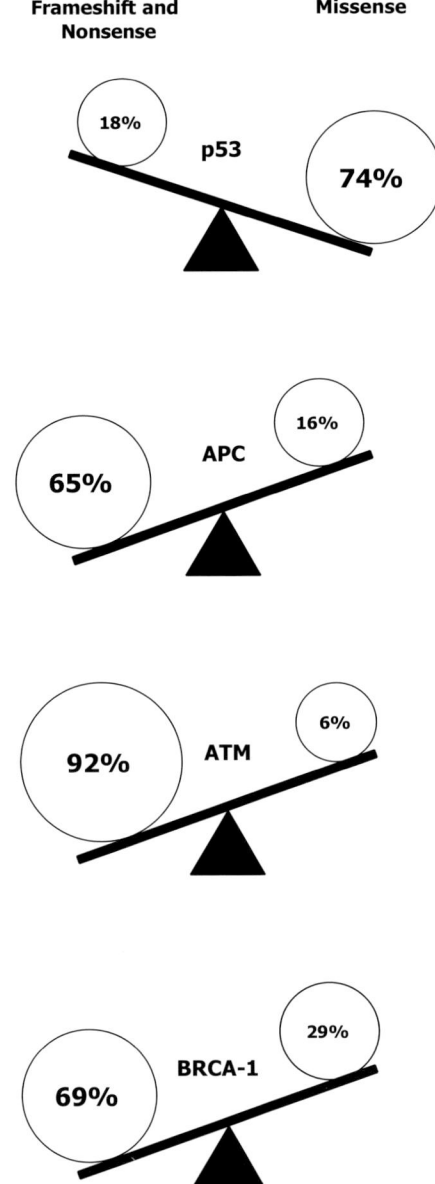

Figure 9.1. Missense mutations are the predominant mutation type in p53. Unlike in other tumor suppressor genes, where the predominant mutation types are nonsense and frameshift, most p53 mutations are missense and result in the production of a full-length mutant protein. This shift in mutation type may indicate that missense p53 mutations offer cancer cells a selective advantage. Data adapted from (Hussain and Harris, 1998b).

of the full-length mutant p53 protein confers some selective advantage to cancer cells.

Moreover, the overwhelming majority of mutations cluster in the DNA binding core domain of the protein (amino acids 102–292; Fig. 9.2). While this domain is large, the absence of mutations at either the N-terminus or C-terminus is striking. Such an absence can reflect the inability of N-terminal or C-terminal mutations to interfere with the wild-type function of p53. Also, it may reflect the inability of the mutated protein to act as an oncogene and aid the selection of cancer cells expressing it. Alternatively, it may at least partly be the result of the preference of some mutagenizing agents for specific sites in the gene segment coding for the DBD.

The DBD is critical for the ability of p53 to bind specific sequences and therefore transactivate target genes. However, few of the DBD residues are involved in direct binding of DNA, and many oncogenic mutations in the DBD occur at residues elsewhere. Furthermore, there are other domains important for p53 function. For example, the N-terminal amino acids 22 and 23 are critical for transactivation (Lin et al., 1994). In addition, the C-terminal tetramerization domain plays a major role in enabling p53 function. Mutations of residues 337 or 344, shown to destabilize or disrupt the domain (DiGiammarino et al., 2002; Varley et al., 1996), were linked to Li-Fraumeni syndrome (Lomax et al., 1997; Varley et al., 1996) and in the case of residue 337, to adrenal cortical carcinoma (Ribeiro et al., 2001). Finally, there are some indications that p53 apoptotic function may be impaired with modifications to the extreme C-terminus (Almog et al., 2000). However, these N and C terminal areas are mutated surprisingly little (Fig. 9.2).

This underlines a distinctive feature of p53 mutations: it is sufficient to modify one amino acid in the core DBD to radically change the function of p53. The mutations at the DNA binding residues (notably at arginine 248 or arginine 273) cause a loss of p53 specific DNA binding, while a second class of mutations cause a pronounced conformational shift of the protein (reviewed in Bullock and Fersht (2001)).

Besides causing a loss of wild-type p53 function, mutations in the DBD offer a selective advantage to cancer cells through the oncogenic activity of the mutated proteins (reviewed in Sigal and Rotter 2000)). This consists of dominant negative activity that neutralizes the p53 from the remaining wild-type allele, and a gain of function (GOF) effect independent of the presence of wild-type p53.

Recent studies have shown that in mutant p53 GOF, there exists a requirement for both the N-terminal transactivation domain and the extreme C-terminus (Matas et al., 2001; Sigal et al., 2001). These areas may still enable the mutant protein to possess abilities, such as transactivation in the case of the N-terminus, though the targets may be different. Therefore, the view that emerges is that oncogenic p53 mutations cause no loss of function, but a shift in function, and the new functions require specific domains to carry them out.

This chapter is an introduction to the oncogenic effects of p53 mutants, their possible mechanisms of action, differences between mutants, and the therapeutic approaches that can be used to combat their activity.

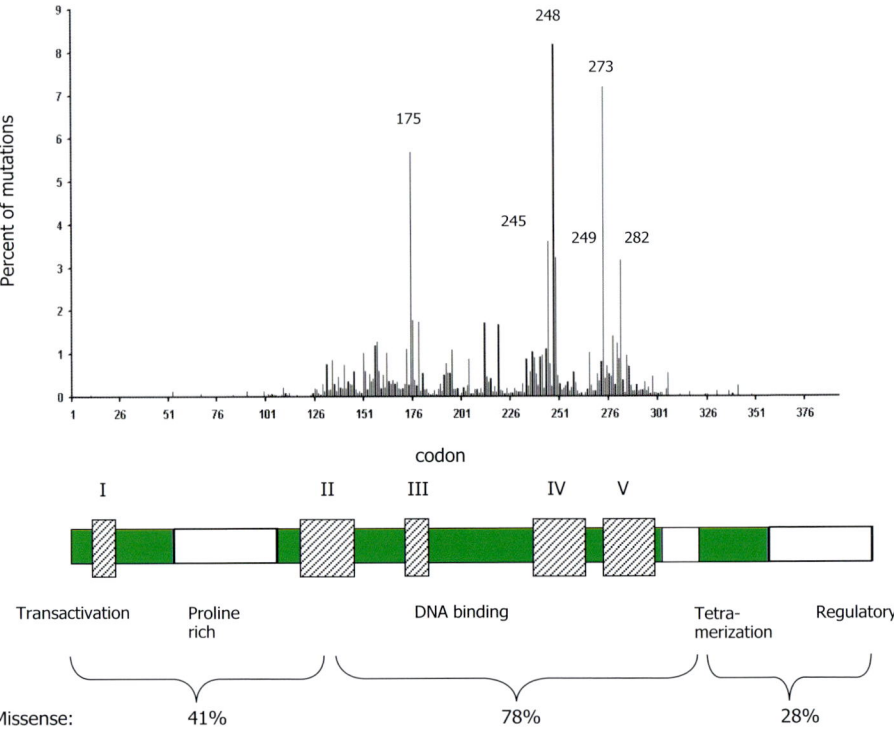

Figure 9.2. Mutation spectrum of the p53 gene in human tumors. The vast majority of p53 mutations in the IARC R5 database [(Hernandez-Boussard et al., 1999), found at http://www.iarc.fr/p53/index.html], cluster in the DBD, especially in conserved regions III to V. The DBD is also enriched for missense mutations, pointing toward a selection for these mutations over other types. A very small proportion of missense mutations is found in the other domains of p53, which indicates that missense mutations outside the DBD are considerably less effective in disrupting p53 function and/or producing proteins with dominant negative and GOF properties. Hatched rectangles denote conserved regions I through V. The percentage of the total mutations that are missense in the regions indicated is taken from Hussain and Harris (1998b).

9.2. ONCOGENIC EFFECTS OF p53 MUTANTS I: DOMINANT NEGATIVE SUPPRESSION

The role of p53 mutants in cancer involves transdominant suppression of wild-type p53, and a wild-type p53-independent oncogenic GOF. The dominant negative effect of p53 mutants has been characterized in a variety of processes that involve wild-type p53 function. Mutant p53 proteins have been found to repress the ability of wild-type p53 to bind to its various specific DNA target sequences (Chene, 1998; Kern et al., 1992; Shaulian et al., 1992b; Srivastava et al., 1993; Unger et al., 1993) and transactivate downstream genes (Kern et al., 1992). They have also been found to repress p53-mediated development and differentiation, (Aloni-Grinstein et al., 1993, 1995; Mazzaro et al., 1999; Soddu et al., 1996; Wallingford et al., 1997) apoptosis (Gottlieb et al., 1994; Lotem and Sachs, 1993), growth arrest (Aloni-Grinstein et al., 1995), constitutive p21 expression (Tang et al., 1998) genomic stability (Liu et al., 1996; Mekeel et al., 1997), resistance to H-1 parvovirus infection (Telerman et al., 1993), immortalization (Cao et al., 1997), and inhibition of *ras* transformation of rat embryo fibroblasts (REFs) (Eliyahu et al., 1988; Hinds et al., 1989; Shaulian et al., 1992a). Instead of full-length mutant p53, Shaulian and coworkers used a fragment named DD, consisting of amino acids 1–4 and 315–390. Transformation by *ras* is thought to be inhibited by wild-type p53 and DD was as effective as the full-length murine p53 135(A to V) mutant in cooperating with *ras* to transform REFs. When part of the tetramerization domain was deleted in DD, it lost its transforming ability.

Evidence of heteromerization between wild-type and mutant p53 species forms the basis for the generally accepted mechanism behind mutant p53 transdominant suppression. Wild-type p53 forms a tetramer to perform its tumor suppresser activity, and this oligomerization is mediated by the tetramerization domain (residues 310–360). This region is fully functional in core domain mutants (Chene, 1998; Unger et al., 1993). The 281 (D to G) mutant, found to inhibit wild-type p53 transactivation of a reporter construct, did not do so when residue 344 in its tetramerization domain was mutated (Chene and Bechter, 1999). Similarly, dominant negative inhibition of transactivation by the 179(H to Q) mutant was lost upon deletion of amino acids 345 to 393 (Unger et al., 1993). It appears that in complex with wild-type p53 the mutant has the ability to drive wild-type p53 into a mutant conformation. Thus, when wild-type and mutant p53 were cotranslated, wild-type p53 lost the epitope recognized by the wild-type p53 specific PAb1620 antibody, and became reactive with the mutant specific PAb240 (Milner and Medcalf, 1991).

While the above conformational change mechanism has strong support, it fails to account for the relatively strong inhibition of wild-type p53 transactivation by the 273(R to H) mutant (Kern et al., 1992), which retains 98% folding of wild-type p53. However, the dominant negative activity of this mutant has been reported to be very weak by other groups (Chene, 1998; Unger et al., 1993).

There is evidence that p53 retains some tumor suppresser activity and transactivation ability as a monomer (Joers et al., 1998; Shaulian et al., 1993; Unger et al., 1993). Some p53 mutants and the fragment DD can repress this transactivation. p53 which is unable to oligomerize due to the deletion in the tetramerization domain, was able to transactivate the p21 promoter. The transactivation was suppressed by core domain mutants (Joers et al., 1998). This may indicate that the dominant negative effect of these mutants is partly due to their ability to sequester factors required for wild-type p53 function, and especially for transactivation (for review, see Blagosklonny (2000)).

9.3. ONCOGENIC EFFECTS OF p53 MUTANTS II: GOF

In the absence of wild-type p53, some p53 mutants are still able to exert oncogenic effects. This was first shown in murine L12 pre B cells, a cell line null for p53 (for a partial list of cell lines null for p53, see the Appendix.) These cells formed tumors in mice that later regressed. However, L12 cells that expressed the murine p53 mutant 132(C to F), caused lethal tumors (Wolf et al., 1984a). Increased tumorigenicity was also demonstrated in (10)3 cells (Dittmer et al., 1993; Lanyi et al., 1998), and T-cell acute lymphoblastic leukemia (T-ALL) cells (Hsiao et al., 1994).

To examine the GOF at the organism level, mice transgenic for the 135(A to V) p53 mutant were made on a p53 null or wild-type background. Mice that expressed 135(A to V) in addition to being null for endogenous p53, did not exhibit decreased survival relative to mice that were null for p53 alone. Both groups did not survive beyond 40 weeks, and both had a similar distribution of tumors, prevalent among which were lymphomas (Harvey et al., 1995). This system shows that the gain of function effect of this p53 mutant is small relative to the effect of wild-type p53 loss. However, the early age at which mice from both groups died may have prevented the detection of GOF. While p53 +/+ mice did not develop tumors during the observation period of 80 weeks, mice that expressed p53 from only one allele, and mice that expressed the 135(A to V) mutant on a p53 +/+ or p53 +/– background, did develop tumors. These mice survived much longer than mice null for p53. Mice that expressed 135(A to V), however, had a frequency of adenocarcinomas of 31% (p53 +/+ and 135(A to V)) and 20% (p53 +/– and 135(A to V)), compared to p53+/– mice, which had a frequency of 2%. This shift in tumor type was not explained by differences in survival, as p53+/– mice survived on average longer than p53 +/– and 135(A to V) mice, but for a shorter period than p53 +/+ and 135(A to V) mice. Therefore, this shift in tumor type seems to be the result of the GOF of the 135(A to V) mutant, and not result of the depletion in active wild-type p53 due to the dominant negative effect of the p53 mutant.

Additional evidence for GOF comes from cellular models. Expression of oncogenic p53 mutants has been shown to mediate increased mutation frequency following irradiation (Iwamoto et al., 1996), genomic instability (Murphy et al., 2000), augmented metastatic potential (Crook and Vousden, 1992; Hsiao et al., 1994), interference with differentiation (Shaulsky et al., 1991), and induction of gene amplification (El-Hizawi et al., 2002).

We and others have found that p53 mutants are able to interfere with the cellular responses of programmed cell death and growth arrest, which play a critical part in chemotherapy and may be important in other contexts such as metastasis. This was first shown in p53 null M1/2 cells, a cell line derived from the murine myeloblastic M1 cell line and selected to be dependent on growth factors secreted by stromal cells. M1/2 cells underwent a slow process of apoptosis when these growth factors were removed, and expression of the 135(A to V) mutant retarded this process (Peled et al., 1996b). This suppression of apoptosis was much more robust when p53-independent apoptosis was induced by DNA-damaging agents (Li et al., 1998). Further studies in M1/2 cells confirmed these observations (Matas et al., 2001; Sigal et al., 2001). Subsequent work in a variety of cell systems has shown that various DBD p53 mutants confer resistance to apoptosis and enhance clonogenic survival in response to the DNA damaging and cytotoxic agents etoposide, cisplatin, 5-flourouracil, mitomycin C, UV, and ionizing radiation. (Atema and Chene, 2002; Blandino et al., 1999; Murphy et al., 2000; Pugacheva et al., 2002).

The interference with DNA-damage-induced apoptosis raises the question of whether mutant p53 interferes with the apoptotic machinery, or the cellular detection of DNA damage. The latter may be the case, since expression of the 135(A to V) mutant in M1/2 cells also increased the concentration of DNA damaging agent needed to initiate growth arrest. Thus, when reversible G2 arrest was triggered by a low concentration of etoposide, cells that expressed the 135(A to V) mutant required a fivefold higher concentration for peak G2 induction, relative to p53 null cells (Sigal et al., 2001).

9.4. PROPOSED MECHANISMS OF GOF

As this chapter is being written, there is no widespread agreement on a mechanism for mutant p53 gain of function. However, several mechanisms have been proposed. The most complete of these was first proposed by Prives and coworkers, and states that mutant p53 has a gain of function because it acts as a dominant negative for the p53 family member p73 (Di Como et al., 1999). Recently, this dominant negative activity of p53 mutants has been shown to repress another p53 family member, p63 (Gaiddon et al., 2001; Strano et al., 2002).

p73 and p63 have been found to have some similar functions to p53, including the ability to induce apoptosis (Jost et al., 1997; Osada et al., 1998; Yang et al., 1998; Zhu et al., 1998). p73 has also been shown to be activated after DNA damage (Agami et al., 1999; Gong et al., 1999; Yuan et al., 1999). Taken together, this may indicate that p73 is involved in the p53-independent DNA damage response pathway, which is suppressed by p53 mutants. p53 mutants interfered with apoptosis, growth suppression, and p21 transactivation induced by p73 (Di Como et al., 1999; Marin et al., 2000; Strano et al., 2000), indicating that they could interfere with the potential DNA damage response mediated by p73.

p53 mutants have been shown to associate with p63 and p73 (Gaiddon et al., 2001; Marin et al., 2000; Strano et al., 2000) lending additional support to this negative dominance model. This association occurs through the DBD (Gaiddon et al., 2001; Strano et al., 2000), and not through the tetramerization domain (Davison et al., 1999), unlike the interaction between mutant and wild-type p53.

Interestingly, the association between a p53 mutant and p73 was shown to be dependent on the p53 isoform. p53 is polymorphic at residue 72 (arginine/proline) (Harris et al., 1986), and the association between p73 and the 143(V to A) mutant was stronger when the isoform was arginine. The arginine isoform also abrogated the growth suppression effect of p73, while the proline isoform did not (Marin et al., 2000). It is postulated that, unlike the proline isoform, the arginine isoform is able to produce p53 mutants that inhibit p73 and cause cancer (Marin et al., 2000). This would predict a bias in cancers toward mutations of the arginine isoform. Such a bias has been found in non-melanoma skin cancers, where 93% of sequenced p53 mutants were of the arginine isoform (McGregor et al., 2002).

While the negative dominance mechanism seems promising, data from mouse models do not fit well with it. p73 deficient mice have been shown to have pronounced neurological, inflammatory, and other defects (Yang et al., 2000), and p63 knockout mice showed defects in limb, craniofacial, and epithelial development (Yang et al., 1999). Such defects were not reported in mice transgenic for mutant p53 (Harvey et al., 1995; Lavigueur et al., 1989). Conversely, the increased tumorigenicity associated with the mutant p53 transgene was not reported in either p63 or p73 deficient mice (Yang et al., 1999, 2000).

A second mechanism proposed for mutant p53 gain of function is mutant p53 transactivation of genes involved in GOF. p53 mutants have been found to transactivate the multidrug resistance 1 (MDR-1) gene (Dittmer et al., 1993; Lin et al., 1995; Sampath et al., 2001), *c-myc* (Frazier et al., 1998; Matas et al., 2001), proliferating cell nuclear antigen (PCNA) (Deb et al., 1992), IL-6 (Margulies and Sehgal, 1993), human heat shock protein 70 (Tsutsumi-Ishii et al., 1995), human epidermal growth factor receptor (EGFR) (Ludes-Meyers et al., 1996) IGF-II (Lee et al., 2000), and dUTPase (Pugacheva et al., 2002). In addition to these, mutant p53 was found to increase the activation of a nonendogenous promoter, the HIV-1 LTR (Subler et al., 1994).

This transactivation can be either direct or indirect. If indirect, it may be the effect of a variety of mutant p53 perturbations to the protein machinery of the cell, including negative dominance over p53 family members. However, there are indications that the transactivation of mutant p53 target genes may be direct. Mutation of residues 22 and 23 (critical to wild-type p53 transactivation ability), in the transactivation domain of the 281(D to G) p53 mutant, inhibited its transactivation of MDR-1 (Lin et al., 1995) and *c-myc* (Frazier et al., 1998; Matas et al., 2001). In addition, deletion of residues 1–58 in 281(D to G) inhibited its transactivation of PCNA, MDR-1, and EGFR (Lanyi et al., 1998). It is still an open question whether there are specific sequences that mutant p53 binds to. It has been shown to bind DNA, in the form of divergent sequences that have in common a tendency to adopt a non B-DNA conformation (Koga and Deppert, 2000).

There appears to be a strong connection between transactivation ability and mutant p53 GOF. Mutation of N-terminal residues 22 and 23 caused a loss of the enhanced tumorigenicity mediated by the 281(D to G) p53 mutant in (10)3 cells (Lin et al., 1995). The murine 135(A to V) p53 mutant interfered with apoptosis induced by the agents cisplatin and etoposide, but not with apoptosis induced by actinomycin D, a powerful transcription inhibitor (Li et al., 1998). Furthermore, mutation of residues 22 and 23 in the 143 (V to A) mutant abolished the suppression of apoptosis mediated by this mutant in M1/2 cells (Matas et al., 2001).

How the upregulation of mutant p53-responsive genes leads to the pathological effects of p53 mutants is not currently clear. Transactivation of MDR-1 seemed a promising candidate as the mechanism for mutant p53 antiapoptotic function, since mutant p53 confers resistance to DNA damaging chemotherapeutic agents such as etoposide, whose toxicity is decreased as a consequence of MDR-1 function. However, mutant p53 mediated resistance to UV, ionizing radiation, and growth factor removal-induced apoptosis is not explained by this mechanism. Neither is mutant p53 mediated resistance to cisplatin-induced apoptosis, since cisplatin toxicity is MDR-1 independent.

Does mutant p53 GOF have a physiological origin? While activated p53 is involved in the induction of apoptosis, very low levels of wild-type p53 have been shown to protect cells from apoptosis in a growth arrest independent manner (Lassus et al., 1996, 1999). It is plausible that nonactivated p53 may have functions different from the activated form, including the suppression of apoptosis, and that oncogenic p53 mutants retain these functions.

9.5. COMBINED EFFECTS OF NEGATIVE DOMINANCE AND GOF

The loss of wild-type p53 activity through mutant p53 transdominance and mutant p53 gain of function occur simultaneously and combine to form the oncogenic effects of p53 mutants. An early study that defined mutant p53 oncogenicity showed that p53 mutants that failed to react with the wild-type p53-specific PAb246 antibody, were tumorigenic in transgenic mice (Lavigueur et al., 1989).

The combined outcome of negative dominance and GOF of p53 mutants were examined in angiogenesis. In these studies there was no separation of GOF and negative dominance, since assays were performed in cells that expressed wild-type p53, such as the murine NIH 3T3 cell line. It was shown that mutant p53 decreased the expression of the angiogenesis inhibitor thrombospondin-1 (Grant et al., 1998), upregulated the basic fibroblast growth factor (bFGF) (Ueba et al., 1994), and the vascular endothelial growth factor (VEGF) (Kieser et al., 1994; Takahashi et al., 1998). Wild-type p53 was shown to be involved in the inhibition of angiogenesis in general (Van Meir et al., 1994) and of VEGF in particular (Mukhopadhyay et al., 1995), so negative dominance of p53 mutants partly accounts for the upregulation.

9.6. THE STABILIZATION OF MUTANT p53 PROTEINS

The optimal mutant p53 protein level for gain of function has not been determined. However, the dominant negative function of mutant p53 is expected to strengthen as the ratio of mutant to wild-type p53 increases, as is indeed the case (Chene, 1998). A hallmark of oncogenic p53 mutants is their high protein levels (Rotter, 1983). The reason is that most p53 mutants are effectively outside the negative feedback loop of Mdm2, which keeps wild-type p53 in check in normal cells. The half-life of wild-type p53 also increases dramatically when it is bound to mutant p53 (Eliyahu et al., 1988; Shaulian et al., 1992a).

There is evidence for several possible mechanisms that attempt to explain what keeps Mdm2 from targeting mutant p53 for degradation. The most accepted is that unlike the wild type, p53 mutants do not transactivate the *mdm2* promoter. Moreover, they inhibit wild-type p53 transactivation of *mdm2* by their dominant negative effect. Hence, insufficient Mdm2 is expressed to target the p53 for degradation (Haupt et al., 1997; Midgley and Lane, 1997). This accounts for the high mutant p53 levels in cells, once the remaining wild-type p53 allele is lost. However, if a wild-type allele is present, the dominant negative activity of the mutant p53 would be expected to cause a buildup of wild-type p53, which may drive Mdm2 expression back up. In fact, exogenous expression of wild-type p53 under a strong promoter in p53 null Saos-2 cells, stably transfected with the 248(R to W) mutant, caused Mdm2 to be induced and the mutant protein levels to drop (Nagata et al., 1999). This incomplete mutant p53 supremacy over wild type, which may be manifested in residual wild-type p53 function and decreased mutant p53 protein levels, is no doubt a source of selective pressure which causes cancer cells to lose the remaining wild-type allele (reviewed in Roemer (1999)). The strength of the selective pressure for this loss of heterozygosity (LOH) should be inversely related to the degree of negative dominance. Interestingly, head and neck squamous cell carcinomas with p53 mutations in DNA contact residues exhibited 100% LOH, while those with mutations in regions important for conformational stability showed 50% LOH (Erber et al., 1998).

There is evidence for a complementary mechanism for mutant p53 stability, which involves its protection from Mdm2-mediated degradation due to mutant p53 binding of heat shock proteins. It has long been known that p53 mutants associate with heat shock proteins (Hinds et al., 1987), including hsp90 (Whitesell et al., 1998). Disruption of hsp90 binding using geldanamycin increased the turnover of the murine 135(A to V) and human 248(R to W) mutants, without restoration of wild-type p53 function (Nagata et al., 1999; Whitesell et al., 1998). How hsp90 protects mutant p53 from degradation was recently addressed (Peng et al., 2001a, b). Mutant p53 was found to enable hsp90 to bind Mdm2. Such binding may neutralize Mdm2, possibly by blocking the ARF binding site on the protein (Peng et al., 2001a, b).

9.7. CLINICAL MANIFESTATIONS OF MUTANT p53 ONCOGENICITY

A disease that has shed much light on the epidemiology of the combined effects of p53 mutants is Li-Fraumeni syndrome (LFS). LFS and Li-Fraumeni like syndrome (LFL) can be broadly defined as familial predisposition to cancer, especially sarcomas, breast cancer, brain tumors, leukemia, and adrenal cortical carcinomas. While the germline changes that predispose individuals to cancer in these syndromes are not always related to mutations in p53, such mutations are exceedingly common. One group observed a 71% p53 mutation frequency in families with LFS and a 22% frequency in families with LFL (Birch et al., 1998). A comparison between families with core domain missense mutations (designated type A) and mutations that led to frameshifts or truncation of p53 (type B), showed that the former had a significantly higher incidence of cancer and the age of cancer onset was earlier. In addition, tumors with type B mutations always proceeded to lose the remaining p53 allele, while only 32% of type A tumors exhibited LOH (Birch et al., 1998).

Several studies have attempted to correlate p53 mutational status with the response of cancer patients to chemotherapy and their overall survival. Missense mutations in p53 are associated with positive immunostaining (Righetti et al., 1996) and tumors with positive p53 immunostaining were found to be resistant to cisplatin chemotherapy (Righetti et al., 1996; Rusch et al., 1995). However, the picture is complicated by the observation that some of the tumors that exhibited positive p53 immunostaining expressed only wild-type p53, and these were resistant to cisplatin as well (Righetti et al., 1996). Examination of the response of breast cancer patients to the chemotherapeutic drug doxorubicin showed that patients with mutations in p53 were much more likely to relapse or never experience remission than patients without p53 mutations (Aas et al., 1996). Another study assessing overall survival of breast cancer patients found that patients with mutations outside conserved regions in p53 had similar survival rates to patients with wild-type p53 tumors. However, mutations in conserved regions II and V (codons 117 to 142, and 270 to 286 respectively, in the DBD (see Fig. 9.2) were associated with worse prognosis (Bergh et al., 1995).

9.8. DIFFERENCES BETWEEN p53 MUTANTS

It has been recognized for some time that different p53 DBD mutations may produce mutant proteins of varying oncogenic potency (Table 9.1). The crystal structure of p53 has revealed that mutations can be classified into two types. The first, termed class I, mutation occurs in residues that come in direct contact with DNA, such as the arginines at positions 248 and 273. These occur on either the L3 loop or the nearby loop-sheet-helix motif of p53 (Cho et al., 1994). The second type of mutation, termed class II, also effectively disrupts DNA binding (see Table 9.1), but does so by disrupting the conformational stability of the p53 protein. Such changes are also detected with conformationally sensitive antibodies. For example, these mutations expose the epitope of the PAb240 antibody and lead to the loss of the wild-type specific epitope detected by PAb1620 (Bartek et al., 1990; Cho et al., 1994; Milner and Medcalf, 1991). An archetypical example of class II is the mutation at arginine 175, which is situated in the L2 loop in the zinc region. Recent data has blurred the sharp distinction between the two classes, and showed that some DNA contact mutations also cause conformational changes (reviewed in Bullock and Fersht (2001); see Table 9.1). However, this categorization is useful because the class II mutations were shown to be more oncogenic than the class I in several systems (Table 9.1).

Bullock and Fersht (2001) further subdivide class II into the temperature sensitive mutations that occur at the β-sandwich, and mutations in the zinc region. They suggest that the wild-type p53 protein is in equilibrium between native and denatured forms, with the equilibrium favoring the native form. β-sandwich mutations, which account for 25% of the total, strongly shift the equilibrium toward the denatured form. They propose that the equilibrium can be shifted back by mass action: small molecules, which bind the native state and not the denatured state, will increase the stability of the native state. Therefore, the temperature at which a given fraction of mutated β-sandwich p53 proteins retain the wild-type conformation will be higher. This approach is predicted not to work for zinc region mutants. Mutations in the zinc region, such as 175(R to H), do not exhibit temperature sensitivity, and so do not show potential for stabilization to the wild-type conformation (Bullock and Fersht, 2001).

The apparent increased tumorigenicity of class II relative to class I mutations would predict that the former would be selected over the latter. However, the most common mutational hotspots are at residues R175, R248, R249, R273, R282, and G245 (Hernandez-Boussard et al., 1999), two of which are class I. The most

Table 9.1. Comparison of structure and some functions of commonly studied p53 mutants.

	Class I		Class II			
	273(R to H)	248(R to W or Q)	249(R to S)	143(V to A)	175(R to H)	Reference
Frequency of mutation at residue	8%	7%	3%	0.5%	6%	Hernandez-Boussard et al. (1999)
Mutation type	DNA Contact	DNA Contact	Conformational	Conformational	Conformational	Cho et al. (1994)
Percentage folded	98	86	85	31	30	Bullock et al. (2000)
Structural changes (NMR)[a]	Local	Local	Local	Extensive	ND	Wong et al. (1999)
Protection from apoptosis[b]	Low	Low	ND	ND	High	Blandino et al. (1999)
Spindle check-point disruption	ND	No	ND	Yes	Yes	Gualberto et al. (1998)
Immortalization of mammary epithelial cells	No	No	ND	Yes	Yes	Cao et al. (1997)
Wt p53 specific DNA binding	No	No	ND	ND	No	Rolley et al. (1995)
Dominant negative inhibition of wt p53 DNA binding	Low	High	ND	ND	High	Chene (1998)
Transformed foci in coop-eration with *ras*[c]	4.7/8	ND/ND	ND/ND	ND/15	11.5/17	Hinds et al. (1990); Slinger-land et al. (1993)

ND: not done.
[a]relative to wild-type p53.
[b]etoposide induced p53-independent apoptosis. All mutants protected to a similar extent against cisplatin induced apoptosis.
[c]average study 1/average study 2 of transformation of rat embryo fibroblasts.

common is the class I mutation at R273, accounting for 8% of all mutations, while the second most common is the class I mutation at R248, accounting for 7%. There are several possible reasons for the high fraction of class I mutations.

First, what mutations are formed may depend on the mode of action of the mutagenizing agent (Denissenko et al., 1997; Hussain and Harris, 1998a; Pfeifer and Holmquist, 1997). A striking example is Aflotoxin B1, associated with mutation in the third base of codon 249 *in vitro* and epidemiologically (Hsu et al., 1991).

Another example is the cigarette smoke carcinogen benzo[*a*]pyrene, which has been shown to preferentially form adducts in codons 157, 248, and 273 of p53 (Denissenko et al., 1996). The relevance of both conclusions *in vivo* has been questioned, however (Denissenko et al., 1998; Rodin and Rodin, 2000), in the case of benzo[*a*]pyrene on the basis of the lack of difference in the mutation frequency at these three codons between smokers and nonsmokers.

Second, since carcinogenesis is a multistep process (Vogelstein and Kinzler, 1993), it is plausible that in some cases, the lower tumorigenicity of class I mutants will lead to a selection of subsequent compensatory steps. An interesting example where this may occur is head and neck squamous cell carcinoma (HNSCC). It was observed that expression of class I mutants was associated with 100% LOH, while expression of class II was associated with only 50% of cancers losing the remaining wild-type p53 allele (Erber et al., 1998). This is consistent with class II mutants exerting a stronger dominant negative (and possibly GOF) effect, thus decreasing the selective pressure for LOH. However, tumors with the class I mutations were shown to be far more aggressive than tumors with class II mutations. They were associated with higher tumor stages, higher incidence of lymph node metastasis (91% vs. 56%), shortened recurrence free survival (8 vs. 23 months), and shorter overall survival (11 months vs. 29 months). Thus, it seems that selection toward LOH greatly increased tumorigenicity, consistent with observations in mice, which show that complete p53 loss has a much more pronounced effect than mutant p53 coexpression (Harvey et al., 1995).

Therefore, while the distinction between class I and class II mutations may be informative for therapy, one must be careful of the generalization that class II mutations make for more "successful" tumors in vivo.

9.9. THERAPEUTIC APPROACHES

The field of mutant p53 has recently been attracting much interest due to studies that have succeeded to restore wild-type p53 functionality to various p53 mutants (Fig. 9.3A). The therapeutic potential of such approaches can be immense. In addition, there have been proposals for other types of approaches that either remove mutant p53 function or use it as a way to remove mutant p53 expressing cells (Fig 9.3B–D).

The reactivation of p53 mutants to a wild-type phenotype relies on interfering with regions and interactions in the mutant p53 protein necessary for it to retain the mutant phenotype. Modification of the extreme C-terminus caused some oncogenic p53 mutants to regain DNA binding to p53 specific elements, and even to regain the ability to transactivate wild-type p53 target genes (Abarzua et al., 1995, 1996; Hupp et al., 1993; Niewolik et al., 1995; Selivanova et al., 1998; Wieczorek et al., 1996). In addition, modification of the extreme C-terminus resulted in a loss of mutant p53 transactivation potential in the 281(D to G) mutant (Frazier et al., 1998) and a loss of antiapoptotic GOF in the murine 135(A to V) mutant (Sigal et al., 2001). The loss of the mutant phenotype is linked to the destabilization of the mutant conformation once the normal C-terminus is removed. Thus, truncation of the C-terminus of the

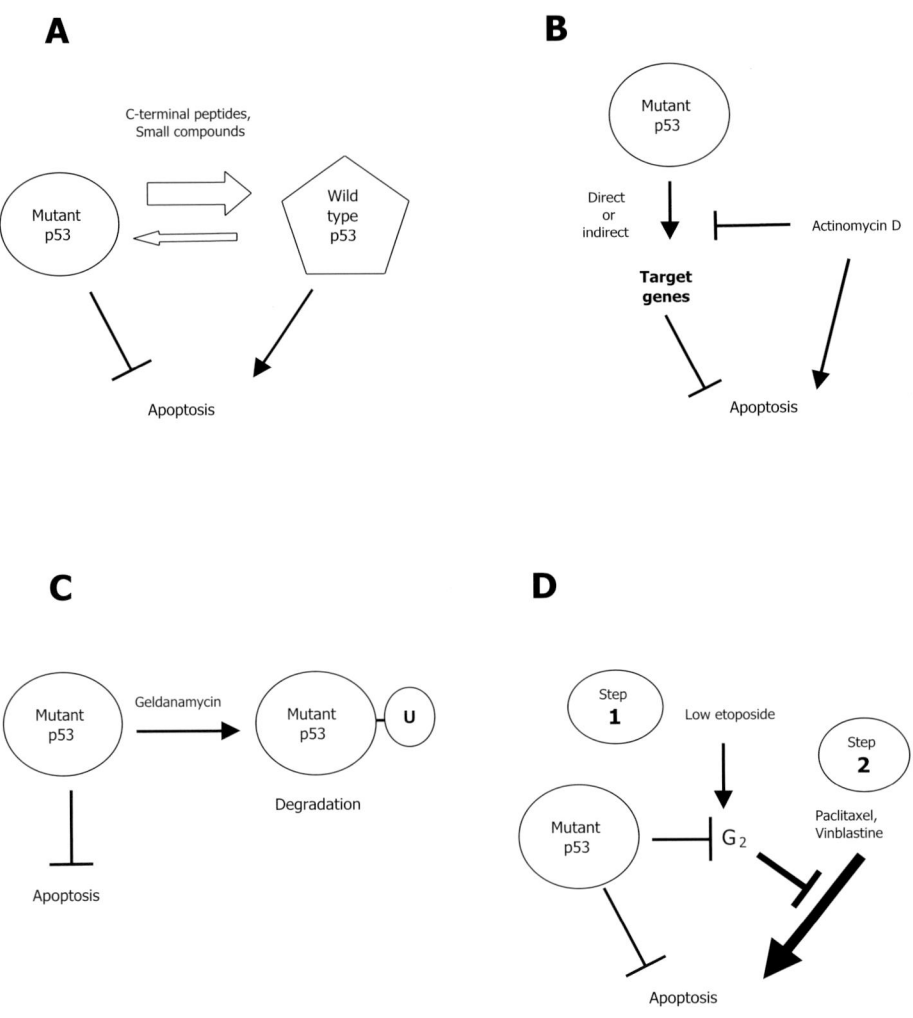

Figure 9.3. Potential therapeutic modalities for tumors that express mutant p53. **A**: Mutant p53 is reactivated to wild-type p53 function by C-terminal peptides or compounds that bind p53 and stabilize it in the wild-type conformation. Due to the cellular milieu and the presence of high p53 levels, such a conversion in cancer cells has been shown to induce growth arrest or apoptosis. **B**: There is evidence that the direct or indirect transactivation of mutant p53 responsive genes is necessary for GOF, including mutant p53 suppression of apoptosis. The use of a chemotherapeutic agent that is also a transcriptional repressor, such as actinomycin D, may therefore reduce GOF and increase the sensitivity of the mutant p53-expressing tumor cells to the initiation of arrest or apoptosis by DNA damaging agents. **C**: GOF may also be reduced by increasing mutant p53 turnover. Targeting of mutant p53 for degradation through Mdm2 mediated ubiquitination (U) was demonstrated with the agent geldanamycin, which disrupts mutant p53–hsp90–Mdm2 complexes. **D**: Mutant p53 has been shown to increase the concentration of the DNA damaging agent etoposide necessary to induce G_2 arrest. Therefore, cells that express mutant p53 could be differentially targeted for apoptosis by first applying a low concentration of DNA damaging agent that only arrests cells not expressing mutant p53. Then, cell death can be induced in dividing cells by the application of agents such as paclitaxel or vinblastine, whose toxicity is specific to mitosis.

murine 135(A to V)mutant caused it to become reactive with antibodies specific for the wild-type conformation (Milner and Medcalf, 1991).

A second target for reactivation studies was the DBD. It was discovered that second, suppressor mutations could be introduced that reverted oncogenic p53 mutants to wild-type function, including the induction of programmed cell death (Brachmann et al., 1998; Nikolova et al., 2000; Wieczorek et al., 1996). The combination of suppressor mutation and C-terminal truncation yielded the best results (Wieczorek et al., 1996).

While suppressor mutations showed the possibility of reactivation, two other approaches showed that this could be achieved by molecules binding to mutant p53, increasing the clinical application of this type of intervention.

The first approach used a peptide derived from the C-terminus (residues 361–382), which was found to bind both to the C-terminus and the DBD. This peptide was found to reinstate wild-type p53 specific DNA binding, and when fused to the Antennapedia homeodomain to facilitate cell entry, it could restore p53 specific transactivation, retard colony formation, and induce apoptosis in cells expressing oncogenic p53 mutants. It had no effect in cells null for p53 (Kim et al., 1999; Selivanova et al., 1997, 1999).

The second approach screened a library of small compounds for molecules that could stabilize the wild-type p53 DBD upon heating to 45°C (Foster et al., 1999) or whose toxicity was specific to mutant p53 expressing cells (Bykov et al., 2002). In the first study, the screen resulted in compounds CP-257042 and CP-31398, shown to stabilize DBDs with oncogenic mutations in the wild-type conformation. These compounds were found to be relatively stable when injected into mice and had no significant toxicity. When administered to nude mice xenografted with a human melanoma cell line with 249(R to S), and a carcinoma cell line with 241(S to F), tumor growth was suppressed (Foster et al., 1999). However, the stabilization of p53 in the wild-type conformation by CP-31398, and its low toxicity in cells not expressing p53 mutants, has since been challenged (Rippin et al., 2002).

The second study, which compared the effect of compounds on Saos-2 cell growth in the presence or absence of the 273(R to H) mutant, identified one compound, designated PRIMA-1, which could suppress growth specifically when the 273(R to H) mutant was expressed. PRIMA-1 was shown to have low toxicity in mice and specifically target Saos-2 tumor xenografts expressing the 273(R to H) mutant (Bykov et al., 2002).

A critical point is that restoration of wild-type p53 function has the potential to be relatively nontoxic to cells expressing normal levels of wild-type p53, as illustrated above. This was also demonstrated by a study that used the C-terminal peptide for the restoration of wild-type p53 function. The peptide induced apoptosis in cell lines expressing mutant p53 or overexpressing wild-type p53, but not in cell lines expressing normal noninduced levels of wild-type p53 (Kim et al., 1999).

There may be several reasons for this selection against cells that express mutant p53, which is crucial for this approach to work. One reason is that, since mutant p53 levels are high because of the breakdown of Mdm2 regulation. Therefore, cells that express mutant p53 will contain, upon reactivation of the protein to wild-type function, high levels of activated p53. This explains why cells like MCF7 with high wild-type p53 expression also undergo apoptosis upon p53 reactivation (Kim et al., 1999). Cells with low p53 levels will be less affected, if at all. Another reason why cells that express mutant p53 will be differentially targeted, is that cancer cells incur changes, such as disregulation of *c-myc,* which can make the cellular environment more permissive for apoptosis or growth arrest.

Restoration of wild-type function to p53 mutants is not expected to work for all mutants, and additional or complementary approaches should be considered. For example, in the event that restoration will not be successful, it may still be possible to eliminate the antiapoptotic GOF of p53 mutants, which, if left unchecked, will lead to resistance to some apoptosis inducing chemotherapy. It was observed that residues 22 and 23, necessary for the ability of p53 mutants to transactivate their target genes, are also necessary for GOF (Lin et al., 1995) and in particular, to the mutant p53 mediated suppression of apoptosis (Matas et al., 2001). Therefore, a transcription inhibitor coadministered with the chemotherapeutic agent is predicted to decrease the resistance of mutant p53 expressing cells to apoptosis (Fig. 9.3B). Mutant p53 inhibition, together with the induction of programmed cell death, can be a feature of the same agent, as is the case with actinomycin D, which induced equal levels of apoptosis in cells expressing mutant p53 and as in null controls (Li et al., 1998).

A second approach that can be used to reduce mutant p53 GOF is to promote degradation of the mutant p53 protein (Fig. 9.3C). As discussed previously, binding of hsp90 to p53 mutants protects the mutant p53 proteins from Mdm2 mediated degradation. Disruption of this binding using geldanamycin destabilizes mutant p53.

Finally, the observation that mutant p53 increases the DNA damage threshold necessary to initiate growth arrest, suggests an additional source of leverage against cancers with oncogenic p53 mutations (Fig. 9.3D). Low concentrations of a DNA damaging agent, such as etoposide, can be calibrated to arrest normal but not mutant p53-expressing cells. A second agent, such as paclitaxel or vinblastine, can then be used to kill the nonarrested cells (Blagosklonny, 2002; Sigal et al., 2001).

These four approaches demonstrate three different principles of mutant p53 targeting: p53 mutants can be induced to convert to wild-type p53 function, eliminated, or cells that express p53 mutants can be selected against based on the phenotype the p53 mutants confer.

With the heightened pace of research into mutant p53, and the increased sophistication of drug screens and clinical techniques, approaches that take advantage of the presence of p53 mutant proteins are expected to gain an important place in the toolkit available to combat cancer.

9.10. CONCLUSION

p53 was discovered in 1979 (Lane and Crawford, 1979; Linzer and Levine, 1979) and classified as an oncogene due to its overexpression in tumors and its transforming ability. Ten years later, a seminal work was published that demonstrated that the oncogenic form of p53 was the mutated form, and the wild-type form was a tumor suppressor (Finlay et al., 1989). This finding has spurred intense research into the functions of wild-type p53 that has for the most part eclipsed the research on the mutant form of the protein. However, with the maturation of the wild-type p53 field, it has come to be recognized that knowledge of how the most common oncogene functions may hold the key to more successful therapy for many cancers.

APPENDIX 9.1

Table A 9.1. Some cell lines that do not express p53

Cell line	Organism	Source	Reference
Caco-2	Human	Colon carcinoma	Djelloul et al. (1997; Sampath et al. (2001)
HCT116 (p53$^{-/-}$)[a]	Human	Colorectal cancer	Bunz et al. (1998)
Hep-3B	Human	Hepatocellular carcinoma	Bressac et al. (1990); Vollmer et al. (1999)
H358	Human	Bronchioalveolar carcinoma	Takahashi et al. (1989); Unger et al., (1993)
H1299	Human	Large cell lung carcinoma	Blandino et al. (1999); Mitsudomi et al. (1992)
Jurkat	Human	Acute T cell leukemia	Iwamoto et al. (1996)
L12	Mouse	Hematopoietic Ab-MuLV transformed of C57L/J origin	Rotter et al. (1983); Wolf et al. (1984b)
MEC	Mouse	Mammary epithelial cell line derived from p53 KO mice	Murphy et al. (2000)
M1 and the M1/2 derivative	Mouse	Myeloid leukemia	Li et al. (1998); Lotem and Sachs (1993); Matas et al. (2001); Peled et al. (1996a); Sigal et al. (2001)
PC-3	Human	Prostate cancer	Carroll et al. (1993); Isaacs et al. (1991)
Saos-2	Human	Osteosarcoma	Dittmer et al. (1993)
10(3)	Mouse	BALB/c murine embryo fibroblast derived	Dittmer et al. (1993)

[a]Derived from the wild-type p53 HCT116 through homologous recombination that disrupted both alleles.

RECENT REVIEWS

Sigal and Rotter (2000)
Roemer (1999)
Bullock and Fersht (2001)
van Oijen and Slootweg (2000)
Cadwell and Zambetti (2001)

USEFUL WEBSITES

The IARC TP53 Mutation Database
   http://www.iarc.fr/p53/index.html
Institut Curie (Terry Soussi's lab)
   http://p53.curie.fr/

REFERENCES

Aas, T., Borresen, A.-L., Geisler, S., Smith-Sorensen, B., Johnsen, H., Varhaug, J. E., Akslen, L. A., and Lonning, P. E. (1996). Specific p53 mutations are associated with de novo resistance to doxorubicin in breast cancer patients. *Nat Med* 2:811–814.

Abarzua, P., LoSardo, J. E., Gubler, M. L., and Neri, A. (1995). Microinjection of monoclonal antibody PAb421 into human SW480 colorectal carcinoma cells restores the transcription activation function to mutant p53. *Cancer Res* 55:3490–3494.

Abarzua, P., LoSardo, J. E., Gubler, M. L., Spathis, R., Lu, Y. A., Felix, A., and Neri, A. (1996). Restoration of the transcription activation function to mutant p53 in human cancer cells. *Oncogene* 13:2477–2482.

Agami, R., Blandino, G., Oren, M., and Shaul, Y. (1999). Interaction of c-Abl and P73alpha and their collaboration to induce apoptosis. *Nature* 399:809–813.

Almog, N., Goldfinger, N., and Rotter, V. (2000). p53-dependent apoptosis is regulated by a C-terminally alternatively spliced form of murine p53. *Oncogene* 19:3395–3403.

Aloni-Grinstein, R., Schwartz, D., and Rotter, V. (1995). Accumulation of wild-type p53 protein upon gamma-irradiation induces a G2 arrest-dependent immunoglobulin kappa light chain gene expression. *EMBO J* 14:1392–1401.

Aloni-Grinstein, R., Zan-Bar, I., Alboum, I., Goldfinger, N., and Rotter, V. (1993). Wild type p53 functions as a control protein in the differentiation pathway of the B-cell lineage. *Oncogene* 8:3297–3305.

Atema, A., and Chene, P. (2002). The gain of function of the p53 mutant Asp281Gly is dependent on its ability to form tetramers. *Cancer Lett* 185:103–109.

Bartek, J., Iggo, R., Gannon, J., and Lane, D. P. (1990). Genetic and immunochemical analysis of mutant p53 in human breast cancer cell lines. *Oncogene* 5:893–899.

Bergh, J., Norberg, T., Sjogren, S., Lindgren, A., and Holmberg, L. (1995). Complete sequencing of the p53 gene provides prognostic information in breast cancer patients, particularly in relation to adjuvant systemic therapy and radiotherapy. *Nat Med* 1:1029–1034.

Birch, J. M., Blair, V., Kelsey, A. M., Evans, D. G., Harris, M., Tricker, K. J., and Varley, J. M. (1998). Cancer phenotype correlates with constitutional TP53 genotype in families with the Li-Fraumeni syndrome. *Oncogene* 17:1061–1068.

Blagosklonny, M. V. (2000). p53 from complexity to simplicity: mutant p53 stabilization, gain-of- function, and dominant-negative effect. *FASEB J* 14:1901–1907.

Blagosklonny, M. V. (2002). P53: an ubiquitous target of anticancer drugs. *Int J Cancer* 98:161–166.

Blandino, G., Levine, A. J., and Oren, M. (1999). Mutant p53 gain of function: differential effects of different p53 mutants on resistance of cultured cells to chemotherapy [In Process Citation]. *Oncogene* 18:477–485.

Brachmann, R. K., Yu, K., Eby, Y., Pavletich, N. P., and Boeke, J. D. (1998). Genetic selection of intragenic suppressor mutations that reverse the effect of common p53 cancer mutations. *EMBO J* 17:1847–1859.

Bressac, B., Galvin, K. M., Liang, T. J., Isselbacher, K. J., Wands, J. R., and Ozturk, M. (1990). Abnormal structure and expression of p53 gene in human hepatocellular carcinoma. *Proc Natl Acad Sci USA* 87:1973–1977.

Bullock, A. N., and Fersht, A. R. (2001). Rescuing the function of mutant p53. *Nat Rev Cancer* 1:68–76.

Bullock, A. N., Henckel, J., and Fersht, A. R. (2000). Quantitative analysis of residual folding and DNA binding in mutant p53 core domain: definition of mutant states for rescue in cancer therapy. *Oncogene* 19:1245–1256.

Bunz, F., Dutriaux, A., Lengauer, C., Waldman, T., Zhou, S., Brown, J. P., Sedivy, J. M., Kinzler, K. W., and Vogelstein, B. (1998). Requirement for p53 and p21 to sustain G2 arrest after DNA damage. *Science* 282:1497–1501.

Bykov, V. J., Issaeva, N., Shilov, A., Hultcrantz, M., Pugacheva, E., Chumakov, P., Bergman, J., Wiman, K. G., and Selivanova, G. (2002). Restoration of the tumor suppressor function to mutant p53 by a low-molecular-weight compound. *Nat Med* 8:282–288.

Cadwell, C., and Zambetti, G. P. (2001). The effects of wild-type p53 tumor suppressor activity and mutant p53 gain-of-function on cell growth. *Gene* 277:15–30.

Cao, Y., Gao, Q., Wazer, D. E., and Band, V. (1997). Abrogation of wild-type p53-mediated transactivation is insufficient for mutant p53-induced immortalization of normal human mammary epithelial cells. *Cancer Res* 57:5584–5589.

Carroll, A. G., Voeller, H. J., Sugars, L., and Gelmann, E. P. (1993). p53 oncogene mutations in three human prostate cancer cell lines. *Prostate* 23:123–134.

Chene, P. (1998). In vitro analysis of the dominant negative effect of p53 mutants. *J Mol Biol* 281:205–209.

Chene, P., and Bechter, E. (1999). p53 mutants without a functional tetramerisation domain are not oncogenic. *J Mol Biol* 286:1269–1274.

Cho, Y., Gorina, S., Jeffrey, P. D., and Pavletich, N. P. (1994). Crystal structure of a p53 tumor suppressor-DNA complex: understanding tumorigenic mutations. *Science* 265:346–355.

Crook, T., and Vousden, K. H. (1992). Properties of p53 mutations detected in primary and secondary cervical cancers suggest mechanisms of metastasis and involvement of environmental carcinogens. *EMBO J* 11:3935–3940.

Davison, T. S., Vagner, C., Kaghad, M., Ayed, A., Caput, D., and Arrowsmith, C. H. (1999). p73 and p63 are homotetramers capable of weak heterotypic interactions with each other but not with p53. *J Biol Chem* 274:18709–18714.

Deb, S., Jackson, C. T., Subler, M. A., and Martin, D. W. (1992). Modulation of cellular and viral promoters by mutant human p53 proteins found in tumor cells. *J Virol* 66:6164–6170.

Denissenko, M. F., Chen, J. X., Tang, M. S., and Pfeifer, G. P. (1997). Cytosine methylation determines hot spots of DNA damage in the human P53 gene. *Proc Natl Acad Sci USA* 94:3893–3898.

Denissenko, M. F., Koudriakova, T. B., Smith, L., O'Connor, T. R., Riggs, A. D., and Pfeifer, G. P. (1998). The p53 codon 249 mutational hotspot in hepatocellular carcinoma is not related to selective formation or persistence of aflatoxin B-1 adducts. *Oncogene* 17:3007–3014.

Denissenko, M. F., Pao, A., Tang, M., and Pfeifer, G. P. (1996). Preferential formation of benzo[a]pyrene adducts at lung cancer mutational hotspots in P53. *Science* 274:430–432.

Di Como, C. J., Gaiddon, C., and Prives, C. (1999). p73 function is inhibited by tumor-derived p53 mutants in mammalian cells. *Mol Cell Biol* 19:1438–1449.

DiGiammarino, E. L., Lee, A. S., Cadwell, C., Zhang, W., Bothner, B., Ribeiro, R. C., Zambetti, G., and Kriwacki, R. W. (2002). A novel mechanism of tumorigenesis involving pH-dependent destabilization of a mutant p53 tetramer. *Nat Struct Biol* 9:12–16.

Dittmer, D., Pati, S., Zambetti, G., Chu, S., Teresky, A., Moore, M., Finlay, C., and Levine, A. (1993). Gain of function mutations in p53. *Nat Genet* 4:42–46.

Djelloul, S., Forgue-Lafitte, M. E., Hermelin, B., Mareel, M., Bruyneel, E., Baldi, A., Giordano, A., Chastre, E., and Gespach, C. (1997). Enterocyte differentiation is compatible with SV40 large T expression and loss of p53 function in human colonic Caco-2 cells. Status of the pRb1 and pRb2 tumor suppressor gene products. *FEBS Lett* 406:234–242.

El-Hizawi, S., Lagowski, J. P., Kulesz-Martin, M., and Albor, A. (2002). Induction of gene amplification as a gain-of-function phenotype of mutant p53 proteins. *Cancer Res* 62:3264–3270.

Eliyahu, D., Goldfinger, N., Pinhasi-Kimhi, O., Shaulsky, G., Skurnik, Y., Arai, N., Rotter, V., and Oren, M. (1988). Meth A fibrosarcoma cells express two transforming mutant p53 species. *Oncogene* 3:313–321.

Erber, R., Conradt, C., Homann, N., Enders, C., Finckh, M., Dietz, A., Weidauer, H., and Bosch, F. X. (1998). TP53 DNA contact mutations are selectively associated with allelic loss and have a strong clinical impact in head and neck cancer. *Oncogene* 16:1671–1679.

Finlay, C. A., Hinds, P. W., and Levine, A. J. (1989). The p53 proto-oncogene can act as a suppressor of transformation. *Cell* 57:1083–1093.

Foster, B. A., Coffey, H. A., Morin, M. J., and Rastinejad, F. (1999). Pharmacological rescue of mutant p53 conformation and function. *Science* 286:2507–2510.

Frazier, M. W., He, X., Wang, J., Gu, Z., Cleveland, J. L., and Zambetti, G. P. (1998). Activation of C-*myc* gene expression by tumor-derived p53 mutants requires a descrete C-terminal domain. *Mol Cell Biol* 18:3735–3743.

Gaiddon, C., Lokshin, M., Ahn, J., Zhang, T., and Prives, C. (2001). A subset of tumor-derived mutant forms of p53 down-regulate p63 and p73 through a direct interaction with the p53 core domain. *Mol Cell Biol* 21:1874–1887.

Gong, J. G., Costanzo, A., Yang, H. Q., Melino, G., Kaelin, W. G., Jr., Levrero, M., and Wang, J. Y. (1999). The tyrosine kinase c-Abl regulates p73 in apoptotic response to cisplatin-induced DNA damage. *Nature* 399:806–809.

Gottlieb, E., Haffner, R., von, R. T., Wagner, E. F., and Oren, M. (1994). Down-regulation of wild-type p53 activity interferes with apoptosis of IL-3- dependent hematopoietic cells following IL-3 withdrawal. *EMBO J* 13:1368–1374.

Grant, S. W., Kyshtoobayeva, A. S., Kurosaki, T., Jakowatz, J., and Fruehauf, J. P. (1998). Mutant p53 correlates with reduced expression of thrombospondin-1, increased angiogenesis, and metastatic progression in melanoma. *Cancer Detect Prev* 22:185–194.

Gualberto, A., Aldape, K., Kozakiewicz, K., and Tlsty, T.D. (1998). An oncogenic form of p53 confers a dominant gain of function phenotype that disrupts spindle checkpoint control. *Natl Acad Sci USA* 95, 5166–5171.

Harris, N., Brill, E., Shohat, O., Prokocimer, M., Wolf, D., Arai, N., and Rotter, V. (1986). Molecular basis for heterogeneity of the human p53 protein. *Mol Cell Biol* 6:4650–4656.

Harvey, M., Vogel, H., Morris, D., Bradley, A., Bernstein, A., and Donehower, L. A. (1995). A mutant p53 transgene accelerates tumour development in heterozygous but not nullizygous p53-deficient mice. *Nat Genet* 9:305–311.

Haupt, Y., Maya, R., Kazaz, A., and Oren, M. (1997). Mdm2 promotes the rapid degradation of p53. *Nature* 387:296–299.

Hernandez-Boussard, T., Rodriguez-Tome, P., Montesano, R., and Hainaut, P. (1999). IARC p53 mutation database: a relational database to compile and analyze p53 mutations in human tumors and cell lines. International Agency for Research on Cancer. *Hum Mutat* 14:1–8.

Hinds, P., W., F., C., A., Frey, A., B., L., and A., J. (1987). Immunological evidence for the association of p53 with a heat shock protein, hsc70, in p53-plus-ras-transformed cell lines. *Mol Cell Biol* 7:2863–2869.

Hinds, P. W., Finlay, C. A., Frey, A. B., and Levine, A. J. (1989). Mutation is required to activate the p53 gene for cooperation with the ras oncogene and transformation. *J Virol* 63:739–746.

Hinds, P. W., Finlay, C. A., Quartin, R. S., Baker, S. J., Fearon, E. R., Vogelstein, B., and Levine, A. J. (1990). Mutant p53 DNA clones from human colon carcinomas cooperate with ras in transforming primary rat cells: a comparison of the "hot spot" mutant phenotypes. *Cell Growth Differ* 1:571–580.

Hsiao, M., Low, J., Dorn, E., Ku, D., Pattengale, P., Yeargin, J., and Haas, M. (1994). Gain-of-function mutations of the p53 gene induce lymphhematopoietic metastatic potential and tissue invasiveness. *Am J Pathol* 145:702–714.

Hsu, I. C., Metcalf, R. A., Sun, T., Welsh, J. A., Wang, N. J., and Harris, C. C. (1991). Mutational hotspot in the p53 gene in human hepatocellular carcinomas. *Nature* 350:427–431.

Hupp, T. R., Meek, D. W., Midgley, C. A., and Lane, D. P. (1993). Activation of the cryptic DNA binding function of mutant forms of p53. *Nucleic Acids Res* 21:3167–3174.

Hussain, S. P., and Harris, C. C. (1998a). Molecular Epidemiology of Human Cancer: Contribution of Mutation Spectra Studies of Tumor Suppressor Genes. *Cancer Res* 58:4023–4037.

Hussain, S. P., and Harris, C. C. (1998b). Molecular epidemiology of human cancer: contribution of mutation spectra studies of tumor suppressor genes. *Cancer Res* 58:4023–4037.

Isaacs, W. B., Carter, B. S., and Ewing, C. M. (1991). Wild-type p53 suppresses growth of human prostate cancer cells containing mutant p53 alleles. *Cancer Res* 51:4716–4720.

Iwamoto, K. S., Mizuno, T., Ito, T., Tsuyama, N., Kyoizumi, S., and Seyama, T. (1996). Gain-of-function p53 mutations enhance alteration of the T-cell receptor following X-irradiation, independently of the cell cycle and cell survival. *Canc Res* 56:3862–3865.

Joers, A., Kristjuhan, A., Kadaja, L., and Maimets, T. (1998). Tumour associated mutants of p53 can inhibit transcriptional activity of p53 without heterooligomerization. *Oncogene* 17:2351–2358.

Jost, C. A., Marin, M. C., and Kaelin, W. G., Jr. (1997). p73 is a human p53-related protein that can induce apoptosis. *Nature* 389:191–194.

Kern, S. E., Pietenpol, J. A., Thiagalingam, S., Seymour, A., Kinzler, K. W., and Vogelstein, B. (1992). Oncogenic forms of p53 inhibit p53-regulated gene expression. *Science* 256:827–830.

Kieser, A., Weich, H. A., Brandner, G., Marme, D., and Kolch, W. (1994). Mutant p53 potentiates protein kinase C induction of vascular endothelial growth factor expression. *Oncogene* 9:963–969.

Kim, A. L., Raffo, A. J., Brandt-Rauf, P. W., Pincus, M. R., Monaco, R., Abarzua, P., and Fine, R. L. (1999). Conformational and molecular basis for induction of apoptosis by a p53 C-terminal peptide in human cancer cells. *J Biol Chem* 274:34924–34931.

Koga, H., and Deppert, W. (2000). Identification of genomic DNA sequences bound by mutant p53 protein (Gly245–>Ser) in vivo. *Oncogene* 19:4178–4183.

Lane, D., and Crawford, L. (1979). T antigen is bound to a host protein in SV40-transformed cells. *Nature* 278:261–263.

Lanyi, A., Deb, D., Seymour, R. C., Ludes-Meyers, J. H., Subler, M. A., and Deb, S. (1998). 'Gain of function' phenotype of tumor-derived mutant p53 requires the oligomerization/nonsequence-specific nucleic acid-binding domain. *Oncogene* 16:3169–3176.

Lassus, P., Bertrand, C., Zugasti, O., Chambon, J. P., Soussi, T., Mathieu-Mahul, D., and Hibner, U. (1999). Anti-apoptotic activity of p53 maps to the COOH-terminal domain and is retained in a highly oncogenic natural mutant. *Oncogene* 18:4699–4709.

Lassus, P., Ferlin, M., Piette, J., and Hibner, U. (1996). Anti-apoptotic activity of low levels of wild-type p53. *EMBO J* 15:4566–4573.

Lavigueur, A., Maltby, V., Mock, D., Rossant, J., Pawson, T., and Bernstein, A. (1989). High incidence of lung, bone, and lymphoid tumors in transgenic mice overexpressing mutant alleles of the p53 oncogene. *Mol Cell Biol* 9:3982–3991.

Lee, Y. I., Lee, S., Das, G. C., Park, U. S., Park, S. M., and Lee, Y. I. (2000). Activation of the insulin-like growth factor II transcription by aflatoxin B1 induced p53 mutant 249 is caused by activation of transcription complexes; implications for a gain-of-function during the formation of hepatocellular carcinoma. *Oncogene* 19:3717–3726.

Li, R., Sutphin, P. D., Schwartz, D., Matas, D., Almog, N., Wolkowicz, R., Goldfinger, N., Pei, H., Prokocimer, M., and Rotter, V. (1998). Mutant p53 protein expression interferes with p53-independent apoptotic pathways. *Oncogene* 16:3269–3277.

Lin, J., Chen, J., Elenbaas, B., and Levine, A. J. (1994). Several hydrophobic amino acids in the p53 amino-terminal domain are required for transcriptional activation, binding to mdm-2 and the adenovirus 5 E1B 55-kD protein. *Genes Dev* 8:1235–1246.

Lin, J., Teresky, A. K., and Levine, A. J. (1995). Two critical hydrophobic amino acids in the N-terminal domain of the p53 protein are required for the gain of function phenotypes of human p53 mutants. *Oncogene* 10:2387–2390.

Linzer, D. I. H., and Levine, A. J. (1979). Characterization of a 54K dalton cellular SV40 tumor antigen present in SV40-transformed cells and uninfected embryonal carcinoma cells. *Cell* 17:43–52.

Liu, P. K., Kraus, E., Wu, T. A., Strong, L. C., and Tainsky, M. A. (1996). Analysis of genomic instablity in Li-Fraumeni fibroblasts with germline p53 mutations. *Oncogene* 12:2267–2278.

Lomax, M. E., Barnes, D. M., Gilchrist, R., Picksley, S. M., Varley, J. M., and Camplejohn, R. S. (1997). Two functional assays employed to detect an unusual mutation in the oligomerisation domain of p53 in a Li-Fraumeni like family. *Oncogene* 14:1869–1874.

Lotem, J., and Sachs, L. (1993). Regulation by bcl-2, c-myc, and p53 of susceptibility to induction of apoptosis by heat shock and cancer chemotherapy compounds in differentiation-competent and -defective myeloid leukemic cells. *Cell Growth Differ* 4:41–47.

Ludes-Meyers, J. H., Subler, M. A., Shivakumar, V., Munoz, R. M., Jiang, P., Bigger, J. E., Brown, D. R., Deb, S. P., and Deb, S. (1996). Transcriptional activation of the human epidermal growth factor receptor promotor by human p53. *Mol Cell Biol* 16:6009–6019.

Margulies, L., and Sehgal, P. B. (1993). Modulation of the human interleukin-6 promoter (IL-6) and transcription factor C/EBPb (NF-IL6) activity by p53 species. *J Biol Chem* 268:15096–15100.

Marin, M. C., Jost, C. A., Brooks, L. A., Irwin, M. S., O'Nions, J., Tidy, J. A., James, N., McGregor, J. M., Harwood, C. A., Yulug, I. G., et al. (2000). A common polymorphism acts as an intragenic modifier of mutant p53 behaviour. *Nat Genet* 25:47–54.

Matas, D., Sigal, A., Stambolsky, P., Milyavsky, M., Weisz, L., Schwartz, D., Goldfinger, N., and Rotter, V. (2001). Integrity of the N-terminal transcription domain of p53 is required for mutant p53 interference with drug-induced apoptosis. *EMBO J* 20:4163–4172.

Mazzaro, G., Bossi, G., Coen, S., Sacchi, A., and Soddu, S. (1999). The role of wild-type p53 in the differentiation of primary hemopoietic and muscle cells. *Oncogene* 18:5831–5835.

McGregor, J. M., Harwood, C. A., Brooks, L., Fisher, S. A., Kelly, D. A., O'Nions, J., Young, A. R., Surentheran, T., Breuer, J., Millard, T. P., et al. (2002). Relationship Between p53 Codon 72 Polymorphism and Susceptibility to Sunburn and Skin Cancer. *J Invest Dermatol* 119:84–90.

Mekeel, K. L., Tang, W., Kachnic, L. A., Luo, C.-M., DeFrank, J. S., and Powell, S. N. (1997). Inactivation of p53 results in high rates of homologous recombination. *Oncogene* 14:1847–1857.

Midgley, C. A., and Lane, D. P. (1997). p53 protein stability in tumour cells is not determined by mutation but is dependent on Mdm2 binding. *Oncogene* 15:1179–1189.

Milner, J., and Medcalf, E. A. (1991). Cotranslation of activated mutant p53 with wild type drives the wild-type p53 protein into the mutant conformation. *Cell* 65:765–774.

Mitsudomi, T., Steinberg, S. M., Nau, M. M., Carbone, D., D'Amico, D., Bodner, S., Oie, H. K., Linnoila, R. I., Mulshine, J. L., Minna, J. D., and et al. (1992). p53 gene mutations in non-small-cell lung cancer cell lines and their correlation with the presence of ras mutations and clinical features. *Oncogene* 7:171–180.

Mukhopadhyay, D., Tsiokas, L., and Sukhateme, V. P. (1995). Wild-type p53 and v-Src exert opposing influences on human vascular endothelial growth factor gene expression. *Cancer Res* 55:6161–6165.

Murphy, K. L., Dennis, A. P., and Rosen, J. M. (2000). A gain of function p53 mutant promotes both genomic instability and cell survival in a novel p53-null mammary epithelial cell model. *FASEB J* 14:2291–2302.

Nagata, Y., Anan, T., Yoshida, T., Mizukami, T., Taya, Y., Fujiwara, T., Kato, H., Saya, H., and Nakao, M. (1999). The stabilization mechanism of mutant-type p53 by impaired ubiquitination: the loss of wild-type p53 function and the hsp90 association. *Oncogene* 18:6037–6049.

Niewolik, D., Vojtesek, B., and Kovarik, J. (1995). p53 derived from human tumour cell lines and containing distinct point mutations can be activated to bind to its consensus target sequence. *Oncogene* 10:881–890.

Nikolova, P. V., Wong, K. B., DeDecker, B., Henckel, J., and Fersht, A. R. (2000). Mechanism of rescue of common p53 cancer mutations by second-site suppressor mutations. *EMBO J* 19:370–378.

Osada, M., Ohba, M., Kawahara, C., Ishioka, C., Kanamaru, R., Katoh, I., Ikawa, Y., Nimura, Y., Nakagawara, A., and Obinata, M., Ikawa, S., (1998). Cloning and functional analysis of human p51, which stracturally and functionally resmbles p53. *Nat Med* 4:839–843.

Peled, A., Zipori, D., and Rotter, V. (1996a). Cooperation between p53-dependent and p53-independent apoptotic pathway of myloid cells. *Cancer Res* 56:2148–2156.

Peled, A., Zipori, D., and Rotter, V. (1996b). Cooperation between p53-dependent and p53-independent apoptotic pathways in myeloid cells. *Cancer Res* 56:2148–2156.

Peng, Y., Chen, L., Li, C., Lu, W., Agrawal, S., and Chen, J. (2001a). Stabilization of the MDM2 oncoprotein by mutant p53. *J Biol Chem* 276:6874–6878.

Peng, Y., Chen, L., Li, C., Lu, W., and Chen, J. (2001b). Inhibition of MDM2 by hsp90 contributes to mutant p53 stabilization. *J Biol Chem* 276:40583–40590.

Pfeifer, G. P., and Holmquist, G. P. (1997). Mutagenesis in the p53 gene. *Biochim Biophys Acta* 1333:M1–M8.

Pugacheva, E. N., Ivanov, A. V., Kravchenko, J. E., Kopnin, B. P., Levine, A. J., and Chumakov, P. M. (2002). Novel gain of function activity of p53 mutants: activation of the dUTPase gene expression leading to resistance to 5-fluorouracil. *Oncogene* 21:4595–4600.

Ribeiro, R. C., Sandrini, F., Figueiredo, B., Zambetti, G. P., Michalkiewicz, E., Lafferty, A. R., DeLacerda, L., Rabin, M., Cadwell, C., Sampaio, G.,et al. (2001). An inherited p53 mutation that contributes in a tissue-specific manner to pediatric adrenal cortical carcinoma. *Proc Natl Acad Sci USA* 98:9330–9335.

Righetti, S. C., Torre, G. D., Pilotti, S., Menard, S., Ottone, F., Colnaghi, M. I., Pierotti, M. A., Lavarino, C., Cornarotti, M., Oriana, S.,et al. (1996). Comparative study of p53 gene mutations, protein accumulation and response to cisplatin-based chemotherapy in advanced ovarian carcinoma. *Canc Res* 56:689–693.

Rippin, T. M., Bykov, V. J., Freund, S. M., Selivanova, G., Wiman, K. G., and Fersht, A. R. (2002). Characterization of the p53-rescue drug CP-31398 in vitro and in living cells. *Oncogene* 21:2119–2129.

Rodin, S. N., and Rodin, A. S. (2000). Human lung cancer and p53: the interplay between mutagenesis and selection. *Proc Natl Acad Sci USA* 97:12244–12249.

Roemer, K. (1999). Mutant p53: gain-of-function oncoproteins and wild-type p53 inactivators. *Biol Chem* 380:879–887.

Rolley, N., Butcher, S., and Milner, J. (1995). Specific DNA binding by different classes of human p53 mutants. *Oncogene* 11:763–770.

Rotter, V. (1983). p53, a transformation-related cellular-encoded protein, can be used as a biochemical marker for the detection of primary mouse tumor cells. *Proc Natl Acad Sci USA* 80:2613–2617.

Rotter, V., Abutbul, H., and Wolf, D. (1983). The presence of p53 transformation-related protein in Ab-MuLV transformed cells is required for their development into lethal tumors in mice. *Int J Cancer* 31:315–320.

Rusch, V., Klimstra, D., Venkatraman, E., Oliver, J., Martini, N., Gralla, R., Kris, M., and Dmitrovsky, E. (1995). Aberrant p53 expression predicts clinical resistance to cisplatin-based chemotherapy in locally advanced non-small-cell lung-cancer. *Cancer Res* 55:5038–5042.

Sampath, J., Sun, D., Kidd, V. J., Grenet, J., Gandhi, A., Shapiro, L. H., Wang, Q., Zambetti, G. P., and Schuetz, J. D. (2001). Mutant p53 cooperates with ETS and selectively up-regulates human MDR1 not MRP1. *J Biol Chem* 276:39359–39367.

Selivanova, G., Iotsova, V., Okan, I., Fritsche, M., Strome, M., Groner, B., Graftstrom, R. C., and Wiman, K. G. (1997). Restoration of the growth suppression function of mutant p53 by a synthetic peptide derived from the p53 C-terminal domain. *Nat Med* 3:632–638.

Selivanova, G., Kawasaki, T., Ryabchenko, L., and Wiman, K. G. (1998). Reactivation of mutant p53: a new strategy for cancer therapy. *Semin Cancer Biol* 8:369–378.

Selivanova, G., Ryabchenko, L., Jansson, E., Iotsova, V., and Wiman, K. G. (1999). Reactivation of mutant p53 through interaction of a C-terminal peptide with the core domain. *Mol Cell Biol* 19:3395–3402.

Shaulian, E., Zauberman, A., Ginsberg, D., and Oren, M. (1992a). Identification of a minimal transforming domain of p53: negative dominance through abrogation of sequence-specific DNA binding. *Mol Cell Biol* 12:5581–5592.

Shaulian, E., Zauberman, A., Ginsberg, D., and Oren, M. (1992b). Identification of minimal transforming domain of p53: negative dominance through abrogation of sequence-specific DNA binding. *Mol Cell Biol* 12:5581–5592.

Shaulian, E., Zauberman, A., Milner, J., Davies, E. A., and Oren, M. (1993). Tight DNA binding and oligomerization are dispensable for the ability of p53 to transactivate target genes and suppress transformation. *EMBO J* 12:2789–2797.

Shaulsky, G., Goldfinger, N., and Rotter, V. (1991). Alterations in tumor development *in vivo* mediated by expression of wild type or mutant p53 proteins. *Cancer Res* 51:5232–5237.

Sigal, A., Matas, D., Almog, N., Goldfinger, N., and Rotter, V. (2001). The C-terminus of mutant p53 is necessary for its ability to interfere with growth arrest or apoptosis. *Oncogene* 20:4891–4898.

Sigal, A., and Rotter, V. (2000). Oncogenic mutations of the p53 tumor suppressor: the demons of the guardian of the genome. *Cancer Res* 60:6788–6793.

Slingerland, J. M., Jenkins, J. R., and Benchimol, S. (1993). The transforming and suppressor functions of p53 alleles: effects of mutations that disrupt phosphorylation, oligomerization and nuclear transloca-tion. *EMBO J* 12, 1029–1037.

Soddu, S., Blandino, G., Scardigli, R., Coen, S., Marchetti, A., Rizzo, M. G., Bossi, G., Cimino, L., Crescenzi, M., and Sacchi, A. (1996). Interference with p53 protein inhibits hematopoietic and muscle differentiation. *J Cell Biol* 134:1–12.

Srivastava, S., Wang, S., Tong, Y. O., Hao, Z. M., and Chang, E. (1993). Dominant negative effect of a germ-line mutant p53: A step fostering tumorigenesis. *Cancer Res* 53:4452–4455.

Strano, S., Fontemaggi, G., Costanzo, A., Rizzo, M. G., Monti, O., Baccarini, A., Del Sal, G., Levrero, M., Sacchi, A., Oren, M., and Blandino, G. (2002). Physical interaction with human tumor-derived p53 mutants inhibits p63 activities. *J Biol Chem* 277:18817–18826.

Strano, S., Munarriz, E., Rossi, M., Cristofanelli, B., Shaul, Y., Castagnoli, L., Levine, A. J., Sacchi, A., Cesareni, G., Oren, M., and Blandino, G. (2000). Physical and functional interaction between p53 mutants and different isoforms of p73. *J Biol Chem* 275:29503–29512.

Subler, M. A., Martin, D. W., and Deb, S. (1994). Activation of the human immunodeficiency virus type 1 long terminal repeat by transforming mutants of human p53. *J Virol* 68:103–110.

Takahashi, T., Nau, M. M., Chiba, I., Birrer, M. J., Rosenberg, R. K., Vinocour, M., Levitt, M., Pass, H., Gazdar, A. F., and Minna, J. D. (1989). p53: a frequent target for genetic abnormalities in lung cancer. *Science* 246:491–494.

Takahashi, Y., Bucana, C. D., Cleary, K. R., and Ellis, L. M. (1998). p53, vessel count, and vascular endothelial growth factor expression in human colon cancer. *Int J Cancer* 79:34–38.

Tang, H. Y., Zhao, K., Pizzolato, J. F., Fonarev, M., Langer, J. C., and Manfredi, J. J. (1998). Constitutive expression of the cyclin-dependent kinase inhibitor p21 is transcriptionally regulated by the tumor suppressor protein p53. *J Biol Chem* 273:29156–29163.

Telerman, A., Tuynder, M., Dupressoir, T., Robaye, B., Sigaux, F., Shaulian, E., Oren, M., Rommelaere, J., and Amson, R. (1993). A model for tumor suppression using H-1 parvovirus. *Proc Natl Acad Sci USA* 90:8702–8706.

Tsutsumi-Ishii, Y., Todakoro, K., Hanaoka, F., and Tsuchida, N. (1995). Response of heat shock element within the human HSP70 promotor to mutated p53 genes. *Cell Growth Differ* 6:1–8.

Ueba, T., Nosaka, T., Takahashi, J. A., Shibata, F., Florkiewicz, R. Z., Vogelstein, B., Oda, Y., Kikuchi, H., and Hatanaka, M. (1994). Transcriptional regulation of basic fibroblast growth factor gene by p53 in human glioblastoma and hepatocellular carcinoma cells. *Proc Natl Acad Sci USA* 91:9009–9013.

Unger, T., Mietz, J. A., Scheffner, M., Yee, C. L., and Howley, P. M. (1993). Functional domains of wild-type and mutant p53 proteins involved in transcriptional regulation, transdominant inhibition and transformation suppression. *Mol Cell Biol* 13:5186–5194.

Van Meir, E. G., Polverini, P. J., Chazin, V. R., Huang, H.-J., de Tribolet, N., and Cavenee, W. K. (1994). Release of an inhibitor of angiogenesis upon induction of wild type p53 expression in glioblastoma cells. *Nat Genet* 8:171–176.

van Oijen, M. G., and Slootweg, P. J. (2000). Gain-of-function mutations in the tumor suppressor gene p53. *Clin Cancer Res* 6:2138–2145.

Varley, J. M., McGown, G., Thorncroft, M., Cochrane, S., Morrison, P., Woll, P., Kelsey, A. M., Mitchell, E. L., Boyle, J., Birch, J. M., and Evans, D. G. (1996). A previously undescribed mutation within the tetramerisation domain of TP53 in a family with Li-Fraumeni syndrome. *Oncogene* 12:2437–2442.

Vogelstein, B., and Kinzler, K. W. (1993). The multistep nature of cancer. *Trends Genet* 9:138–141.

Vollmer, C. M., Ribas, A., Butterfield, L. H., Dissette, V. B., Andrews, K. J., Eilber, F. C., Montejo, L. D., Chen, A. Y., Hu, B., Glaspy, J. A., et al. (1999). p53 selective and nonselective replication of an E1B-deleted adenovirus in hepatocellular carcinoma. *Cancer Res* 59:4369–4374.

Wallingford, J. B., Seufert, D. W., Virta, V. C., and Vize, P. D. (1997). p53 activity is essential for normal development in Xenopus. *Curr Biol* 7:747–757.

Whitesell, L., Sutphin, P. D., Pulcini, E. J., Martinez, J. D., and Cook, P. H. (1998). The physical association of multiple molecular chaperone proteins with mutant p53 is altered by geldanamycin, an hsp90-binding agent. *Mol Cell Biol* 18:1517–1524.

Wieczorek, A. M., Waterman, J. L. F., Waterman, M. J. F., and Halazonetis, T. D. (1996). Structure-based rescue of common tumor-derived p53 mutants. *Nat Med* 2:1143–1146.

Wolf, D., Harris, N., and Rotter, V. (1984a). Reconstitution of p53 expression in a nonproducer Ab-MuLV-transformed cell line by transfection of a functional p53 gene. *Cell* 38:119–126.

Wolf, D., Harris, N., and Rotter, V. (1984b). Reconstitution of p53 expression in a nonproducer Ab-MuLV-transformed cell line by transfection of a functional p53 gene. *Cell* 38:119–126.

Wong, K. B., DeDecker, B. S., Freund, S. M., Proctor, M. R., Bycroft, M., and Fersht, A. R. (1999). Hot-spot mutants of p53 core domain evince characteristic local structural changes. *Proc Natl Acad Sci USA* 96:8438–8442.

Yang, A., Kaghad, M., Wang, Y., Gillett, E., Fleming, M. D., Dotsch, V., Andrews, N. C., Caput, D., and McKeon, F. (1998). p63, a p53 homolog at 3q27-29, encodes multiple products with transactivating, death-inducing, and dominant-negative activities. *Mol Cell* 2:305–316.

Yang, A., Schweitzer, R., Sun, D. Q., Kaghad, M., Walker, N., Bronson, R. T., Tabin, C., Sharpe, A., Caput, D., Crum, C., and McKeon, F. (1999). p63 is essential for regenerative proliferation in limb, craniofacial and epithelial development. *Nature* 398:714–718.

Yang, A., Walker, N., Bronson, R., Kaghad, M., Oosterwegel, M., Bonnin, J., Vagner, C., Bonnet, H., Dikkes, P., Sharpe, A., et al. (2000). p73-deficient mice have neurological, pheromonal and inflammatory defects but lack spontaneous tumours. *Nature* 404:99–103.

Yuan, Z. M., Shioya, H., Ishiko, T., Sun, X., Gu, J., Huang, Y. Y., Lu, H., Kharbanda, S., Weichselbaum, R., and Kufe, D. (1999). p73 is regulated by tyrosine kinase c-Abl in the apoptotic response to DNA damage. *Nature* 399:814–817.

Zhu, J., Jiang, J., Zhou, W., and Chen, X. (1998). The potential tumor suppressor p73 differentially regulates cellular p53 target genes. *Cancer Res* 58:5061–5065.

10

Therapeutic Strategies Based on Pharmacological Modulation of p53 Pathway

Andrei V. Gudkov

SUMMARY

p53 plays a dual role in cancer treatment being, on one hand, a major cancer preventive factor, which can kill or sensitize tumors to radio- and chemotherapy and, on the other hand, a determinant of cancer treatment side effects by inducing apoptosis in normal tissues during cancer therapy. This dualism defines two major therapeutic applications targeting p53: p53 activation to reduce viability of tumor cells and p53 inhibition to increase the viability of normal cells thereby reducing treatment side effects. Prospective new anticancer agents are being developed that recover p53 function in tumor cells by disrupting its interactions with natural inhibitors, such as MDM2 or E6, or restore wild-type conformation of mutant p53. In parallel, p53 inhibitory strategy is being developed to protect normal tissues from chemo- and radiotherapy and to treat other pathologies associated with stress-mediated activation of p53.

10.1. WHY p53 IS A THERAPEUTIC TARGET

p53 is a tumor suppressor that is lost or inactivated in the majority of tumors (Soussi, 2000). Moreover, many tumors respond to ectopic expression of wild-type

---

A. V. GUDKOV • Department of Molecular Biology, Lerner Research Institute, Cleveland Clinic Foundation, Cleveland, OH 44195, USA

The p53 Tumor Suppressor Pathway and Cancer, edited by Zambetti.
Springer Science+Business Media, New York, 2005.

p53 by rapid apoptosis or irreversible growth arrest (Vousden and Lu, 2002) thereby defining gene therapy applications of p53. This therapeutic strategy has been aggressively explored by many with modest success explained by challenges of effective delivery of p53-expressing vectors into tumor cells, reviewed by Willis and Chen (2002) and Fang and Roth (2003). Tumor suppressor genes are not viewed as promising therapeutic targets because they have not much to offer, in terms of therapeutic strategies, besides gene therapy. However, there are several reasons why p53 is different from other tumor suppressors.

The first reason is that functional inactivation of p53 in tumors is rarely reached by its physical loss. In the majority of tumors p53 either mutated or inactivated by inhibitory mechanisms, involving known, such as Mdm2 or E6 (see Prives and Hall, 1999; Soussi, 2000; Woods and Vousden, 2001), and unknown inhibitory factors, thereby making the situation potentially reversible (Bullock and Fersht, 2001; Lain and Lane, 2003). Although the majority of conventional anticancer treatments by drugs and radiation can effectively induce p53 in normal cells, none of them is capable of "waking up" mutated or inhibited p53 in cancer cells. This possibility indicates an attractive untouched therapeutic resource of specific killing tumor cells by pharmacological rescue of inactive p53. It is noteworthy that overexpression of mutant p53 is also used as a basis for the biological strategy of activating immune response for selective elimination of tumor cells, reviewed by (Offringa et al., 2000; Chene, 2001).

Another reason making p53 an important therapeutic target is its role as a determinant of sensitivity of some normal tissues to genotoxic stress associated with cancer treatment and other p53-activating conditions. Ironically, this major cancer preventive factor can complicate cancer treatment by triggering massive programmed cell death in certain normal (but not in cancerous) tissues during systemic genotoxic stress associated with chemo- and radiotherapy. This makes p53 a target for therapeutic suppression—an approach to reduce side effects associated with treatment of p53-deficient cancers (Komarova and Gudkov, 1998, 2001; Gudkov and Komarova, 2003).

Remarkably, inhibition of p53 in some instances may have a direct antitumor effect, not necessarily through reducing the side effects. There is accumulating evidence that in some tissues p53 can play a role of a survival factor under conditions of severe genotoxic stress. This is true to those tissues in which apoptosis is not the major outcome of p53 activation but which mostly undergo p53-dependent growth arrest, thereby increasing chances for successful repair and survival, reviewed by (Gudkov and Komarova, 2003).

Hence, both activation and suppression of p53 can be beneficial for cancer treatment outcome (Fig. 10.1). The first approach is expected to contribute to more efficient tumor cell killing by restoring p53 function to the level of a normal tissue, while the other one should improve treatment outcome by reducing tissue injury by temporary reversible conversion of normal tissues to a p53-deficient state characteristic for tumors. Many laboratories are currently exploring these approaches, both of which offer some additional promises outside of the cancer treatment field.

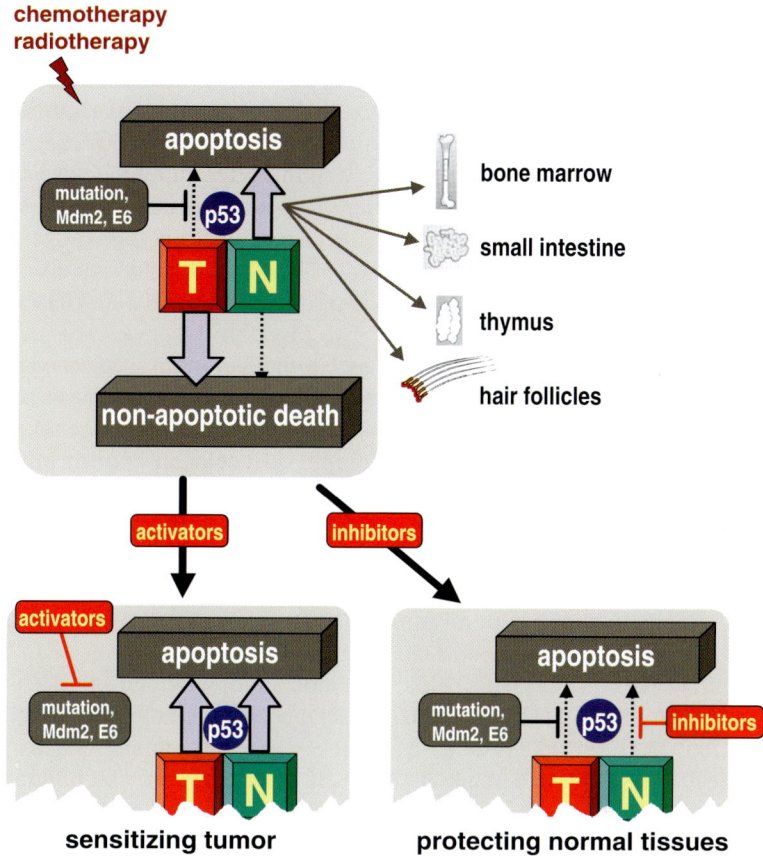

Figure 10.1. Rationale for pharmacological modulation of p53 in cancer treatment. Therapeutic index of cancer treatment is a ratio between antitumor effect of therapy and the side effects. Killing of tumor cells (indicated as "T") as a result of treatment largely goes through nonapoptotic mechanisms since these mechanisms (including p53-mediated apoptosis) are frequently repressed in cancer. At the same time, systemic genotoxic stress associated with chemo- and radiotherapy induces p53-dependent apoptotic response in sensitive normal tissues (indicated as "N") leading to severe side effects. Improving of the therapeutic index can be reached either (i) by reactivating the apoptotic program in tumor cells by using p53 activators or (ii) by reversible temporary inhibition of p53-mediated death of normal cells. Both goals can be reached through pharmacological modulation of p53.

10.2. THERAPIES BASED ON PHARMACOLOGICAL ACTIVATION OF p53

10.2.1. Restoration of Wild-Type Function of Mutant p53 in Tumors

Of more than 10,000 different human tumors analyzed so far, 45–50% contained inactivating mutations within the p53 gene (Landis et al., 1999) making it one of the most frequently mutated gene in cancer. Frequencies of mutations vary from one tumor type to another (Chow et al., 1999) ranging from 60–65% of lung and colon cancer, 40–45% in stomach, esophagus, and bladder cancers, 25–30% in breast, liver, prostate cancer and in lymphomas, and only 10–15% in leukemia (Landis et al., 1999). Interestingly, the nature of mutations acquired by p53 in the tumors clearly differentiates this gene from other tumor suppressors that are inactivated predominantly by deletions or transcriptional silencing, resulting in the lack of functional protein. On the contrary, 80–85% of tumor-derived alterations in p53 gene are missense point mutations localized within the DNA-binding domain of the protein (Lowe, 1995; Chow et al., 1999; Landis et al., 1999; Bast et al., 2000). Such mutations result in the accumulation of large amounts of abnormally stable p53 protein that loses specific DNA-binding and transactivation functions. The fact that tumor-derived mutations in p53 preserve expression of full-length, though altered, proteins suggests that mutant p53 could provide some selective advantages for tumor cells. At the same time, this situation opens an attractive possibility of converting mutant p53 into a wild-type killing conformation: importantly, supplying exogenous wild-type p53 in cancer cells by gene delivery is effective in suppressing tumor growth of both mutant and wild-type p53-containing tumors.

The principal possibility of such restoration was supported by observations that the wild-type structure of some of the tumor-derived p53 mutants can be restored (structurally and functionally) both in vitro and in the cells by glycerol that is known to stabilize folding of destabilized temperature sensitive proteins (Ohnishi et al., 1999).

Attempts to develop more p53 specific small molecules with similar properties resulted so far in two publications. Foster et al. (1999) reported the isolation of compound CP-31398 picked from the in vitro screening system allowing to distinguish between wild type and denatured conformations of p53 in a high throughput manner. CP-31398 was shown to have antitumor effect in human tumor xenografts expressing mutant p53. Later reports by other groups questioned the p53-restoring ability of CP-31398 (Rippin et al., 2002). A successful attempt to isolate a p53 rescuing molecule was made by a Swedish group led by G. Selivanova who isolated the compound named PRIMA using cell-based readout system allowing for monitoring p53-dependent transactivation (Bykov et al., 2002). Again, the antitumor effect of PRIMA, dependent on mutant p53, was demonstrated in vivo in human xenografts growing in nude mice. Remarkably, the effect of PRIMA was not limited to the mutant that was expressed in the cells used as a readout system but was also effective against other tumor-derived mutants.

The well determined folding structure of p53 protein opens the opportunity for a rational design of compounds that could stabilize the wild-type-like conformation of mutant p53 that is unstable under regular conditions. This approach taken by the group led by A. Fersht and G. Selivanova resulted in designing and successful testing of a 9-amino-acid-long peptide CDB3 that effectively converted several tumor-derived p53 mutants into the wild-type conformation in vivo causing restoration of its ability to transactivate its gene targets, induce apoptosis, and sensitize tumor cells to gamma radiation-induced apoptosis. Recently, similar observations were published by Peng et al. (2003) for experimental anticancer drug ellipticine that was shown capable of restoring function to a broad spectrum of p53 tumor-derived mutants (175 H, 248W, 249S, 273 H, 281G, 194F, 233L, 241F, and 273C), thus broadening the list of molecules of that functional class.

All these data indicate that pharmacological rescue of mutant p53 in tumors is possible and that a single agent can be effective against numerous mutants making this approach even more attractive. It obviously has some limitations and is probably inapplicable to those classes of p53 mutants whose loss of function does not result from destabilized conformation (Bullock and Fersht, 2001; Friedler et al., 2003). As in many other cases, the success of rational design of p53 rescuing peptides will provide a solid basis for the development of drugs after it becomes possible to generate cell penetrating effective peptide mimetics.

10.2.1.1. Activation of Wild-Type p53 in Tumors by Targeting p53 Inhibitory Proteins: Mdm2

While p53 is mutated in more than half of human tumors, it is still either completely or partially nonfunctional in the majority of remaining cases (Soussi, 2000; Woods and Vousden, 2001). The known mechanisms of such repression involve either overexpression of natural inhibitory proteins of cellular (Mdm2, MdmX; Michael and Oren, 2002) or viral (E6; Tommasino et al., 2003) nature or the loss of components of p53 pathway essential for its activity (Arf; Sherr and Weber, 2000). Inactivation of p53 by dominant inhibitors makes the situation "druggable" since it opens the opportunity to release functional p53 by disrupting its interactions with the inhibitors.

The most common among these is Mdm2, an ubiquitin ligase that physically binds to p53, mediates its nuclear export, and promotes proteasomal degradation of p53 (Vargas et al., 2003). Normally, Mdm2 is not abundant enough to constitutively prevent p53 activity; however, it is encoded by a p53-regulated gene and is activated as a result of p53 response, playing the role of a feedback regulator. Some tumors express constitutively high levels of Mdm2 that represses p53 activation; this mechanism is active in a variety of cancers making this protein an attractive target for selective killing of tumor cells (Chene, 2003; Zheleva et al., 2003).

The development of inhibitors of p53–Mdm2 interactions is facilitated by accurate functional mapping of protein interaction domains and by the availability of structures of both proteins resolved by X-ray crystallography and NMR, reviewed by Chene (2003) and in Chapter 2. Precise localization of interacting domains allowed the development of a series of short peptides and peptide mimetics imitating the p53-protein region interacting with Mdm2 and interfering with this interaction. It was shown that the minimal Mdm2 binding site in p53 molecule could be reduced to p53 fragment between the 18th and 23rd amino acids. Further improvement of structure and composition of this peptide, based on a combination of functional selection in phage display libraries, use of nonnatural amino acids, and computational analysis, allowed to increase the strength of peptide-Mdm2 binding almost 2,000-fold of that of the original p53-derived peptide (Garcia-Echeverria et al., 2000). These peptides demonstrated expected p53-activating abilities in vitro (Kanovsky et al., 2001; Chene et al., 2002) and could form the basis for creation of small molecules with similar structure and effect suitable for in vivo delivery (Zhao et al., 2002).

Development of small molecules inhibiting p53–Mdm2 interaction has not yet resulted in isolation of a clinically proven drug, although several chemicals with the desirable activity have been reported. These include 1,2-benzodiazopine-2-one, predicted by a computational approach based on the known shape of the p53-interacting region of Mdm2 molecule (see Chene, 2003). Chalcones (1,3-diphenyl-2-propen-1-ones) were identified as MDM2 inhibitors that bind to a subsite of the p53-binding cleft of human MDM2 (Stoll et al., 2001). Biochemical experiments showed that these compounds could disrupt the MDM2/p53 protein complex, releasing p53 from both the p53/MDM2 and DNA-bound p53/MDM2 complexes. Chlorofusin was isolated by screening of microbial products in vitro (Duncan et al., 2001). Recently, two groups reported identification of small molecules disrupting p53/Mdm2 intercation. Vassilev et al. (2004) used structure-based rational drug design to generate compounds named nutlins, potent and selective small-molecule antagonists of MDM2, and confirmed their mode of action through the crystal structures of complexes. These compounds bind MDM2 in the p53-binding pocket and activate the p53 pathway in cancer cells, leading to cell cycle arrest, apoptosis, and growth inhibition of human tumor xenografts in nude mice. A few months later Issaeva et al. (2004) described successful application of chemical library screening approach in a cell-based readout system resulted in isolation another compound with similar properties named RITA, which changes conformation of p53 in the way that retains its function but prevents its binding to Mdm2. Strong anti-tumor effect of both compounds in mouse models with no detectable general toxicity one could expect from the results of genetic knockout of Mdm2 (Montes de Oca Luna et al. 1995) is very encouraging and stimulates rapid development of a new type of anticancer drugs targeting p53/Mdm2 interaction.

Another member of the Mdm2 gene family, MdmX, shares structural and many functional features of its better-studied relative (Michael and Oren, 2002). Not surprisingly, some of the isolated p53–Mdm2-disrupting peptides are active against p53–MdmX interaction (Bottger et al., 1999; Chene, 2003).

10.2.1.2. Activation of Wild-Type p53 in Tumors by Targeting p53 Inhibitory Proteins: E6 of Papilloma Virus

Papilloma virus protein E6 is another example of a natural dominant inhibitor of p53 that is involved in cancer development. Human papilloma viruses types 16, 18, and several others are etiological agents of several human malignant diseases (Bast et al., 2000). More than 95% of cervical carcinoma, the second most common malignant disease among women worldwide (Parkin et al., 1999), are associated with infection of high-risk types of human papilloma viruses (HPV) (Bosch et al., 2002). In addition, HPV infection is linked to more then 50% of other anogenital cancers, and approximately 20% of oral carcinomas (Doorbar et al., 1997).

Two gene products of papilloma viruses, E6 and E7, are responsible for the initiation of all events leading to cancer development by blocking two major tumor suppressor pathways. E7 protein inactivates the tumor suppressor pRB thus eliminating control over the cell cycle (Munger et al., 2001), while the product of E6 gene inhibits p53 (Mantovani and Banks, 2001). Physical binding of E6 to p53 promotes its degradation by cellular U3-ubiquitin-ligase E6AP (Huibregtse et al., 1993). Other functions of the E6 and E7 proteins interfere with differentiation, intercellular communications, and apoptosis (Mantovani and Banks, 2001; Munger et al., 2001).

The virus-imposed inhibition of products of two major tumor suppressor genes is so efficient, that apparently there is no need for further inactivation of these genes during the carcinogenesis. This explains why, unlike most other malignant diseases, both the pRB and the p53 tumor suppressor pathway stay structurally intact in HPV-expressing cervical carcinomas (Scheffner et al., 1991; Goodwin and DiMaio, 2000). Many approaches have been tested to block expression of viral oncogenes in HPV-positive cervical cancer cells, such as inhibiting the E6/E7 gene transcription by the expression of the E2 gene product (Goodwin and DiMaio, 2000), or using antisense constructs (von Knebel Doeberitz et al., 1992; Hamada et al., 1996), ribozymes (Alvarez-Salas et al., 1998), siRNA (Jiang and Milner, 2002) directed against the polycistronic E6/E7 mRNA, or by inhibition of HPV transcription with 2-deoxyglucose (Maehama et al., 1998) and antiviral drug cidofovir (Abdulkarim et al., 2002). Indeed, the above treatments resulted in reactivation of the pRB and p53 tumor suppressor pathways, leading to significant growth suppression, radio sensitization, and apoptosis. The structural integrity of tumor-suppressor pathways in cervical carcinomas, along with the distinctly foreign nature of viral E6 and E7 proteins, makes HPV a perfect potential target for pharmacological suppression. Several attempts have been made to develop drugs that interfere with functions of HPV in cervical carcinoma cells. Peptide aptamers developed to the HPV16 E6 gene have been shown to induce apoptotic response in cervical carcinoma cells (Butz et al., 2000). Application of zinc-ejecting compounds, such as 4,4′-dithiodimorpholine, inhibited the interaction of E6 with the E6AP protein, leading to accumulation of p53 in HPV-containing cells, accompanied by the induction of apoptosis (Beerheide et al., 1999).

All this evidence proves that targeting the E6–p53 interaction to treat (or to prevent from) cervical cancer is a feasible approach. Nevertheless, it is still unclear how the ultimate drug acting through this mechanism will look like and what will be the regimen of its application.

Inhibition of p53 in tumors is not limited to Mdm2 family members and E6 of HPV. There are cancers that almost never acquire mutations in p53 but in which p53 is functionally repressed. Tumors of these type include melanoma, renal cell carcinoma (RCC), neuroblastoma, and others (Tweddle et al., 2003). In some of them, like in melanoma, p53 function is repressed by a recessive mechanism (loss of Apaf-1 expression resulting in lack of apoptosis, Soengas et al., 2001) and therefore is not well suitable for drug development. However, in some other tumors, such as renal cell carcinoma, the mechanism of repression is dominant but not involving Mdm2 or MdmX (Gurova et al., 2004). Although the exact molecular target for rescuing p53 in RCC remains unknown, the dominant nature of p53 repression creates a "druggable" situation making it possible to screen chemical libraries for p53-reactivating drugs using transactivation function as a readout in RCC. This approach has resulted in the identification of a series of small molecules capable of p53 activation in this practically so far untreatable form of cancer (Gurova et al., submitted). In this scenario, target identification will follow drug discovery.

10.2.2. Can Cancer-Preventing Agents Act through Activating p53?

p53 is believed to perform its tumor suppressor function by controlling genetic stability through self-elimination of cells with damaged DNA. This broadly accepted paradigm favors development of therapeutic approaches involving restoration of the lost p53 function in the tumors, leading to growth inhibition of cancer cells. However, it leaves no room for cancer prevention through activation of p53 since increased activity of this protein should be detrimental for growth of normal tissues. Interestingly, the paper by Seo et al. (2002) suggested a revision of this traditional thinking. It indicated that maintenance of genomic stability by p53 can be separated from its growth suppressive or proapoptotic functions and may involve direct activation of DNA repair machinery.

They found that incubation with selenomethionine (SeMet), the major source of selenium in our diet, results in an unusual activation of p53 in cultured cells: a reduction of two specific cysteine residues within p53 molecule leading to a conformational shift in the p53 molecule and induction of its DNA binding activity. This reduction is neither the result of a direct interaction of p53 with SeMet nor is it a consequence of DNA damage, that SeMet does not induce. It requires the cellular protein Ref1, a known redox factor that was previously shown to physically interact with p53; inactivation of this protein blocks p53 modification by the selenium-containing compound. However, the most unusual property of p53 activated by SeMet-induced Ref1-mediated reduction is that it becomes capable of activating DNA repair machinery without affecting cell growth. As a result, the cells with wild-type p53 can tolerate higher doses of UV irradiation, while p53-deficient cells could not benefit from the presence of the selenium-containing compound. Hence, p53 can contribute to genomic stability not only by eliminating damaged cells from the population, but also through a direct activation of DNA repair system.

Thus, p53 modified by SeMet can induce DNA repair through a specific modification that is distinct from its DNA damage-responsive form. These observations provide plausible explanation for cancer preventive activity of SeMet that is likely to reduce accumulation of mutations by somatic cells causing permanent (as long as the compound is present) p53-mediated activation of DNA repair.

It would be somewhat premature to generalize this model before the described phenomenon is confirmed in different cell types and tested in vivo. However, if proven right, it would provide a mechanistic explanation for the activity of one of the most promising cancer preventive agents and define the way toward new cancer preventive drugs.

10.2.3. Search for p53 Activating Agents as an Approach
to New Anticancer Drugs

The majority of anticancer agents, including drugs and radiation, are potent inducers of p53 in normal cells. This is not considered surprising because p53 is known to be a key mediator of cellular responses to a variety of stresses and conventional cancer treatment is stressful by definition. However, if this explanation were correct, one could expect many p53-inducing agents among generally toxic compounds. This question was experimentally addressed in the work of Sohn et al. (2002) who screened a chemical library consisting of a range of conventional chemotherapeutic agents as well as over 16,000 diverse small compounds in a cell-based readout system for detecting activators of p53-dependent transcription of surrogate p53 responsive reporter. While two-thirds of conventional chemotherapeutic agents were found capable of activating p53 activity by twofold or greater, only 0.2% of diverse compounds showed this property. Cyto-toxicity was independent of p53 genetic status as judged by their testing in syngenic wild-type p53 and p53-null cells. Hence, there is enormous enrichment of p53 activators among established anticancer agents. This result is most surprising, taking into account that anticancer effect of conventional chemotherapeutic drugs does not depend on p53 activation because the p53 pathway is inactive in the majority of tumors. This phenomenon validates the approach of searching for new anticancer drugs among p53 inducers.

10.3. PROSPECTIVE THERAPEUTIC APPLICATIONS OF p53 INHIBITORS

10.3.1. p53 and Cancer Treatment Side Effects

Although activation of p53 is generally viewed as a most direct and promising anticancer strategy, it is not a favorable one for normal tissues. Cancer treatment with radiation and cytostatic drugs is associated with the induction of genotoxic stress resulting from either direct DNA damage (radiation, antitopoisomerase drugs, nucleotide analogs, etc.) or the inability to undergo normal mitosis (antimicrotubule agents: Vinca alkaloids, taxol, etc.). Cell reaction to genotoxic stress in vitro involves activation of p53 that initiates a cascade of events leading to growth arrest or apoptosis. By analyzing mice expressing the *lacZ* reporter gene from the p53-responsive promoters (Komarova et al., 1997), it was found that whole-body gamma irradiation or treatment with high dosages of DNA-damaging chemotherapeutic drugs led to a pronounced activation of the transgene, indicative of the p53 activity, in the most obvious areas of radiation or drug-induced apoptosis that was not seen in p53-deficient mice (Lowe et al., 1993). These areas, in turn, coincided with the sites affected by anticancer treatment, suggesting p53 involvement in the treatment-induced damage of sensitive tissues. Sites of apoptosis match a tissue-specific pattern of p53 mRNA expression, indicating that p53 regulation at the mRNA level is a determinant of acute radiosensitivity of tissues. Comparison of wild-type and p53-knockout mice showed that acute apoptotic response to gamma irradiation in the hematopoietic system (Cui et al., 1995; Wang et al., 1996), in hair follicles (Song and Lambert, 1999), in oligodendroblasts of spinal cord (Chow et al., 2000), and, in part, in epithelia of digestive tract (Merritt et al., 1994) is p53 dependent. All these facts indicate that p53 plays a key role in radiation and chemo-sensitivity of tissues, thus contributing to general radiosensitivity of the organism. Consistently, p53-deficient mice survive high doses of radiation that are lethal for the wild-type animals (Westphal et al., 1997).

10.3.2. Pharmacological Suppression of p53 May Reduce Cancer Treatment Side Effects

Based on these observations, we hypothesized that p53 is a mediator and a determinant of radiation and drug toxicity and, therefore, could be considered a target for therapeutic suppression to reduce cancer treatment side effects (Komarova and Gudkov, 1998). Obviously, such an approach should be applicable to the treatment of p53-deficient tumors that are a major portion of all cancers. To prove this principle, a chemical inhibitor of p53 named PFT was isolated which rescued p53 wild-type cells from apoptotic death induced by DNA damage in vitro and reduced lethality of mice from gamma radiation in vivo without a detectable increase in tumor incidence (Komarov et al., 1999), and did not cause a protective effect on treatment sensitivity of p53-deficient tumors (unpublished observations). PFT had a protective effect in experimental chemotherapy models (Zhang et al., 2003) as well as in other pathological conditions involving p53 activation (see below). These results indicate that reversible repression of p53 is a valid approach to reduce cancer treatment side effects and that p53 inhibitors could be useful drugs to be applied in combination with chemo- or radiation therapy.

10.3.3. Applications of p53 Inhibitors Outside of Cancer Treatment

p53 can trigger apoptotic cell death in response to a variety of other stresses besides cancer therapy. There is an accumulating bulk of experimental evidence supporting the involvement of p53 apoptosis in pathological consequences of such frequent natural stresses as hypoxia and hyperthermia, including such common diseases as heart and brain ischemia. Thus, p53 was shown to be involved in HIF1α-mediated cell response to hypoxia (An et al., 1998; Blagosklonny et al., 1998). It was found that HIF-1α stabilizes p53 through the formation of hypoxic complex, which in turn enhances the transcription of known p53 targets (Halterman et al., 1999). p53 can promote MDM2-mediated ubiquitination and proteasomal degradation of the HIF-1α, leading to the suppression of angiogenic stimulus (Ravi et al., 2000). We have summarized and discussed these facts in our recent review (Komarova and Gudkov, 2001). Unlike genotoxic stress, there is no unequivocal proof that p53 is the major determinant of tissue sensitivity to hypoxia and heat shock. Additional experiments with p53-deficient mice and p53 inhibitors are required before we can conclude that p53 suppression will really make a difference and reduce the rate of fatalities in, for example, ischemic diseases. Nevertheless, the collected information provides, in our opinion, strong rationale for testing this possibility experimentally.

During the last two years, there have been a number of publications indicating that PFT can in fact be useful in rescuing cells and organisms from pathological conditions involving p53 activation, other than anticancer treatment (Culmsee et al., 2001; Lakkaraju et al., 2001; Alves da Costa et al., 2002; Pani et al., 2002). These include the reduction of neuronal death after treatment with hypoxia or dopamine (Culmsee et al., 2001; Lakkaraju et al., 2001), and even the rescue of p53-dependent embryonic lethality associated with PAX3 gene knockout (Pani et al., 2002). PFT was also found active against a mouse model of Parkinson disease (Pirrkala et al., 2001; Duan et al., 2002), endotoxin-mediated liver toxicity (Schafer et al., 2003) and hypoxia-induced renal failure (Kelly et al., 2003). These observations add support to our hypothesis that p53 inhibitors can be useful for the treatment of a variety of pathological conditions.

10.3.4. Wild-Type p53 can be a Treatment Resistance Factor in Tumors

p53 is known as a major determinant of DNA damage-induced apoptosis, and loss of p53 in tumors is associated with an unfavorable prognosis in many forms of cancer (Cordon-Cardo et al., 1994; Falette et al., 1998; Molina et al., 1998). Wild-type p53 is therefore thought to make tumors more sensitive to treatment through the induction of apoptosis and p53 inactivation is thought to lead to treatment resistance. However, this model is only applicable to those tumor cells that are capable of p53-dependent apoptosis, a property that is frequently lost in tumors. What is the role of p53 in those tumors that lack apoptotic program?

As p53 is responsible for the prolonged arrest after IR treatment, it is expected to facilitate DNA repair in the absence of apoptotic response. Therefore, tumors that inactivate p53 during progression should be less capable of DNA repair and more sensitive to DNA damage-induced mitotic catastrophe. Hence, in the absence of apoptosis p53 might act as a survival factor. This was shown to be true in several tumor cell models in which inactivation of p53 function has no effect on radio sensitivity (Brachman et al., 1993; Slichenmyer et al., 1993; Bunz et al., 1999; Roninson et al., 2001).

If p53 is, indeed, a survival factor in tumors, why is the loss of p53 associated with a poor prognosis (Cordon-Cardo et al., 1994; Falette et al., 1998; Molina et al., 1998; Nieder et al., 2000)? This apparent controversy probably reflects the role of p53 in maintaining genomic stability. It has been demonstrated in several tumor models that cell variants that either express Bcl-2 or lack p53 have similar growth advantages in vivo (Graeber et al., 1996; Schmitt and Lowe, 2001; Gurova et al., 2002). However, these two traits have an opposite prognostic value: whereas p53 inactivation is associated with an unfavorable prognosis, paradoxically, Bcl-2 expression could be a favorable prognostic marker in different types of cancer (Gurova and Gudkov, 2003). The reason for this difference is that whereas loss of p53 or expression of Bcl-2 both prevent apoptosis, only loss of p53 makes cells genetically unstable thereby promoting rapid progression (Schmitt and Lowe, 2001; Gurova et al., 2002). Moreover, Bcl-2-positive tumors tend to maintain wild-type p53 simply because they provide no selective advantages for the p53-deficient variants (Schmitt and Lowe, 2001; Gurova et al., 2002).

Studies from Judah Folkman's lab indicated that experimental chemotherapy of mouse tumors targeting tumor vascular endothelium was more effective in p53-null mice suggesting that p53 can play a protective role in tumor endothelium under the conditions of genotoxic stress (Browder et al., 2000). This observation defines a new potential application of p53 inhibitors as antiangiogenic factors, an approach that is now supported by experimental data (Burdelya and Gudkov, in preparation). Although the mechanism of this phenomenon is yet to be understood, it might be similar to the p53-mediated protection of the epithelium of small intestine from gamma radiation. In this latter case, p53 plays the role of a survival factor by allowing cells to reside in growth arrest thereby reducing the risk of a mitotic catastrophe (Komarova et al., 2004).

The role of p53 in tumor susceptibility to treatment is therefore not as simple as was originally thought. Its impact could vary from negative to positive depending on its ability to perform distinct functions (apoptosis, temporary growth arrest, irreversible growth arrest, control of genomic stability). Hence, the diagnostic and prognostic value of p53 depends on many additional factors and should be evaluated in connection with a specific tumor context. Nevertheless, it is clear that loss of p53 in many cases does not lead to an increased resistance of tumor to treatment and, on the contrary, can be a factor contributing to chemo- and radiation sensitivity.

10.3.5. Is There a Risk Associated with the Use of p53 Inhibitors?

As in the case of cancer treatment, safety is an obvious concern in potential clinical applications of p53 inhibitors. p53 suppression could result in the survival of genetically altered cells (which otherwise would have been eliminated by apoptosis) that potentially may form a high risk subpopulation from which tumorigenic cells could eventually be recruited. The fact that radioprotection by the p53 inhibitor was not associated with a detectable induction of tumor occurrence in mice indicates that temporary reversible inhibition of p53 can be relatively safe compared to total p53 deficiency, which is associated with a high incidence of cancer in p53-knockout mice (Donehower et al., 1992; Jacks et al., 1994). However, recent in vitro studies indicated that the rescuing effect of PFT on p53 wild-type cells treated with chemotherapeutic drugs was accompanied by a higher rate of chromosomal abnormalities (Bassi et al., 2002). Thus, this issue requires more attention, both in statistical and pharmacological aspects, to evaluate the interdependence between prolonged applications of the inhibitors (imitating future clinical applications) and cancer frequency.

The risk/benefit ratio for the use of p53 inhibitors could vary greatly for different diseases. While the risk is worth taking in life-threatening diseases in adults (cancer, stroke, severe burns), the use of similar approaches to prevent embryos from maternal fever seems less attractive due to the heightened risk of developmental malformations. However, any conclusions would be premature at our current level of knowledge.

10.4. CONCLUDING REMARKS: PERSPECTIVES OF PHARMACOLOGICAL MODULATORS OF p53

In summary, there are a number of clinical conditions under which p53 modulation could be considered as a beneficial therapeutic approach, making the isolation of small molecules targeting p53 function a desirable task. Such molecules could target p53 protein, p53 gene regulators, or other components of the p53 pathway. On the other hand, p53 tumor suppressor function is so vitally important for the organism that it is essential to carefully verify the risk of cancer development associated with the use of this new class of prospective pharmaceuticals. The role of a drug resistance gene and a survival factor that p53 might play in some tumors is a relatively new concept. Interestingly, this seems to be true for both wild-type p53 and p53-deficient tumors; in the latter case, p53 can presumably serve as a protector of tumor vascular endothelium. Importantly, the above-mentioned negative roles of p53 in cancer treatment are presumably exerted through different mechanisms: cancer treatment side effects result from p53-mediated apoptosis while drug and radiation resistance function of p53 occurs in tumors—through p53-mediated control of growth arrest (see Fig. 10.2).

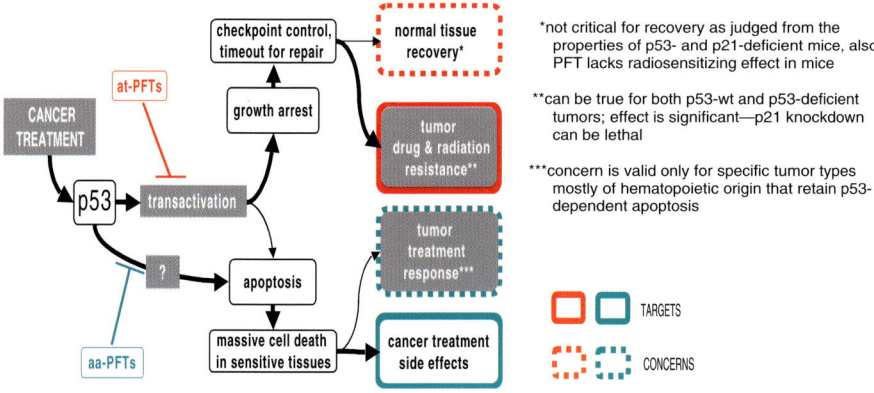

Figure 10.2. p53 as a target for pharmacological suppression in cancer treatment. In tumor-bearing organisms, p53 gets activated as a result of systemic genotoxic stress both in normal tissues and in the tumor, if the tumor retains wild-type p53. But even if it does not, p53 is activated in tumor stroma. Depending on the cell type and severity of stress, p53 activation results in activation of apoptosis or growth arrest. Growth arrest is mediated by activation of transcription of p53-responsive genes involved in cell cycle checkpoint control such as p21 and 14-3-3-sigma. Apoptosis is also induced in part through transactivation of proapoptotic genes, such as Bax or PUMA, but it can also be induced by a different mechanism that presumably involves direct interaction of p53 with mitochondria. Although p53-dependent apoptosis seems to be a major determinant of cancer treatment side effects, such as hematopoietic syndrome or hair loss; it is rarely involved in tumor response to treatment. On the contrary, p53-mediated growth arrest facilitates recovery of damaged cells by preventing their entrance into mitotic catastrophe; it can contribute to drug resistance phenotype of tumor cells. These considerations suggest that (i) temporary inhibition of p53 during acute phase of chemo- or radiotherapy could be beneficial for the treatment outcome, (ii) selective inhibitors of p53-dependent apoptosis are likely to be useful against cancer treatment side effects, while (iii) selective inhibitors of transactivation (that mainly target growth arrest function of p53) could be used to sensitize tumors to treatment.

p53 controls growth arrest through the transactivation of checkpoint control genes, such as p21 or 14-3-3-sigma (Prives and Hall, 1999). However, control of apoptosis by p53 only in part goes through the transactivation of proapoptotic p53-responsive genes (e.g., Bax, PUMA, NoxA, etc.); p53 can induce apoptosis directly through an alternative mechanism that might involve its interaction with mitochondria and does not require transactivation (Jeffers et al., 2003; Mihara et al., 2003; Schuler et al., 2003). In fact, p53 inhibitors that have been isolated for their ability to suppress p53-dependent transactivation (atPFTs): their strength as transactivation inhibitors does not correlate with their antiapoptotic effect (Gudkov et al., unpublished observations; Bonini et al., 2004). Such inhibitors are expected to be effective primarily against the growth arrest function of p53 and therefore be considered as potential tumor sensitizing agents. Compounds targeting specifically antiapoptotic function of p53 (aaPFTs) are expected to be more effective against treatment side effects. New readout systems for chemical screening are currently being used for the isolation of new classes of p53 inhibitors and the above expectations will be experimentally tested in the near future.

ACKNOWLEDGMENTS

The author would like to thank Elena Komarova, Peter Chumakov, and Katerina Gurova for their stimulating discussions and years of fruitful collaboration. The author's work on p53 modulators, in his laboratory, was supported by grants from NIH (CA60730 and CA75179), Quark Biotech, Inc., and Cleveland BioLabs, Inc.

REFERENCES

Abdulkarim B, Sabri S, Deutsch E, Chagraoui H, Maggiorella L, et al. (2002). Antiviral agent Cidofovir restores p53 function and enhances the radiosensitivity in HPV-associated cancers. *Oncogene* 21:2334–2346.

Alvarez-Salas LM, Cullinan AE, Siwkowski A, Hampel A, DiPaolo JA. (1998). Inhibition of HPV-16 E6/E7 immortalization of normal keratinocytes by hairpin ribozymes. *Proc Natl Acad Sci USA* 95:1189–1194.

Alves da Costa C, Paitel E, Mattson MP, Amson R, Telerman A, et al. (2002). Wild-type and mutated presenilins 2 trigger p53-dependent apoptosis and down-regulate presenilin 1 expression in HEK293 human cells and in murine neurons. *Proc Natl Acad Sci USA* 99:4043–4048.

An WG, Kanekal M, Simon MC, Maltepe E, Blagosklonny MV, Neckers LM. (1998). Stabilization of wild-type p53 by hypoxia-inducible factor 1alpha. *Nature* 392:405–408.

Bassi L, Carloni M, Fonti E, Palma de la Pena N, Meschini R, Palitti F. (2002). Pifithrin-alpha, an inhibitor of p53, enhances the genetic instability induced by etoposide (VP16) in human lymphoblastoid cells treated in vitro. *Mutat Res* 499:163–176.

Bast R, Kufe D, Pollock R, Weichselbaum R, Holland J, Frei E. (2000). *Cancer Medicine*. Hamilton: B C Decker.

Beerheide W, Bernard HU, Tan YJ, Ganesan A, Rice WG, Ting AE. (1999). Potential drugs against cervical cancer: zinc-ejecting inhibitors of the human papillomavirus type 16 E6 oncoprotein. *J Natl Cancer Inst* 91:1211–1220.

Blagosklonny MV, An WG, Romanova LY, Trepel J, Fojo T, Neckers L. (1998). p53 inhibits hypoxia-inducible factor-stimulated transcription. *J Biol Chem* 273:11995–11998.

Bonini P, Cicconi S, Cardinale A, Vitale C, Serafino AL, et al. (2004). Oxidative stress induces p53-mediated apoptosis in glia: p53 transcription-independent way to die. *J Neurosci Res* 75:83–95.

Bosch FX, Lorincz A, Munoz N, Meijer CJ, Shah KV. (2002). The causal relation between human papillomavirus and cervical cancer. *J Clin Pathol* 55:244–265.

Bottger V, Bottger A, Garcia-Echeverria C, Ramos YF, van der Eb AJ, et al. (1999). Comparative study of the p53-mdm2 and p53-MDMX interfaces. *Oncogene* 18:189–199.

Brachman DG, Beckett M, Graves D, Haraf D, Vokes E, Weichselbaum RR. (1993). p53 mutation does not correlate with radiosensitivity in 24 head and neck cancer cell lines. *Cancer Res* 53:3667–3669.

Browder T, Butterfield CE, Kraling BM, Shi B, Marshall B, et al. (2000). Antiangiogenic scheduling of chemotherapy improves efficacy against experimental drug-resistant cancer. *Cancer Res* 60:1878–1886.

Bullock AN, Fersht AR. (2001). Rescuing the function of mutant p53. *Nat Rev Cancer* 1:68–76.

Bunz F, Hwang PM, Torrance C, Waldman T, Zhang Y, et al. (1999). Disruption of p53 in human cancer cells alters the responses to therapeutic agents. *J Clin Invest* 104:263–269.

Butz K, Denk C, Ullmann A, Scheffner M, Hoppe-Seyler F. (2000). Induction of apoptosis in human papillomaviruspositive cancer cells by peptide aptamers targeting the viral E6 oncoprotein. *Proc Natl Acad Sci USA* 97:6693–6697.

Bykov VJ, Issaeva N, Shilov A, Hultcrantz M, Pugacheva E, et al. (2002). Restoration of the tumor suppressor function to mutant p53 by a low-molecular-weight compound. *Nat Med* 8:282–288.

Chene P. (2001). Targeting p53 in cancer. *Curr Med Chem Anti-Canc Agents* 1:151–161.

Chene P. (2003). Inhibiting the p53-MDM2 interaction: an important target for cancer therapy. *Nat Rev Cancer* 3:102–109.

Chene P, Fuchs J, Carena I, Furet P, Garcia-Echeverria C. (2002). Study of the cytotoxic effect of a peptidic inhibitor of the p53-hdm2 interaction in tumor cells. *FEBS Lett* 529:293–297.

Chow BM, Li YQ, Wong CS. (2000). Radiation-induced apoptosis in the adult central nervous system is p53-dependent. *Cell Death Differ* 7:712–720.

Chow WH, Devesa SS, Warren JL, Fraumeni JF, Jr. (1999). Rising incidence of renal cell cancer in the United States. *Jama* 281:1628–1631.

Cordon-Cardo C, Dalbagni G, Sarkis AS, Reuter VE. (1994). Genetic alterations associated with bladder cancer. *Important Adv Oncol*:71–83.

Cui YF, Zhou PK, Woolford LB, Lord BI, Hendry JH, Wang DW. (1995). Apoptosis in bone marrow cells of mice with different p53 genotypes after gamma-rays irradiation in vitro. *J Environ Pathol Toxicol Oncol* 14:159–163.

Culmsee C, Zhu X, Yu QS, Chan SL, Camandola S, et al. (2001). A synthetic inhibitor of p53 protects neurons against death induced by ischemic and excitotoxic insults, and amyloid beta-peptide. *J Neurochem* 77:220–228.

Donehower LA, Harvey M, Slagle BL, McArthur MJ, Montgomery CA, Jr., et al. (1992). Mice deficient for p53 are developmentally normal but susceptible to spontaneous tumours. *Nature* 356:215–221.

Doorbar J, Foo C, Coleman N, Medcalf L, Hartley O, et al. (1997). Characterization of events during the late stages of HPV16 infection in vivo using high-affinity synthetic Fabs to E4. *Virology* 238:40–52.

Duan W, Zhu X, Ladenheim B, Yu QS, Guo Z, et al. (2002). p53 inhibitors preserve dopamine neurons and motor function in experimental parkinsonism. *Ann Neurol* 52:597–606.

Duncan SJ, Gruschow S, Williams DH, McNicholas C, Purewal R, et al. (2001). Isolation and structure elucidation of Chlorofusin, a novel p53-MDM2 antagonist from a Fusarium sp. *J Am Chem Soc* 123:554–560.

Falette N, Paperin MP, Treilleux I, Gratadour AC, Peloux N, et al. (1998). Prognostic value of P53 gene mutations in a large series of node-negative breast cancer patients. *Cancer Res* 58:1451–1455.

Fang B, Roth JA. (2003). Tumor-suppressing gene therapy. *Cancer Biol Ther* 2:S115–S121.

Foster BA, Coffey HA, Morin MJ, Rastinejad F. (1999). Pharmacological rescue of mutant p53 conformation and function. *Science* 286:2507–2510.

Friedler A, Veprintsev DB, Hansson LO, Fersht AR. (2003). Kinetic instability of p53 core domain mutants: implications for rescue by small molecules. *J Biol Chem* 278:24108–24112.

Garcia-Echeverria C, Chene P, Blommers MJ, Furet P. (2000). Discovery of potent antagonists of the interaction between human double minute 2 and tumor suppressor p53. *J Med Chem* 43:3205–3208.

Goodwin EC, DiMaio D. (2000). Repression of human papillomavirus oncogenes in HeLa cervical carcinoma cells causes the orderly reactivation of dormant tumor suppressor pathways. *Proc Natl Acad Sci USA* 97:12513–12518.

Graeber TG, Osmanian C, Jacks T, Housman DE, Koch CJ, et al. (1996). Hypoxia-mediated selection of cells with diminished apoptotic potential in solid tumours. *Nature* 379:88–91.

Gudkov AV, Komarova EA. (2003). The role of p53 in determining sensitivity to radiotherapy. *Nat Rev Cancer* 3:117–129.

Gurova KV, Hill JE, Razorenova OV, Chumakov PM, Gudkov AV. (2004). p53 pathway in renal cell carcinoma is repressed by a dominant mechanism. *Cancer Res* 64:1951–1958.

Gurova KV, Kwek SS, Koman IE, Komarov AP, Kandel E, et al. (2002). Apoptosis inhibitor as a suppressor of tumor progression: expression of Bcl-2 eliminates selective advantages for p53-deficient cells in the tumor. *Cancer Biol Ther* 1:39–44; discussion 5–6.

Halterman MW, Miller CC, Federoff HJ. (1999). Hypoxia-inducible factor-1alpha mediates hypoxia-induced delayed neuronal death that involves p53. *J Neurosci* 19:6818–6824.

Hamada K, Sakaue M, Alemany R, Zhang WW, Horio Y, et al. (1996). Adenovirus-mediated transfer of HPV 16 E6/E7 antisense RNA to human cervical cancer cells. *Gynecol Oncol* 63:219–227.

Huibregtse JM, Scheffner M, Howley PM. (1993). Cloning and expression of the cDNA for E6-AP, a protein that mediates the interaction of the human papillomavirus E6 oncoprotein with p53. *Mol Cell Biol* 13:775–784.

Issaeva N, Bozko P, Enge M, Protopopova M, Verhoef LG, et al. (2004). Small molecule RITA binds to p53, blocks p53-HDM-2 interaction and activates p53 function in tumors. *Nat Med* 10:1321–1328.

Jacks T, Remington L, Williams BO, Schmitt EM, Halachmi S, et al. (1994). Tumor spectrum analysis in p53-mutant mice. *Curr Biol* 4:1–7.

Jeffers JR, Parganas E, Lee Y, Yang C, Wang J, et al. (2003). Puma is an essential mediator of p53-dependent and -independent apoptotic pathways. *Cancer Cell* 4:321–328.

Jiang M, Milner J. (2002). Selective silencing of viral gene expression in HPV-positive human cervical carcinoma cells treated with siRNA, a primer of RNA interference. *Oncogene* 21:6041–6048.

Kanovsky M, Raffo A, Drew L, Rosal R, Do T, et al. (2001). Peptides from the amino terminal mdm-2-binding domain of p53, designed from conformational analysis, are selectively cytotoxic to transformed cells. *Proc Natl Acad Sci USA* 98:12438–12443.

Kelly KJ, Plotkin Z, Vulgamott SL, Dagher PC. (2003). P53 mediates the apoptotic response to GTP depletion after renal ischemia-reperfusion: protective role of a p53 inhibitor. *J Am Soc Nephrol* 14:128–138.

Komarov PG, Komarova EA, Kondratov RV, Christov-Tselkov K, Coon JS, et al. (1999). A chemical inhibitor of p53 that protects mice from the side effects of cancer therapy. *Science* 285:1733–1737.

Komarova EA, Chernov MV, Franks R, Wang K, Armin G, et al. (1997). Transgenic mice with p53-responsive lacZ: p53 activity varies dramatically during normal development and determines radiation and drug sensitivity in vivo. *EMBO J* 16:1391–1400.

Komarova EA, Gudkov AV. (1998). Could p53 be a target for therapeutic suppression? *Semin Cancer Biol* 8:389–400.

Komarova EA, Gudkov AV. (2001). Chemoprotection from p53-dependent apoptosis: potential clinical applications of the p53 inhibitors. *Biochem Pharmacol* 62:657–667.

Komarova EA, Kondratov RV, Wang K, Christov K, Golovkina TV, et al. (2004). Dual effect of p53 on radiation sensitivity in vivo: p53 promotes hematopoietic injury, but protects from gastro-intestinal syndrome in mice. *Oncogene* 23:3265–3271.

Lain S, Lane D. (2003). Improving cancer therapy by non-genotoxic activation of p53. *Eur J Cancer* 39:1053–1060.

Lakkaraju A, Dubinsky JM, Low WC, Rahman YE. (2001). Neurons are protected from excitotoxic death by p53 antisense oligonucleotides delivered in anionic liposomes. *J Biol Chem* 276:32000–32007.

Landis SH, Murray T, Bolden S, Wingo PA. (1999). Cancer statistics, 1999. *CA Cancer J Clin* 49: 8–31, 1.

Lowe S, Schmitt E, Smith S, Osborne B, Jacks T. (1993). p53 is required for radiation-induced apoptosis in mouse thymocytes. In *Nature*, pp. 847–849.

Lowe SW. (1995). Cancer therapy and p53. *Curr Opin Oncol* 7:547–553.

Maehama T, Patzelt A, Lengert M, Hutter KJ, Kanazawa K, et al. (1998). Selective down-regulation of human papillomavirus transcription by 2-deoxyglucose. *Int J Cancer* 76:639–646.

Mantovani F, Banks L. (2001). The human papillomavirus E6 protein and its contribution to malignant progression. *Oncogene* 20:7874–7887.

Merritt AJ, Potten CS, Kemp CJ, Hickman JA, Balmain A, et al. (1994). The role of p53 in spontaneous and radiation-induced apoptosis in the gastrointestinal tract of normal and p53-deficient mice. *Cancer Res* 54:614–617.

Michael D, Oren M. (2002). The p53 and Mdm2 families in cancer. *Curr Opin Genet Dev* 12:53–59.

Mihara M, Erster S, Zaika A, Petrenko O, Chittenden T, et al. (2003). p53 has a direct apoptogenic role at the mitochondria. *Mol Cell* 11:577–590.

Molina R, Segui MA, Climent MA, Bellmunt J, Albanelll J, et al. (1998). p53 oncoprotein as a prognostic indicator in patients with breast cancer. *Anticancer Res* 18:507–511.

Montes de Oca Luna R, Wagner DS, Lozano G. (1995). Rescue of early embryonic lethality in mdm2-deficient mice by deletion of p53. *Nature* 378:203–206.

Munger K, Basile JR, Duensing S, Eichten A, Gonzalez SL, et al. (2001). Biological activities and molecular targets of the human papillomavirus E7 oncoprotein. *Oncogene* 20:7888–7898.

Nieder C, Petersen S, Petersen C, Thames HD. (2000). The challenge of p53 as prognostic and predictive factor in gliomas. *Cancer Treat Rev* 26:67–73.

Offringa R, Vierboom MP, van der Burg SH, Erdile L, Melief CJ. (2000). p53: a potential target antigen for immunotherapy of cancer. *Ann NY Acad Sci* 910:223–33; discussion 33–36.

Ohnishi T, Ohnishi K, Wang X, Takahashi A, Okaichi K. (1999). Restoration of mutant TP53 to normal TP53 function by glycerol as a chemical chaperone. *Radiat Res* 151:498–500.

Pani L, Horal M, Loeken MR. (2002). Rescue of neural tube defects in Pax-3-deficient embryos by p53 loss of function: implications for Pax-3-dependent development and tumorigenesis. *Genes Dev* 16:676–680.

Parkin DM, Pisani P, Ferlay J. (1999). Estimates of the worldwide incidence of 25 major cancers in 1990. *Int J Cancer* 80:827–841.

Peng Y, Li C, Chen L, Sebti S, Chen J. (2003). Rescue of mutant p53 transcription function by ellipticine. *Oncogene* 22:4478–4487.

Pirkkala L, Nykanen P, Sistonen L. (2001). Roles of the heat shock transcription factors in regulation of the heat shock response and beyond. *Faseb J* 15:1118–1131.

Prives C, Hall PA. (1999). The p53 pathway. *J Pathol* 187:112–126.

Ravi R, Mookerjee B, Bhujwalla ZM, Sutter CH, Artemov D, et al. (2000). Regulation of tumor angiogenesis by p53-induced degradation of hypoxia-inducible factor 1alpha. *Genes Dev* 14:34–44.

Rippin TM, Bykov VJ, Freund SM, Selivanova G, Wiman KG, Fersht AR. (2002). Characterization of the p53-rescue drug CP-31398 in vitro and in living cells. *Oncogene* 21:2119–2129.

Roninson IB, Broude EV, Chang BD. (2001). If not apoptosis, then what? Treatment-induced senescence and mitotic catastrophe in tumor cells. *Drug Resist Updat* 4:303–313.

Schafer T, Scheuer C, Roemer K, Menger MD, Vollmar B. (2003). Inhibition of p53 protects liver tissue against endotoxin-induced apoptotic and necrotic cell death. *Faseb J* 17:660–667.

Scheffner M, Munger K, Byrne JC, Howley PM. (1991). The state of the p53 and retinoblastoma genes in human cervical carcinoma cell lines. *Proc Natl Acad Sci USA* 88:5523–5527.

Schmitt CA, Lowe SW. (2001). Bcl-2 mediates chemoresistance in matched pairs of primary E(mu)-myc lymphomas in vivo. *Blood Cells Mol Dis* 27:206–216.

Schuler M, Maurer U, Goldstein JC, Breitenbucher F, Hoffarth S, et al. (2003). p53 triggers apoptosis in oncogene-expressing fibroblasts by the induction of Noxa and mitochondrial Bax translocation. *Cell Death Differ* 10:451–460.

Seo YR, Kelley MR, Smith ML. (2002). Selenomethionine regulation of p53 by a ref1-dependent redox mechanism. *Proc Natl Acad Sci USA* 99:14548–14553.

Sherr CJ, Weber JD. (2000). The ARF/p53 pathway. *Curr Opin Genet Dev* 10:94–99.

Slichenmyer WJ, Nelson WG, Slebos RJ, Kastan MB. (1993). Loss of a p53-associated G1 checkpoint does not decrease cell survival following DNA damage. *Cancer Res* 53:4164–4168.

Soengas MS, Capodieci P, Polsky D, Mora J, Esteller M, et al. (2001). Inactivation of the apoptosis effector Apaf-1 in malignant melanoma. *Nature* 409:207–211.

Sohn TA, Bansal R, Su GH, Murphy KM, Kern SE. (2002). High-throughput measurement of the Tp53 response to anticancer drugs and random compounds using a stably integrated Tp53-responsive luciferase reporter. *Carcinogenesis* 23:949–957.

Song S, Lambert PF. (1999). Different responses of epidermal and hair follicular cells to radiation correlate with distinct patterns of p53 and p21 induction. *Am J Pathol* 155:1121–1127.

Soussi T. (2000). The p53 tumor suppressor gene: from molecular biology to clinical investigation. *Ann NY Acad Sci* 910:121–37; discussion 37–39.

Stoll R, Renner C, Hansen S, Palme S, Klein C, et al. (2001). Chalcone derivatives antagonize interactions between the human oncoprotein MDM2 and p53. *Biochemistry* 40:336–344.

Tommasino M, Accardi R, Caldeira S, Dong W, Malanchi I, et al. (2003). The role of TP53 in Cervical carcinogenesis. *Hum Mutat* 21:307–312.

Tweddle DA, Pearson AD, Haber M, Norris MD, Xue C, et al. (2003). The p53 pathway and its inactivation in neuroblastoma. *Cancer Lett* 197:93–98.

Vargas DA, Takahashi S, Ronai Z. (2003). Mdm2: A regulator of cell growth and death. *Adv Cancer Res* 89:1–34.

Vassilev LT. (2004). Small-molecule antagonists of p53-MDM2 binding: research tools and potential therapeutics. *Cell Cycle* 3:419–421.

von Knebel Doeberitz M, Rittmuller C, zur Hausen H, Durst M. (1992). Inhibition of tumorigenicity of cervical cancer cells in nude mice by HPV E6-E7 anti-sense RNA. *Int J Cancer* 51:831–834.

Vousden KH, Lu X. (2002). Live or let die: the cell's response to p53. *Nat Rev Cancer* 2:594–604.

Wang L, Cui Y, Lord BI, Roberts SA, Potten CS, et al. (1996). Gamma-ray-induced cell killing and chromosome abnormalities in the bone marrow of p53-deficient mice. *Radiat Res* 146:259–266.

Westphal CH, Rowan S, Schmaltz C, Elson A, Fisher DE, Leder P. (1997). atm and p53 cooperate in apoptosis and suppression of tumorigenesis, but not in resistance to acute radiation toxicity. *Nat Genet* 16:397–401.

Willis AC, Chen X. (2002). The promise and obstacle of p53 as a cancer therapeutic agent. *Curr Mol Med* 2:329–345.

Woods DB, Vousden KH. (2001). Regulation of p53 function. *Exp Cell Res* 264:56–66.

Zhang M, Liu W, Ding D, Salvi R. (2003). Pifithrin-alpha suppresses p53 and protects cochlear and vestibular hair cells from cisplatin-induced apoptosis. *Neuroscience* 120:191–205.

Zhao J, Wang M, Chen J, Luo A, Wang X, et al. (2002). The initial evaluation of non-peptidic small-molecule HDM2 inhibitors based on p53-HDM2 complex structure. *Cancer Lett* 183:69–77.

Zheleva DI, Lane DP, Fischer PM. (2003). The p53-Mdm2 pathway: targets for the development of new anticancer therapeutics. *Mini Rev Med Chem* 3:257–270.

Index